Lecture Notes in Computer Science 1288

Edited by G. Goos, J. Hartmanis a[nd J. van Leeuwen]

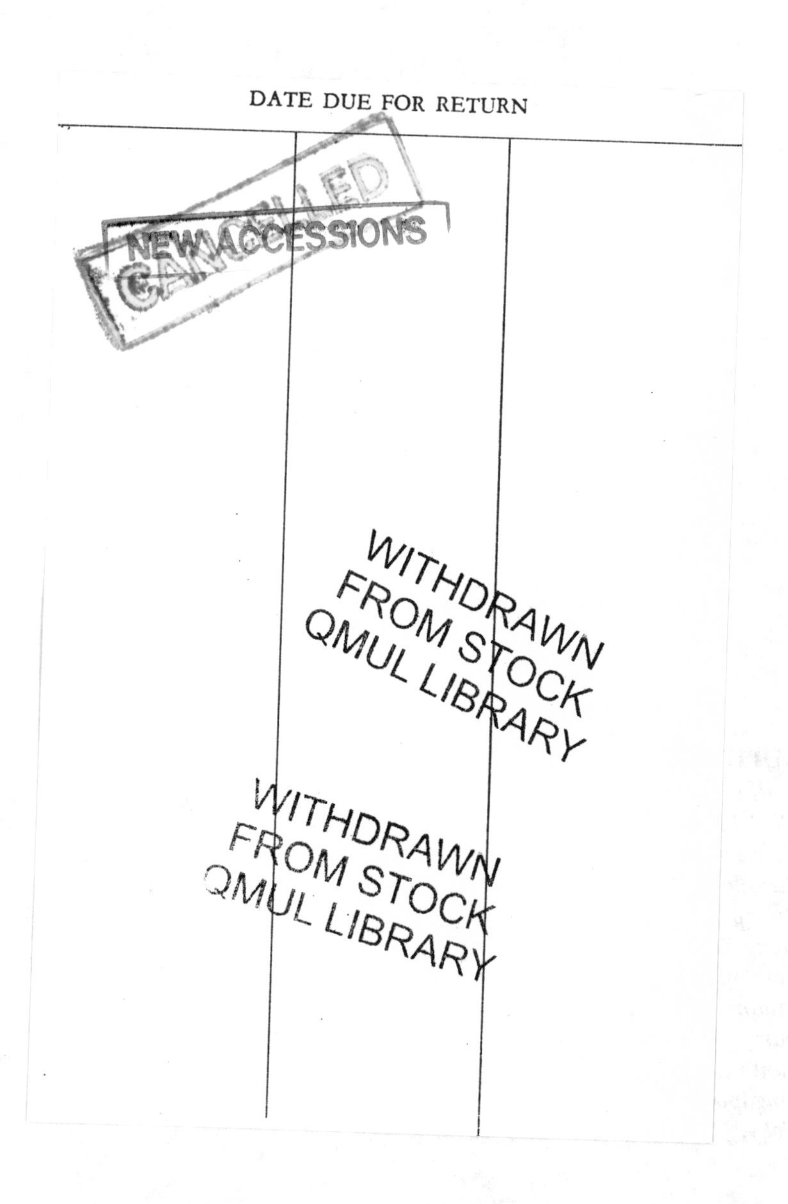

Springer
Berlin
Heidelberg
New York
Barcelona
Budapest
Hong Kong
London
Milan
Paris
Santa Clara
Singapore
Tokyo

Markus Schneider

Spatial Data Types for Database Systems

Finite Resolution Geometry for Geographic Information Systems

Author

Markus Schneider
FernUniversität Hagen, Fachbereich Informatik
Postfach 940, D-58084 Hagen, Germany
E-mail: markus.schneider@fernuni-hagen.de

Cataloging-in-Publication data applied for

Die Deutsche Bibliothek - CIP-Einheitsaufnahme

Schneider, Markus:
Spatial data types for database systems : finite resolution geometry for geographic information systems / Markus Schneider. - Berlin ; Heidelberg ; New York ; Barcelona ; Budapest ; Hong Kong ; London ; Milan ; Paris ; Santa Clara ; Singapore ; Tokyo : Springer, 1997
(Lecture notes in computer science ; Vol. 1288)
ISBN 3-540-63454-1

CR Subject Classification (1991): H.2, E.1-2, H.4, J.2

ISSN 0302-9743
ISBN 3-540-63454-1 Springer-Verlag Berlin Heidelberg New York

Typesetting: Camera-ready by author
SPIN 10546430 06/3142 – 5 4 3 2 1 0 Printed on acid-free paper

Foreword

Database research in the last decade has focused on developing support for so-called *non-standard applications*. One important area is the representation of *spatial* information, needed, for example, in *Geographic Information Systems*. Database systems extended by capabilities for managing spatial information are called *spatial database systems*. This book contributes to the design and implementation of geometry in spatial database systems.

The fundamental question in a spatial database is how to represent geometry. Attempts to use a relational database with the standard set of data types (integer, real, string, etc.) had serious shortcomings. Representing a simple polygon in a table of x and y coordinates is conceptually inadequate, leads to complex queries and is, last but not least, hopelessly inefficient. To determine which polygon in the database contains a given point, it is first necessary to reconstruct the polygon from the set of tuples representing it.

The fundamental idea is to introduce abstract data types for spatial objects with the corresponding operations. For example, in two dimensions, there may be types to represent points, (poly-)lines, or polygons in the plane with operations such as testing whether two polygons overlap or whether a point lies within a polygon, computing the intersection of a line and a polygon, etc. The relational model, or in fact any DBMS data model, can be extended by such types in the role of *attribute data types*. Hence we may now have, for example, a relation describing countries with an attribute of type *polygon* representing the country area.

This leads directly to the question how geometric constructions, as defined by the Euclidean axioms, can be represented with the finite approximations available in a computer system. Several systems of spatial data types (or *spatial algebras*) had been proposed in the literature, but these formal designs were based on exact Euclidean geometry. For example, one assumed that the intersection point of two line segments can be computed precisely. However, computers work with finite representations and can represent coordinates only approximately. It is usually necessary to round the coordinates of the intersection point to the nearest *grid point* where the grid corresponds to the resolution of the number system used. This introduces numerous types of errors, e.g., a subsequent test will tell you that the intersection point does not lie on either of the two line segments that created it in the first place! These errors were not only known theoretically, but hounded the daily practice of Geographic Information Systems: the most important "overlay" function of most commercial sytems failed sometimes even for simple inputs. It became crucial to find a solution for a robust implementation of geometric algorithms.

The essential new idea developed in this thesis work was not to base the design of a spatial algebra directly on Euclidean geometry but to use an underlying discrete basis, a so-called *realm*. A realm is a set of points and line segments defined over a discrete grid, with the additional property that no two realm segments intersect (except at their end points), and that no realm point lies on the interior of a realm segment. In other words, a realm is a planar graph embedded into the plane. It can be represented with finite coordinate values. A realm is intended to represent the entire geometry of an application, regardless of how the application is structured into different classes of geometric objects. The values of spatial data types are composed of points and line segments existing in the underlying realm. New geometries entered into the database are first put into the realm – so any intersections are computed at this level – and then propagated to the spatial data type values defined over the realm in the database. As a result, one can define a spatial algebra whose operations obey the algebraic laws that one would expect not only in theory, but also in the implementation. For example, computing the overlay of parcels with soil types gives the same result as the overlay of soil types over parcels. This fundamental law is generally violated in today's Geographic Information System software.

This book develops this strategy in detail. It describes the concept of realm and shows how updates can be performed in the realm layer so that the realm properties are maintained. This results in the design of the *ROSE algebra*, a comprehensive and coherent system of spatial types that can be implemented in (extensible) database systems without incurring the risk of computational problems due to the finiteness of computer algebra.

The algorithms and data structures designed are very efficient, in many cases even more efficient than the corresponding standard, floating point based algorithm. This is the rare case where the correct solution is also more efficient. The algorithms have been implemented and tested. They are available as a software package and we hope that they can be integrated by the industry into Geographic Information Systems. This book describes all the pieces of the design and implementation in a coherent manner, and should be a valuable resource for industry to help transfer these concepts into practice.

It has been a pleasure for me (RHG) to work with Markus, from the early days when he was a student member of a project group at the University of Dortmund that implemented a first prototype of a spatial database system called *Gral*, until now. His research results have been well received in the spatial database community and the concept of realms is well known now. The ROSE algebra, which is the central part of his Ph.D. thesis and described here, is used in some projects at other universities. It is a new approach to this challenging problem and has been an inspiration for others to focus on similar questions.

May 1997 *Ralf Hartmut Güting*

Andrew U. Frank

Preface

In various application fields there is a need to manage and process *geometric* or *spatial* data, that is, data related to *space*. For this purpose *spatial database systems* are designed which are full-fledged database systems with additional capabilities for storing, retrieving, manipulating, and querying spatial data. Such systems provide the underlying database technology needed to support applications such as *geographical information systems*. *Spatial data types* like *point*, *line*, and *region* provide a fundamental abstraction for modelling the structure of geometric entities, their relationships, properties, and operations. Their definition and implementation is probably the most fundamental issue in the development of spatial database systems and has influence on multiple disciplines like computer science, computer aided design, multimedia systems, VLSI design, (applied) computational geometry, geography, cartography, surveying engineering, linguistics, and psychology.

In this book we first give a comprehensive survey of the modelling approaches for spatial data types and operations so far discussed in the literature. We then introduce a system of spatial data types, or a *spatial algebra*, called *ROSE* (*RObust Spatial Extension*) *algebra*. The ROSE algebra avoids many of the deficiencies of current approaches and provides a satisfactory solution for a formal definition and robust implementation of a spatial type system in a single model. Current approaches to spatial data types primarily suffer from shortcomings regarding generality, formal definitions, closure properties, finite representations, geometric consistency, efficiency, extensibility, and data model independence. The development of the ROSE algebra consists of three steps.

The first step introduces the concept of a *realm* as a *discrete* geometric basis underlying one or more spatial data types. A realm is a finite set of points and non-intersecting line segments over a discrete grid and conceptually describes the complete underlying geometry of a particular application space in two dimensions. The idea is to construct the geometries of spatial objects by composing them from points or segments in the realm. As a basis for the design of spatial data types, very often Euclidean space is used or implicitly assumed. A point in the plane is then represented by a pair of real numbers. Unfortunately, in practice there are no real numbers available in computers but only finite approximations. This leads to many problems in geometric computation. The realm concept solves these problems of numerical robustness and topological correctness below and within the realm layer. In particular, it solves the line segment intersection problem over a discrete grid and enforces geometric consistency of related spatial objects. It also enables one to formally define general spatial data types or algebras on top of realms. These types enjoy nice closure properties not only in theory but also in computational practice.

The second step deals with the formal definition of the *ROSE algebra* itself which is defined on top of realms and offers *realm-based* spatial data types to represent point, line, and region features in the plane together with a comprehensive set of spatial operations. The ROSE algebra has a number of interesting features: it (i) offers (values of) data types of a very general structure, (ii) has a complete formal definition of the semantics of types and operations, (iii) has realms as a discrete geometric basis which allows for a numerically correct and robust implementation of types and operations in terms of integer arithmetic, (iv) treats consistency between distinct spatial objects with common parts, and (v) has a general object model interface which allows it to cooperate with different kinds of database systems.

The third step relates to the *ROSE system* as an implementation of the ROSE algebra. This system realizes the algebra's types and operations by providing efficient data structures and new *realm-based geometric algorithms* defined over a discrete grid. The main techniques used are parallel traversal of objects, plane-sweep, and graph algorithms. All algorithms are analysed with respect to their worst-case time and space requirements. Due to the realm properties, these algorithms are relatively simple and efficient, and numerically completely robust. In contrast to traditional work on algorithms, the focus is not on finding the most efficient algorithm for each single problem, but rather on considering a spatial algebra as a whole, and on reconciling the various requirements posed by different algorithms within a single data structure for each type. As an important result, a comparison with the algorithms of classical computational geometry reveals that realm-based geometric algorithms are much simpler and more efficient than their Euclidean counterparts.

This book is a revised and slightly extended version of my doctoral thesis (Ph.D. thesis) [Sc95a] submitted to the Computer Science Department of the Fernuniversität Hagen in Hagen, Germany, in December 1995. All parts of this book are worked out in considerable detail. This may be seen as a negative consequence of this text being essentially my thesis. However, it may also be interesting in its own right, in particular for readers who like to go into further detail or intend to construct and implement a spatial database system, a geographical information system, or a CAD system. Since detailed expositions are always preceded by conceptual overviews, it is left to the reader to omit detailed explanations and to focus on the book parts he or she is interested in.

Acknowledgments

I am indebted to several people who assisted me in one way or another to realize the original thesis which led to this book. First of all, I would like to express my esteem and my sincere thanks to my supervisor Prof. Dr. Ralf Hartmut Güting of the Fernuniversität Hagen, Germany, for his guidance and support throughout the development of the thesis. Furthermore, my gratitude extends to Prof. Dr. Andrew U. Frank of the Technical University of Vienna, Austria, for his interest in the subject of my thesis and his willingness to be its co-referee.

Prof. Güting suggested dealing with the subject of spatial data types for database systems and gave me the opportunity and the time to work on this topic and to pursue my ideas. He accompanied and supported my work by stimulating discussions and constructive criticism and helped me to structure my concepts and to focus on essential ideas. Prof. Frank made many valuable comments on a first draft and helped me to improve the contents and the presentation of my work.

Furthermore, I would like to thank Thomas de Ridder and Andreas Rohrbach for their excellent implementation work. Thanks also to the German Research Community (Deutsche Forschungsgemeinschaft (DFG)) for sponsoring a four-year research project entitled "Geo-Database Systems". Without its financial support, neither my publications in this period nor my thesis nor this book would have been possible. Special thanks go to Alfred Hofmann of Springer-Verlag for his editorial help.

Last but not least, I would like to thank my wife Annette, my son Florian, and my parents Hans and Christel Schneider for their love, support, encouragement, and patience. This book is dedicated to them.

June 1997 *Markus Schneider*

Table of Contents

1 Introduction **1**

1.1 Background and Motivation 1
1.2 Focus 3
1.3 Organization of the Book 6

2 Spatial Data Types - A Survey **11**

2.1 Spatial Data Modelling 12
2.1.1 A Three-Level Model for Phenomena in Space 12
2.1.2 Spatial Modelling 14
2.1.3 Properties of Spatial Data 16
2.1.4 Spatial Concepts 18
2.1.5 Spatial Data Models and Spatial Data Structures 20
2.2 Spatial Data Types and Spatial Operations 22
2.2.1 Conceptual Views of Spatial Objects 22
2.2.2 Spatial Operations 33
2.2.3 Design Criteria for Modelling Spatial Data Types 43
2.3 Formal Definition Methods 48
2.3.1 Introduction 48
2.3.2 Mathematical Concepts of Space 50
2.3.3 Point-Set Theory 55
2.3.4 Point-Set Topology 59
2.3.5 Algebraic Topology 68
2.3.6 Order Theory 73
2.4 Numerical Robustness and Topological Correctness of Geo-Algorithms 77
2.5 Finite Precision Computational Geometry 83

3 Realms: A Foundation for Spatial Data Types in Database Systems **85**

3.1 The Realm Concept 85
3.2 Integer Arithmetic and Robust Geometric Primitives 88
3.3 Redrawing as a Solution of the Discrete Segment Intersection Problem 90
3.3.1 The Segment Intersection Problem Defined over a Discrete Space 90
3.3.2 Concepts and Properties of the Redrawing Process 92

3.3.3 Number-Theoretical Foundations for the Redrawing of a Hooked Segment 97
3.3.4 The Restricted Redrawing Algorithm 103
3.3.5 The Generalized Redrawing Algorithm 116
3.4 Realms 122
3.5 Operations on Realms: The Realm Interface 122
3.5.1 Modelling the Realm Interface 122
3.5.2 Algorithms of the Realm Interface 126
3.6 Realm-Based Structures and Primitives 130
3.7 Related Work 139

4 Realm-Based Spatial Data Types: The ROSE Algebra 141
4.1 Realm-Based Spatial Data Types 141
4.2 The Type System of the ROSE Algebra 144
4.2.1 Second-Order Signature 144
4.2.2 The Type of a Partition 146
4.3 The Object Model Interface 148
4.3.1 OMI Concepts for Defining the ROSE Algebra 148
4.3.2 OMI Concepts for Embedding the ROSE Algebra into a DBMS Query Language 149
4.4 The ROSE Algebra 153
4.4.1 Spatial Predicates Expressing Topological Relationships 154
4.4.2 Spatial Operators Returning Spatial Data Type Values 155
4.4.3 Spatial Operators Returning Numbers 156
4.4.4 Spatial Operators on Sets of Spatially-Referenced Objects 158
4.5 Integration with a DBMS Query Language: O_2SQL/ROSE 161
4.6 Related Work 167

5 Efficient Algorithms for Realm-Based Spatial Data Types 169
5.1 Descriptive and Executable ROSE Algebra 169
5.2 Data Structures for the Realm-Based Spatial Data Types 172
5.3 Algorithms of the Executable ROSE Algebra 176
5.3.1 Algorithms with Simple or Parallel Object Traversal 177
5.3.2 Algorithms Using the Plane Sweep Paradigm 182
5.3.3 Graph Algorithms 196
5.3.4 Special Algorithms for Distance Problems 208
5.3.5 Using Filter Techniques: The Bounding Box 209
5.3.6 Algorithms for Operations on Sets of Database Objects 212
5.4 Related Work 214

6 Implementing Concepts: Realm System and ROSE System **215**
6.1 Providing Realm System and ROSE System as Modular Components 215
6.2 The Realm System 216
6.2.1 Modularization and Architecture 216
6.2.2 Implementation Aspects 217
6.2.3 The Realm Editor 220
6.3 The ROSE System 226
6.3.1 Modularization and Architecture 226
6.3.2 Implementation Aspects 227
6.3.3 The ROSE Editor 230
6.4 Integration of Realm System and ROSE System 230
6.5 Related Work 231

7 Conclusions, Open Problems, and Future Work **233**
7.1 Conclusions 233
7.2 Open Problems 236
7.3 Future Work 237
7.3.1 Spatial Type Extension Packages 237
7.3.2 Vague Spatial Objects 240
7.3.3 Spatiotemporal Objects 244

Bibliography **245**

Appendix A: Definition of Robust Geometric Primitives **259**

Appendix B: Definition Layers for Realm-Based Spatial Data Types **263**

Appendix C: Translation of the Descriptive ROSE Operators into Executable Operators **265**

Index **269**

Chapter 1
Introduction

1.1 Background and Motivation

In the past, conventional database systems were developed and successfully employed predominantly for the organization of large data sets in business management and administrative applications, so-called "standard" applications. But in recent years much of the database research has increasingly concentrated on database support for technical-scientific applications, so-called "non-standard" applications. It is widely recognized that conventional data models and database systems are not particularly qualified to meet the requirements of non-standard application areas such as computer aided design, image processing, multimedia systems, geographic information systems, VLSI design, office information systems, or applications in medicine and biology. Hence, these new requirements have stimulated the interest of research in extending database technology to the needs of the new application areas and have led to the design of *non-standard database systems*.

In an increasing number of application fields, database support is needed to manage *geometric* or *spatial* data, which means data related to *space*. Several terms have been coined for non-standard database systems that offer support for this kind of data like *pictorial*, *image*, *geographic*, *spatial*, or *geometric* database systems. Image or pictorial database systems manage data that are available as digital raster images which can be stored, manipulated, and retrieved as discrete entities and from which meaningful objects in space can be extracted by sophisticated analysis techniques. Examples are remote sensing by satellites or computer tomography in medical applications. Spatial or geometric database systems view sets of objects in space (rather than images of a space) which have features like identity, location, extent, and relationships to other objects. Because of the different nature of the data the requirements and techniques for dealing with objects in space are rather different from those for dealing with raster images.

In this book we will focus only on spatial database systems. A spatial database system is a full-fledged database system with *additional* capabilities for representing, querying, and manipulating objects in space. This implies that the user can treat spatial and non-spatial (e.g., alphanumeric) data in a homogeneous way and that the treatment of standard data modelling and querying tasks is also included. A special purpose system which is not able to handle standard tasks would not find acceptance by the user. *Geographical information systems* (GIS) are the most important applications on top of spatial (and image) database systems. They contain components to process *geographical data*, that is, spatial data occurring in a geographical context, to display these data in the form of maps and to perform analytical tasks like overlay and buffering. The aspect of analysing geographical data is one of the main purposes of a GIS.

Conventional data models, query languages, and access methods were designed to deal with simple data types representing alphanumerical data like integers, floating point numbers, and strings as they are used in standard applications. In geometric application areas, however, it is necessary not only to manage alphanumerical data but at the same time to treat spatial data having both complex structure and complex semantics. Conventional database systems are not suitable to store, retrieve, and manipulate spatial data, and in addition to this, they are unable to efficiently support the type of operations that are required for geometric applications. Another essential feature of spatial database systems is that they cover an extremely wide and diverse range of applications. The special requirements, which to a large extent determine the database design and the available set of functionalities, usually vary from one given application to another. Hence, *extensibility* at all levels of the spatial database design is a central concept.

The diversity of geometric applications has led to several proposals both for the modelling of spatial data and for the design of new data models and query languages integrating traditional alphanumerical data as well as geometric data. Frequently, the approach has been pursued to suitably extend an existing data model or query language. In the literature the general opinion prevails that special data types are necessary to model geometry and to suitably represent geometric data in database systems. These data types are commonly called *spatial* or *geometric*[1] *data types*, such as *point*, *line*, or *region*. Spatial data types provide a fundamental abstraction for modelling the geometric structure of objects in space as well as their relationships, properties, and operations. We speak of *spatial objects* as occurrences of spatial data types. Thus, we take an entity-oriented view of spatial phenomena and advocate a concept of "identifiable objects in space", that is, objects can be handled individually.

---

[1] In the literature, the use of the two notions *spatial* and *geometric* is not uniform. Either no difference is made and both notions denote the pure geometry (location, shape, extension) of objects in space, or the notion *spatial* denotes the data which describe meaningful objects located in space (e.g., highway, city), and the notion *geometric* relates to their pure geometry. We decide to conform to the first alternative and use both notions as synonyms. For meaningful objects located in space we use the term "spatially-referenced objects" (see Section 2.1.1).

Which data types are needed depends on the kind of application to be supported (e.g., rectangles in VLSI design, surfaces and volumes in 3D-applications). The definition of spatial data types and operations as well as the mechanisms for providing them to the user are to a large degree responsible for the design of a spatial data model and the performance of a spatial database system and have great influence on the expressive power of spatial query languages. This is true regardless whether a DBMS uses a relational, complex object, object-oriented, or some other data model. Hence, *the definition and implementation of spatial data types is probably the most fundamental issue in the development of spatial database systems.*

In some sense, this work reflects the learning process, experience, and knowledge the author has gained from participating in the implementation of a prototype for a spatial database system called the *Gral system* [BG92, Gü88a, Gü88b, Gü89]. Its theoretical basis, the *geo-relational algebra*, is a formal extension of the relational data model and algebra by geometric data types and operations. At the same time it represents a query language for a spatial database system.

1.2 Focus

The subject of this book is a *formal* definition and a numerically and topologically *robust* implementation of spatial data types based on general design criteria which will be introduced in this section. Spatial data types (SDTs) are used routinely in the description of spatial query languages (e.g., [CW87, Eg89a, Gü88a, JC88, LN87, SH91, To90]) and have been implemented in some prototype systems (e.g., [Gü89, OM88, RFS88]). Furthermore, some formal definitions have been given [Gü88a, GNT91, SV89]. But there is still no completely satisfactory solution available according to the following aspects which are raised to general *design criteria* for the modelling of spatial data types:

- *Generality and versatility.* It should be feasible to model spatial objects that are the occurrences of SDTs as generally as possible. For example, a line object should be able to model a network like the ramification of the Nile delta. A region object should be able to represent a collection of disjoint areas each of which may have holes. This allows one, for instance, to model the German state of Niedersachsen enclosing the state of Bremen and having offshore islands in the North Sea as one object. Generality is the prerequisite for the versatility of spatial data types in many different kinds of applications.

- *Closure properties*. The domains of SDTs like *point*, *line*, and *region* should be closed under union, intersection, and difference of their underlying point sets. That is, the result of such an operation must be a well-defined object, too, and correspond to the definition of the SDTs. This makes it possible to define powerful data type operations with nice closure properties. By observing this criterion, geometric anomalies are avoided which can occur when, for instance, set-theoretic operations are carried out on point sets.

- *Rigorous definition.* The semantics of SDTs, that is, the possible values for the types and the functions associated with the operations, require a *formal* and clear definition to avoid ambiguities both for the user and the implementor.
- *Finite resolution, numerical robustness, and topological correctness.* The formal definitions must take into account *the finite representations available in computers*. This has so far been neglected in definitions of SDTs. It is still left to the programmer to close this gap between theory and practice which leads inevitably not only to numerical but also to topological errors.
- *Geometric consistency.* Distinct spatial objects may be related through geometric consistency constraints (e.g., adjacent regions have a common boundary, or two lines meet in a point). The definition of SDTs must offer facilities to enforce such consistency.
- *Efficiency.* Spatial operations needed in spatial query languages should be evaluated as fast as possible. Hence, geometric algorithms realizing spatial operations should be processed as fast as possible and make extensive use of the well-known and efficient algorithms, data structures, and methods of Computational Geometry.
- *Extensibility.* Even though the designer of a spatial database system may provide a good collection of spatial data types and operations, there will always be applications requiring additional operations on available types or requiring new types with new operations. A type system for SDTs should therefore be extensible for new data types.
- *Data model independence.* Spatial data types as such are rather useless; they need to be integrated into a DBMS data model and query language. However, a definition of SDTs should be valid regardless of a particular DBMS data model and therefore not depend on it.[2] Instead, the SDT definition should be based on a general abstract interface to the DBMS data model which we call the *object model interface*.

[2] This also holds for the implementation level: A *spatial type extension package* (*STEP*) should be able to cooperate with any extensible DBMS offering a suitable interface regardless of its data model.

Within the framework of this book we develop a formal definition and robust implementation of spatial data types taking these design criteria into account and offering a satisfactory solution in one single model. A central idea is to introduce into the DBMS the concept of a *realm* as a foundation for the definition of spatial data types. A realm is, in general, a finite, user-defined structure that is used as a basis for one or more system data types. Thus, a realm is somewhat similar to an enumeration type in programming languages. A realm used as a basis for spatial data types is essentially a finite set of points and *non-intersecting* line segments. Intuitively, it models the entire underlying geometry for a collection of spatially-referenced objects of an application. All spatial objects like points, lines, and regions can then be defined in terms of realm objects, that is, in terms of points and line segments present in the realm. In fact, spatial objects are never created directly but only by selecting some realm objects. They are never updated directly. Instead, updates are performed on the realm and from there propagated to the dependent spatial objects. Hence, all spatial objects occurring in a database or another environment are *realm-based.*

The concept of realms enables us to formally define the semantics of the proposed spatial data types points, lines, and regions together with their operations. The definition of algebraic operations for the types guarantees that the geometric objects constructed are realm-based as well. So the spatial algebra which is the formal framework for the spatial data types and which is called *ROSE* (*RObust Spatial Extension*) *algebra* is closed with respect to a given realm.

No two SDT objects occurring in geometric computation have "proper" intersections of line segments. Instead, two initially intersecting segments have already been split at the intersection point when they were entered into the realm. One could say that any two intersecting SDT objects (e.g., lines or regions) "have become acquainted" already when they were entered into the realm. This is a crucial property for the correct and efficient implementation of geometric operations. During query processing no new intersection points have to be computed.

Realm objects and consequently spatial objects are defined not in abstract continuous Euclidean space but in terms of finite representations over a discrete space, i.e., a grid. All geometric primitives and realm operations (e.g., updates) are defined in error-free integer arithmetic. numerical robustness and topological correctness problems are treated below and within the realm level. Geometric data of an application are, in general, not free from intersections as required for a realm. For mapping an application's set of intersecting line segments into a realm's set of non-intersecting segments the concept of *redrawing* and *finite resolution computational geometry* from Greene and Yao [GY86] is used. Although intersection points computed with finite resolution, in general, move away from their exact Euclidean position, this concept ensures that the unavoidable distortion of geometry (that is, the numerical error) remains bounded and very small and that essentially[3] no topological errors occur. This means that a programmer has a precise specification that directly lends itself to a correct implementation. It also means that the *spatial algebra obeys algebraic laws precisely in theory as well as in practice*. Furthermore, it is rather obvious that realms also solve the geometric consistency problem.

The realization of spatial operations of the ROSE algebra is carried out by geometric algorithms, methods, and data structures as they are subject of Computational Geometry [Me84, PS85]. One of the fundamental methods of this research field is the plane-sweep paradigm [BO79, NP82] which in variations is used for the geometric algorithms. An important aspect is that all these algorithms are realm-based. They require realm-based spatial objects as operands, i.e., no two segments properly intersect. This property can be exploited for the design of efficient algorithms. Since numerical and topological problems have already been solved below and within the realm level and are invisible at the algebra level, the design of *realm-based geometric algorithms* can focus on efficiency and algorithmical correctness. Plane-sweep algorithms are now much simpler and more efficient, since all intersection points are explicitly known, during a sweep new intersection points of segments can never be discovered, and special cases do not exist. Sweep stations are only isolated points or end points of segments that are all known in advance. We call this subfield of Computational Geometry or, more precisely, of finite resolution computational geometry *realm-based computational geometry.*

In summary, the ROSE algebra provides rather a complete theoretical foundation for the development of spatial data types which fulfills the design criteria mentioned above and which allows for and integrates the interaction with realms, robust arithmetic, and the interfaces to a DBMS data model in the same modelling process.

1.3 Organization of the Book

The book is organized into seven chapters. Chapter 2 gives a detailed survey of modelling approaches for spatial data types and operations in the literature and additionally investigates whether and how the design criteria of Section 1.2 have been taken into account by these approaches. Special interest relates to approaches for a *formal* definition of the semantics of spatial data types and operations. Another objective is to show the position of this book in the research field of spatial database systems.

The next two chapters present the formal definition of realm-based spatial data types which is organized as a series of layers that are described bottom-up. Each layer defines its own structures and primitives and uses the notions of the layers below. An

[3] See the discussion in Section 3.3.

overview of the different definition layers for realm-based spatial data types is given in Appendix B.

Chapter 3 introduces the realm concept. It begins with the two lowest layers of *integer arithmetic* and *robust geometric primitives*. The integer arithmetic is error-free with respect to overflow and offers the usual standard integer operations as well as additional operations like conversion routines. The layer of robust geometric primitives, which is formally defined in Appendix A, describes a discrete space (i.e., a grid) $N \times N$ where $N = \{0, ..., n-1\}$ is a subset of the natural numbers. The objects in this space are points and line segments with coordinates in N, called *N-points* and *N-segments*. A number of operations (predominantly predicates) are defined such as whether an N-point lies *on* an N-segment or whether two N-segments *intersect*, and which N-point is the result of intersecting two N-segments. The crucial point is that these definitions are given in terms of the integer arithmetic; hence they are directly implementable. Based on the integer arithmetic and the robust geometric primitives, a *redrawing* algorithm following [GY86] is presented and generalized which solves the problem of intersecting line segments defined over a discrete grid and the problem of topological correctness.

Next, geometric *realms* are described and formally defined as a set of points and non-intersecting line segments over the discrete space $N \times N$. The realm objects are called *R-points* and *R-segments*. Basic operations on realms are insertion and deletion of N-points and N-segments which may trigger the redrawing of segments. All necessary operations are summarized in the *realm interface* which is able to cooperate with a database system. For example, the operation of inserting an N-segment returns besides a modified realm a redrawing of the inserted segment and a set of redrawings of segments in the database that need to be modified together with logical pointers to database representations of these segments.

The third layer defines *structures present in a realm* that serve as a basis for the definition of SDTs. A realm can be viewed as a spatially embedded planar graph. An *R-cycle* is a cycle of this graph. An *R-face* is an R-cycle possibly enclosing some other disjoint R-cycles corresponding to a region with holes. An *R-unit* is a minimal R-face. These three notions support the definition of a <u>*regions*</u> data type. An *R-block* is a connected component of the realm graph; it supports the definition of a <u>*lines*</u> data type. For all of these structures there are also predicates defined to describe their possible relationships.

In Chapter 4, based on realm-based structures and primitives, the fourth layer introduces the *spatial data types* <u>*points*</u>, <u>*lines*</u>, and <u>*regions*</u> and defines the structure of corresponding objects. A <u>*points*</u> object is a set of R-points. There are two alternative views of <u>*lines*</u> and <u>*regions*</u>. The first is that a <u>*lines*</u> object is a set of R-segments and a <u>*regions*</u> object a set of R-units. The other view is equivalent but "semantically richer": A <u>*lines*</u> object is a set of disjoint R-blocks and a <u>*regions*</u> object a set of (edge-)disjoint *R*-faces. Moreover, *spatial algebra primitives* are defined on objects of these types.

The next two concepts prepare the definition of the fifth and final layer. A flexible type system is introduced that allows one to precisely describe polymorphic operations that are central to the ROSE algebra. In this type system it is also possible to cleanly model partitions of the plane so that operations can be constrained to be applicable to partitions or regions of partitions only. A *partition* is a set of spatially-referenced objects whose regions attribute values are disjoint.

Afterwards, the *object model interface* (*OMI*) is defined. We identify a number of concepts that need to be present in the DBMS data (or object) model to allow it to cooperate with our spatial algebra. The OMI has two parts. The first part is needed to define the semantics of operations of the ROSE algebra, in particular for complex operations that manipulate sets of database objects. The second part is needed to embed the ROSE algebra into a query language. It consists of a number of facilities within the query language that are required to make full use of the ROSE algebra possible. The corresponding idea at the system level is that any extensible database system offering an OMI implementation can cooperate with a *spatial type extension package* (*STEP*) realizing the spatial algebra.

Then as a top layer the *ROSE algebra* is described, and the semantics of all operations are formally defined. The operations are classified into four classes:

- spatial predicates expressing topological relationships (e.g., **area-disjoint**, **adjacent**, **inside**)
- spatial operations returning spatial data type values (e.g., **intersection**, **contour**, **common_border**)
- spatial operations returning numbers (e.g., **length**, **dist**, **area**)
- spatial operations on sets of spatially-referenced objects (e.g., **overlay**, **decompose**, **fusion**)

The last group of operations manipulates not only SDT objects but also the spatially-referenced objects they are associated with. We then show how the ROSE algebra can be integrated with a given DBMS data model and query language by choosing O_2 as an example. This illustrates the object model interface. Example queries in O_2SQL/ROSE demonstrate the "expressive power" of this spatial algebra.

Chapter 5 deals with the implementation of the ROSE algebra by providing data structures for its types and new *realm-based geometric algorithms* for its operations. Implementing the ROSE algebra means the transition of a *descriptive* to an *executable* ROSE algebra (this transition is summarized in Appendix C). The main algorithmic techniques used are based on parallel traversal of objects, plane-sweep, and graph algorithms. For each algorithmic technique example algorithms are described using a high-level notation. All algorithms are analysed with respect to their worst case time and space requirements. Due to the realm properties, these algorithms are relatively simple, efficient, numerically completely robust, and in particular more efficient than their Euclidean counterparts. Additionally, the use of filter techniques preceding the actual geometric algorithms is investigated by approximating spatial objects through bounding boxes.

The concept of realms and the algorithms and data types of the executable ROSE algebra have been implemented within the framework of two software systems called the *Realm system* and the *ROSE system*. Chapter 6 gives an overview of their architecture, functionality, and modularization. Both systems are designed as modular components which can be used either separately or connected together (applying the second alternative the ROSE system is put on top of the Realm system). Two so-called *application environment interfaces* and two window- and graphic-oriented editors called the *Realm editor* and the *ROSE editor* support the handling of both systems.

Chapter 7 summarizes the main results of the book, discusses some open problems, and gives a prospect of future work.

Chapter 2

Spatial Data Types - A Survey

This chapter gives a survey of modelling approaches for *spatial data types* in the literature and in particular investigates whether and how the design criteria of Section 1.2 have been taken into account by these approaches. Intuitive explanations of the meaning of spatial data types are insufficient both for the user and the programmer because of their ambiguity and vagueness. Hence, we are in particular interested in approaches for a *formal* definition of the semantics of spatial data types. Spatial information can be viewed and modelled on different context levels. An objective is to categorize this survey and this book as a whole with respect to these levels and to show the position of this book in the fields of spatial database systems and geographic information systems.

The first section deals with *spatial data modelling* and summarizes the modelling process for phenomena in space in a three-level model. Furthermore, it categorizes the contents of this book with regard to this model and describes the properties of spatial data and their differences in comparison to alphanumerical data. The three important notions of *spatial concepts*, *spatial data models*, and *spatial data structures* are introduced that exert a great influence on the design and implementation of spatial data types.

The second section informally describes the different proposals in the literature for the modelling of spatial data types from the user's perspective and gives classifications for them. It then investigates from the designer's perspective whether and how the design criteria have been realized by these approaches.

The third section presents different formal definition methods for spatial data types which are based on *point set theory*, *point set topology*, *algebraic topology*, and *order theory*.

The fourth section deals with the problems of *numerical robustness* and *topological correctness* of geometric algorithms.

The fifth section considers a small subfield of Computational Geometry called *finite precision computational geometry* which deals with computational geometry performed on a discrete domain.

2.1 Spatial Data Modelling

2.1.1 A Three-Level Model for Phenomena in Space

A study of the literature (e.g., [Al90, CG82, CL90, ELNR87, Fr91, LN87, Sm90, SV91]) shows that the modelling process of phenomena in space amounts to a three-level model (Figure 2-1) and that often the top level is missing in these approaches. Each level represents and describes a different context in which spatial information with different complexity is considered. Hence, we may speak of *context levels*. Each higher level is based on, uses, and contains the next lower level. A level comprises a number of proposals of models, each describing certain kinds of entities with respective operations. The term *spatial data modelling* is used to summarize the complete modelling process for phenomena in space and thus relates to the whole three-level model.

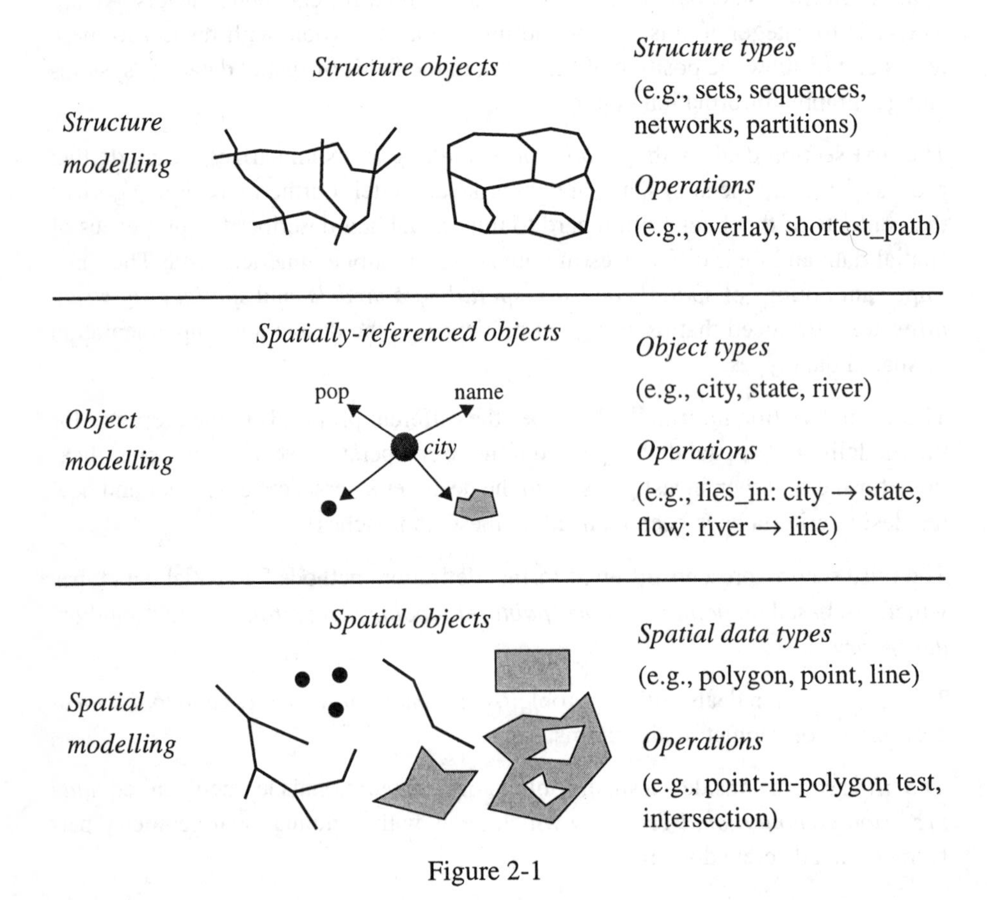

Figure 2-1

This survey and this book will predominantly focus on the lowest context level, namely that of spatial modelling. *Spatial modelling* (or *geometric modelling*) in the sense of this book refers to the design process for abstract representation objects describing the geometric structure, extent, and shape of objects in space, for their relationships to other representation objects, and for their relationships to the space in which they are located. As already mentioned, these representation objects are called *spatial objects*. Examples of spatial objects are points, lines, and regions. Operations at this level are geometric ones which for instance calculate the intersection of two polygons or check whether a point or a line lies inside a polygon.

The middle context level named *object modelling* describes the modelling and organization of spatial information from the real-world and DBMS point of view. A spatial database consists of several classes of real-world objects which are related to specific locations in space and which are associated with spatial and non-spatial (e.g., alphanumerical) objects (Figure 2-2). Following the terminology in [Wo92] these objects are called *spatially-referenced objects*. Only if the geometry and the location of the real-world object is taken into account in the database, the object becomes a spatially-referenced object. Thus, it is always associated with a spatial object in the database. Geographical objects are a special kind of spatially-referenced objects relating to phenomena of the earth.

An example of a spatially-referenced object is an object "London" of a type "city" which is associated with a region object and with non-spatial objects describing its name and population. Note that such an object can be associated with more than one spatial object. A city, for example, may be geometrically modelled as a point or a region which represent different abstractions of the object, a feature which is known as generalization in geography. Operations at this level, for example, yield the geometry of a river object which is a line object or examine whether a city object lies inside a state object, an information which is either explicitly stored or internally computed by reducing the question to the geometries of both objects.

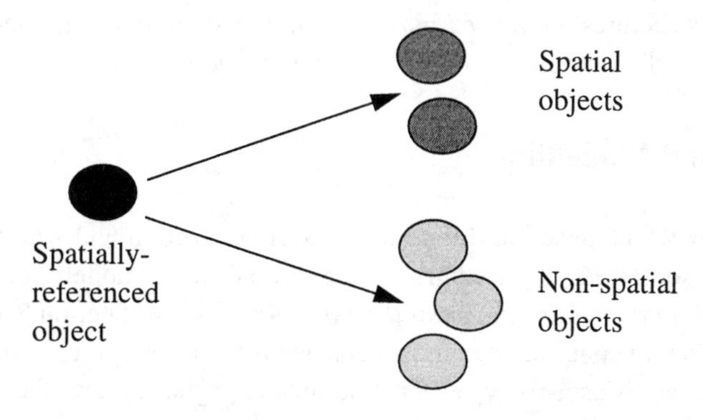

Figure 2-2

The top context level termed *structure modelling* is deeply interconnected with object modelling. It takes into account that spatially-referenced objects need "containers" in which they are kept and which we call (*connectivity*) *structures*. A connectivity structure describes a collection of spatially-referenced objects with explicitly or implicitly modelled connectivity properties or constraints (e.g., topological ones) between these objects. Well-known examples of implicit connectivity structures are sets, relations, and object classes. They represent simple collections of objects and model either no specific (sets, relations) or hierarchical (object classes) connectivity properties.

More complex examples are explicit structures like *networks* and *partitions* of the plane. A *network* can be viewed as a spatially embedded graph which consists of a set of point objects representing its nodes and a set of line objects describing the geometry of its edges. Networks are widespread in geography; examples are highways, rivers, public transport systems, power and phone lines. In a highway network that is viewed as a graph, two kinds of spatially-referenced objects can be distinguished, namely highway junctions and highway sections that are associated with the nodes and with the edges of the graph, respectively, and geometrically modelled by point and line objects, respectively. An example for an operation on networks is to find the shortest path between a source and a destination, e.g., from Berlin to Frankfurt. Current spatial database models and systems support well enough the handling of the geometry of such networks but have no concept of their connectivity, that is, of their underlying graph structure. In the literature, explicit graphs [EG94, Er94, Gü91, Gü94a] as a modelling concept for database systems and their integration into a data model and query language have been proposed.

A *partition* can be considered as a set of region objects together with the constraint that each pair of region objects is disjoint. The neighbourhood relationship is of particular interest here, since region objects may have a common boundary. A classical example for using partitions are thematic maps. An example for an operation on partitions is the overlay of two partitions producing a new partition. Other interesting connectivity structures are *nested partitions* (e.g., a country partitioned into states partitioned into districts etc.) and *digital terrain models*.

2.1.2 Spatial Modelling

Three notions which describe different levels of abstraction and which have to be conceptually separated play a central role in spatial (data) modelling: *spatial concepts*, *spatial data models*, and *spatial data structures* (see Section 2.1.4 and Section 2.1.5). The diversity of spatial concepts leads to various and non-uniform data models. This fact is especially responsible for the vagueness and the multitude of existing definitions of spatial data types and spatial operations as the components of spatial data models. Our main interest focuses on spatial data types describing the structure of spatial objects and on spatial operations representing geometric functions that can be applied to these objects. More precisely, we are interested in the definition of their syntax and semantics.

Spatial data types can be viewed from three different points of view (Figure 2-3). The *conceptual* or *designer's model* describes the designer's logical view of spatial data and determines the degree of abstraction and the concepts the user can employ to model reality. The *implementation model* yields implementation strategies and describes the representation of spatial objects (i.e., their data structures) and operations (i.e., their algorithms). The *user's model*, which is often neglected, describes how the user perceives and handles spatial data. This model may be equal to the conceptual model[4]. One of the user's main requirements is, of course, that he is furnished with highly expressive, versatile, well-defined, and consistent data types and operations. A user's model of spatial objects and operations should not be too complex, so that a user can easily learn and understand it and foresee the consequences of his actions.

User's *Model / Level*	User's conceptual view of spatial objects and operations
Conceptual *Model / Level*	Designer's logical view of spatial data Design methods Formal definition methods
Implementation *Model / Level*	Implementation aspects e.g., data structures, algorithms, efficiency

Figure 2-3

The designer of a spatial database system in particular decides whether the structure of spatial objects is made visible or hidden at the user's level. This decision on *visibility* determines whether the internal structure of objects is made accessible to the general facilities of a query language or only to domain-specific operations. Furthermore, the designer should attach importance to a formal frame for the correct

4 This three-model view of spatial data types is analogous to the three-layered architecture of modern DBMS consisting of *internal*, *conceptual*, and *external* layers.

specification of the structure and the precise definition of the semantics of spatial data types in order to guarantee clarity and consistency both at the user's and the implementation level. Hence, an important issue relates to formal methods and principles for the modelling of spatial data. The diversity and vagueness of proposed spatial data types reveals that there is an urgent need for their formal definition. Implementation aspects like data structures and algorithms are beyond the scope of this survey and not treated.

In most programming languages and database systems numerical, boolean, and string types have turned out to be widely accepted as fundamental data types for standard applications. In this context these types can be regarded as atomic and standardized types. But on the one hand these data types have been designed for applications which are much simpler than geometric applications, and on the other hand (representations of) spatial data types and (algorithms for) spatial operations differ greatly from one application to another.

Two main modelling strategies can be pursued when devising a collection of spatial data types. The first modelling strategy assumes the existence of a fixed and standardized set of spatial data types (on the analogy of standard applications) which is universal enough to satisfy the requirements of any geometric application. This strategy poses the question if such a set can exist, or in other words, it poses the question of *completeness*. The second modelling strategy assumes the existence of a kernel set of largely application-independent spatial data types which always has to be provided and which is suitable to support most geometric applications. In order to be applicable to new geometric application fields, this strategy must contain *extensibility* for new spatial data types as a central concept. Consequently, the specific extensions added are application-specific and thus restricted in generality.

Another problem is given by the two concepts of *minimality* and *convenience*. On the one hand the designer should aim at furnishing the user with a minimal set of data types and operations in order not to overtax him. On the other hand such a minimal set does not necessarily contain the most convenient data types and operations, and the user expects some comfort when using them. Frequently applied, complex operations which have to be constructed from elementary operations every time are very troublesome to use. Thus, a reasonable compromise between both concepts has to be found.

2.1.3 Properties of Spatial Data

The design of spatial data types should be based on an understanding of the properties of spatial data in order to accommodate the design process as far as possible to the nature of spatial phenomena in reality. But surprisingly very little is known about the inherent characteristics of and the relationships between spatial phenomena. Reasons for this lacking knowledge are the complexity and diversity of spatial data.

The fundamental properties of spatially-referenced objects can be classified in *non-geometric* (non-spatial, aspatial, thematic) and *geometric* (spatial) properties [Bu86, Fr87, OSM89, Oo90]. Non-geometric properties describe attribute-based data which are usually expressed by standard alphanumerical data, e.g., the population or name of a city, or the costs and the energy consumption of a house. Geometric properties can be distinguished in *metric* (locational) and *topological* features.

Metric features describe shape (geometry, structure), location, extent and measurements of spatial objects in a *reference system* or *reference frame*. A reference system is the standardized spatial background into which a set of spatial objects is conceptually embedded, e.g., the cartesian coordinate system. Each point is attached to a certain reference (coordinate) point. The shape of a spatially-referenced object describes an abstraction of its geometric structure like point, line, rectangle, or polygon. For example, the shape of a city can be modelled as a polygon, the shape of a river as a line. The location of an object indicates the position of the object with regard to the selected reference system. According to the spatial extent of an object, we distinguish 0-, 1-, 2-, and 3-dimensional objects. The introduction of a metric allows to derive further metric information and to compute measurements like the distance between two objects, perimeter and area of a region, or the length of a line.

Topological features characterize relationships between spatial objects that refer to statements concerning adjacency, connectivity, inclusion, and similar relationships of objects. These relationships are independent from the used reference system and invariant under topological transformations like translation and rotation.

Another distinctive property of spatial data is for what Anselin [An89] and Goodchild [Go90] coin the term *spatial dependence* and what Güting [Gü88a] characterizes as *connectedness*. Both terms describe the same property, namely the observation that nearby locations in a continuous space influence each other and possess equal or similar attributes. Spatial dependence primarily makes it possible to define spatial objects in a finite way (e.g., in the form of a boundary representation) and to map continuous space to a discrete and finite computer, even if the mapping only yields an approximate view of reality.

Spatially-referenced objects can be subject to *temporal changes* [Eg93, Fr91] and thus be dynamic. Changes may relate to geometry (the course of a street was modified) and to thematic values (a new street name). Temporal changes[5] may be temporary (a river overflows its banks) or permanent (construction of a new highway).

[5] The treatment of temporal aspects in spatial data models leads to spatio-temporal data models and is out of the scope of this book.

A comparison of spatial with alphanumerical data shows the great conceptual differences between both kinds of data. All spatial data are related to objects in space and have thus a position in space [Fr82]. Alphanumerical data are very simply structured (e.g., a floating point number is composed of a mantissa and an exponent, a fact which is uninteresting from the user's point of view), and operations on them are also very simple. By contrast, spatial data usually possess a complex and meaningful structure (e.g., a polygon is composed of line segments describing its surrounding boundary line), and operations on them have a complex semantics. That is, from the user's perspective, spatial objects are viewed as *complex objects*, and special operations are needed to access (components of) them. The use of spatial operations, however, is very similar to alphanumerical operations. The specific properties of spatial data are also reflected by their representation at the user interface. While alphanumerical data are usually represented in textual or tabular form, these kinds of representation play a minor role for spatial data which are normally represented by graphical drawings like maps in geographical applications.

2.1.4 Spatial Concepts

The complexity and diversity of spatial data, their occurrence in different contexts and situations, and humans' different experiences with the treatment of these data seem to be in charge of the fact that humans intuitively use several different methods to conceptualize space. For example, when arguing spatial arrangements in their immediate neighbourhood, people use Euclidean geometry; otherwise when planning a trip by railway or when navigating a car, they employ a network-topology point of view. Furthermore, humans are able to simultaneously apply several different methods, to select the most favourable one for solving a specific task and to switch between them whenever necessary.

Data modelling for standard applications relates to data that represent artifacts and that are dominated by business practice, rules, and regulations which define how things should be understood and done. The scope for modelling data is very small compared to geometric applications. Spatial data modelling, spatial database systems, and GIS enable the user to compare the data model and the facilities of the real system with (the user's concepts of) reality. Unfortunately, the user's concepts of reality and the facilities of the real system do not always agree.

All methods which serve to organize and structure our perception of space describe spatial concepts [EH91, Fr90, MF89] and refer to all levels of the three-level model in Section 2.1.1. A *spatial concept* is either an informal or a formal description of human's understanding of space, objects within space, and relationships between these objects. It forms an abstraction of reality in the sense that humans have to abstract from reality and to concentrate on the relevant aspects in order to cope with the complexity of a real geometric situation. These abstractions depend upon the respective situation and can thus vary a lot, so that we can imagine a multitude of abstractions of reality. Traditionally, two major groups of spatial concepts are distinguished: (1) *entity-oriented* concepts (in [He91] called *feature-based*), where space is determined by the objects of interest filling space and (2) *space-oriented* concepts (in [He91] called *position-based*), where each point in space is associated with some properties or attributes. This distinction has both theoretical and practical relevance, since it leads to different categories of data models, data types, operations, data structures, algorithms, and systems.

Because of the importance of spatial concepts for the design of spatial data types a few essential examples of entity-oriented and space-oriented concepts are informally given with some explanations. A more comprehensive but still incomplete set of spatial concepts is given in [Fr90].

Examples of entity-oriented spatial concepts are

- *Boundary representation*
  Space is regarded as being composed of an infinite collection of dimensionless points forming a continuum. Each point is characterized and identified by its coordinate values which in two-dimensional space correspond to the elements of $\mathbf{R} \times \mathbf{R}$. Infinite point sets over this space can be finitely specified as entities by boundary representations through different basic types which are usually points, lines, and polygons.

- *Euclidean geometry*
  Euclidean geometry is very popular for solving geometric problems. The basic objects are points and infinite lines, and the operations on them are defined axiomatically. One of its most important functions is a metric describing the Pythagorean distance between two points in Euclidean space. Euclidean geometry is based on a continuous space being composed of an infinite set of points. Its mathematics depends on real numbers where between any two numbers another one exists. This property is necessary, since in Euclidean geometry between any two points another one can be inserted.

- *Graphs*
  Graphs consist of two sets of objects: vertices (nodes) representing spatial information and edges (arcs) representing connections between two vertices and indicating the adjacency of them. An edge can also be interpreted geometrically and carry spatial information. Many algorithms are known to compute properties of graphs. Graphs have a great practical meaning and are often employed for network problems (e.g., traffic networks, transportation networks).

- *Object-based partitions*
  An object-based partition describes a complete subdivision of space into pair-

wise disjoint spatial objects. Adjacency of objects is a crucial property of partitions.

Examples of space-oriented spatial concepts are

- *Point sets*
  Space is regarded as being composed of an infinite collection of dimensionless points forming a continuum. Each point can be characterized and identified by its coordinate values which in two-dimensional space correspond to the elements of $\mathbf{R} \times \mathbf{R}$. For each point (at least theoretically) a vector of attribute values exists describing its properties.

- *Cell-based partitions*
  A cell-based partition describes a complete subdivision of space into pairwise disjoint small cells. Attached to each cell is information about its contents, e.g., that a cell is part of a polygon. Partitions can be based on attribute values, that is, composed of a set of connected cells with the same value (e.g., the same land use).

- *Layers*
  Spatial data are described as a set of layers, each layer representing a single property distributed over a region. For each point of a layer one can ascertain if the point has the layer feature or not. The same point of the space can have a meaning on different layers.

Although frequently employed in everyday life, most spatial concepts are only informally defined. Abler [Ab87] emphasizes the absence and necessity of mathematical formalisms of spatial concepts. An essential requirement is that these formalisms have to correspond to the spatial concepts humans have in mind. Otherwise, they will be of no use for geographers and GIS users.

The examples show that the few mathematically based spatial concepts rest on certain assumptions, such as infinite numbering systems, infinite point sets, or some other form of continuum. Because of the finiteness of computers these assumptions prevent spatial concepts from being directly implemented. An appropriate representation of spatial concepts must take the finiteness of computers into account and be based on finite descriptions [EH91].

2.1.5 Spatial Data Models and Spatial Data Structures

Since spatial concepts are usually not directly implementable, abstractions from reality must be employed which include discretization of spatial reality as the central concept and which fulfil the requirement of being implementable on a computer system. *Spatial data models* form these abstractions. They provide formalizations of spatial concepts and at an abstract level define those objects needed for handling geometric data as well as the operations being applicable to them and describing their behaviour. This corresponds to the definition of an algebra. Although fulfilling the requirement of being implementable, spatial data models do not describe implementation details, so that the geometry modelled in a spatial data model is independent of implementation aspects. The formalization of spatial concepts requires well-understood mathematical methods to describe the structure of spatial data and the semantics of operations that are performed on these data.

At the levels of structure and object modelling, the advanced data models currently used can be classified into *extended relational models*, *N1NF models*, *complex object models*, and *object-oriented models*. It is beyond the scope of this survey to delineate classifications for spatial data models at these levels. But considering existing data models shows that a "semantic" gap exists between spatial concepts and their transformations into spatial data models. An example is the lack of modelling concepts for graphs (networks) in spatial databases, although recently there has been some work on this topic [EG94, Er94, Gü91, Gü94a]. Reasons for the semantic gap are again the lack of formal definitions for describing spatial concepts and the difficulties of suitably mapping a spatial concept to a spatial data model, a task which is complicated and has not yet been satisfactorily achieved in existing systems.

Current GIS and spatial database systems usually support only one single data model. But since different spatial concepts exist and are simultaneously employed by the user, such systems should be able to integrate several concepts and provide the corresponding spatial data models. The integration aspect is very important, because besides the co-existence of several data models with distinct features it gives a chance to realize interconnections between data of different models and the facility to switch between different data models in order to solve a task. A theoretical approach for such a system has been suggested by using *category theory* for establishing a unifying theory integrating different spatial concepts. Category theory studies structure-preserving morphisms between algebraic structures [HEF90, He89, He91].

Realizing a spatial concept as implementation on a computer is carried out by designing *spatial data structures*. These are based on the formal definitions of the corresponding spatial data model and together with appropriate algorithms provide the details how the desired behaviour of the operations specified in the data model can be achieved. A spatial data model defines semantics, which includes the description of spatial data types, operations, and constraints, and does not take implementation details into account. A spatial data structure, however, contains all details necessary for implementation and allows for performance aspects like storage utilization and response time. Of course, a spatial data structure describes only one of many possible implementations. The goal of spatial data structures is to support the design of algorithms for spatial operations by taking into consideration the principles of robustness, correctness, and efficiency.

On the analogy of the classification of spatial concepts, spatial data structures can be grouped into two well-known categories. Data structures supporting the entity-oriented concept are usually called *vector-oriented*, while data structures supporting the space-oriented concept are usually called *raster-oriented* [Bu86, EH91, Fr90].

2.2 Spatial Data Types and Spatial Operations

2.2.1 Conceptual Views of Spatial Objects

Spatial data types have been recognized as suitable abstractions of spatial data and as fundamental for modelling geometry, efficiently representing spatial objects in database systems, and augmenting the expressive power of spatial query languages. They are to reflect humans' perception of objects in space (i.e., their spatial concepts) in an adequate way and are normally used to describe a spatial attribute of a class of spatially-referenced objects. Their definition has a great influence on the design of a spatial data model and on the power of a spatial database system. The development of a user's model of spatial data in the form of spatial data types implies the definition of complex objects representing the geometric structure of spatial phenomena and the definition of corresponding operations.

This section deals with the user's and the designer's conceptual model of spatial data at the level of spatial modelling and describes the basic approaches for the design of spatial data types that are proposed in the literature. Closely connected to the design of spatial data types are the question of *(in)visibility* of the inner structure of spatial objects and the question of the consequences of this design decision with regard to spatial query languages.

When working with a concrete spatial database system, the user has to cope with the questions which conceptual view of spatial objects is presented to him, how spatial objects are structured, how they are distinguished, which operations can be applied to them, and many more. The acceptance of spatial data types depends on their correspondence with the spatial concepts the user has in mind or is able to comprehend. The conceptual view of spatial objects is seldom presented formally at the user's level, but rather informally, colloquially, and based on intuition. One cannot expect from a user that he deals with formal descriptions which are mostly formulated mathematically, unless it is known that he has a corresponding educational background. Hence, formal specifications should restrict themselves to the designer's level and the implementation level. At the user's level informal but precise explanations should suffice.

Designers of spatial data models, developers of spatial database systems, and geographers have made many proposals for the modelling of spatial data. Not all of these proposals amount to the introduction of spatial data types. From the user's point of view they can be classified into the following categories:

- Modelling spatial data as relations
- Universal spatial data types
- Spatial data types modelling spatial objects as point sets
- Spatial data types modelling spatial objects by using half planes
- Data types associated with spatial properties
- Structure-oriented spatial data types

In the following, this classification will be explained in more detail, and each category will be illustrated by some examples. Note that a few approaches can be assigned to different categories. Formal definitions of spatial data types will be given in Section 2.3 and Chapter 3.

Modelling spatial data as relations

The success and efficiency of relational database technology for standard applications, which is rooted in its simple data model, its high-level interactive query languages, and its well-understood underlying theory, has led to many proposals (surveyed in [CF81]) to transfer this technology directly to geometric applications and to explicitly model the structure of spatial data as relations. The consequence is that the user conceives spatial data in tabular form, just the same as standard data, and that a spatial object is represented by several tuples. Following this approach, both the user's and the designer's model coincide, since both the user and the designer have the same conceptual view of spatial data and (have to) think of them in tables.

One of the earliest attempts to employ relational database technology for geometric applications is proposed in [BS77]. Here, a geometric front-end called GEO-QUEL is implemented on top of the relational database system INGRES. This approach stands as a typical representative for a whole class of spatial database systems implementing a geometric component on top of a relational system. All geometric data in a database are treated as relations. A tuple represents either a point or a line segment. A relation is given by the scheme *RelName* (X_1, Y_1, X_2, Y_2, *PLZTYPE*, *<other information>*). X_1, Y_1, X_2 and Y_2 are the coordinates of a point or a line segment. *PLZTYPE* indicates whether the tuple describes a point, a single line, a line segment of a line or a line segment of a polygon. The other attributes are for displaying purposes. A concept of spatial objects is not visible at the user's level, since all lines and polygons are decomposed into a sequence of line segments (tuples) scattered over a relation. That is, a spatial object is not treated as an entity but only *corresponds* to several tuples. Modelling spatial data as relations does therefore not imply the existence of spatial data types.

Another example pursuing a similar strategy is proposed in [CF80, CF81] with the query language QPE (Query-by-Pictorial-Example). An improvement of this approach is the introduction of some spatial modelling concepts in the form of points, lines, and regions, and the provision of a small collection of geometric operations. But geometric objects are still represented in relations and thus decomposed into a set of line segments as the basic units so that an operator's operand of a region type must be a relation. Within the system GADS (Geo-data Analysis and Display System) [MC80] the *map* has been recognized as a fundamental geometric type. A map represents a partition of the plane, that is, a subdivision of the plane into disjoint regions, and is also modelled in relational form. Other approaches have exploited the more subtle features of standard relational database technology. Examples are the system GEO [Ap85], the works of van Roessel [Ro84, Ro87], and the system GEOVIEW [WH87]. But they all have their origin in modelling spatial data as relations.

The drawbacks of this modelling approach predominate. The designer of a database has to be skilful to represent a spatial data model by means of a suitable set of relation schemes and to keep this representation consistent. At a fundamental level the relational data model lacks the facilities which are needed to model the characteristic properties of spatial objects, because it enforces the user to model complex spatial objects in flat, independent relations. Since the representation of spatial data occurs on a very low level and is exclusively based on standard domains like integers, strings, and reals (while the user has originally intended to deal with points, lines, or polygons), an adequate treatment of spatial data is impeded. Apart from that, tables commonly do not correspond to spatial concepts the user has in mind.

Although the facilities of the query language of a relational DBMS are available, they are only of limited use. Since such a language is based on standard domains and has no concept of spatial data types, it cannot provide any meaningful geometric operations. The query language QUEL only allows to select the tuples belonging to a geometric object and to display them on a map. The support of geometric operations by the special query language extension GEO-QUEL is very poor. For instance, retrievals that are based on spatial relationships are impossible. The only chance is to precompute such relationships and to store them explicitly. This all leads to the insight that a high-level view of spatial data is indispensable.

A general and certainly correct estimation in the literature with respect to the visibility of the inner structure of spatial data is that the advantages of a high-level query language are compensated if the user has to deal with low-level concepts like internal codings, such as the representation of spatial objects in the form of (x, y)-coordinates. The modelling approach presented here requires that the user has an understanding of implementation details concerning spatial data which should be better hidden at the user's level. Without going into detail, it is obvious that our design criteria are not fulfilled.

The numerous deficiencies of the approach of modelling spatial data as relations have led to the assessment that this approach is unsuitable to manage spatial data in a conceptually clean and efficient manner and that a high-level view of spatial objects is essential.

Universal spatial data types

In [Eg89a, Eg91a, Eg94] Egenhofer describes the design of a spatial query language called Spatial SQL for a relational database system. Analysing the requirements and establishing the design criteria for a spatial query language, he comes to the conclusion that without a high-level treatment of spatial objects the efficiency of such a query language cannot be guaranteed. Therefore, he proposes to extend the relational data model by an abstract data type *spatial* with corresponding operations. This type has to be seen at the same level as a type for integers, floating point numbers, or character strings. Thus, a value of type *spatial* is just the same a value as a value of type integer. Both values differ, of course, in the complexity of their structure. The internal structure of a value of type *spatial* is hidden at the user's level to make the user independent from internal codings of spatial objects. A value is only accessible by special operations. Since the aspects of dimensionality and shape of spatial objects are not considered, we can regard the data type *spatial* as a *universal*, dimension-independent spatial data type.

Within the data definition language four different kinds of spatial data types are distinguished, namely *spatial_0*, *spatial_1*, *spatial_2*, and *spatial_3* for 0-, 1-, 2-, and 3-dimensional spatial objects. In this way, the data type *spatial* can be regarded as a generalization of the four dimension-dependent types to a dimension-independent domain. The type can be used as an attribute type of a relation scheme, i.e., a spatial object describes an attribute value of a tuple in a relation. The four spatial data types themselves can be regarded as universal, dimension-dependent types. A value of type *spatial_2* can, for instance, be a polygon, a rectangle, a triangle or even a mixed collection of them.

A similar proposal of a universal spatial data type has been made in [GNT91]. In a relational setting a non-standard domain called *Geometry* is defined as an abstract data type with corresponding operations. Each element of *Geometry* represents either a spatial object or a part of it (e.g., a point, a line segment, a line, or a polygon) in the form of an atomic value which can be used as an attribute value of a tuple.

What is represented can only be discerned by the name of the attribute. In [LPV93], also in a relational setting, a universal spatial data type *Geometry* is provided which is defined as a collection of independent geometric values and which can model complex heterogeneous geometries that are composed of points, lines, and polygons possibly with holes.

In [CAR80] a universal, single "geographical data type" is advocated, since in the authors' opinion data types for points, lines, and polygons do not form a completely satisfactory basis for geographical data processing but are too restrictive. Their second argument is that less problems with closure properties of spatial operations arise. They consider a co-existence of the geographical data type and of additional point, line, and polygon types as worth discussing. In such a case, appropriate conversion and projection operators must be provided that map an object of some additional type to an object of the geographical data type, and vice versa.

Another example of a universal spatial data type is the POINT-SET type of the Probe data model which will be discussed from a different point of view in the following category.

From the user's perspective the question has to be posed if universal data types are sufficient to represent the user's perception of spatial concepts, because the aspect of modelling the shape of spatial objects has been neglected but seems to be an important criterion for the user's understanding of space. This is analogous to the question whether a system should provide different types *integer* and *real*, or just a single type *number*. An advantage of a single type may be that closure properties, one of our design criteria, are easier to achieve. On the other hand, a collection of several data types that takes shape and dimension into account is more expressive and allows to define and apply operations that are tailored to the respective type and perhaps not appropriate for other types. Hence, a specialization of universal spatial data types should be possible.

Spatial data types modelling spatial objects as point sets

Another conceptual model for spatial data types at the user's level is based on mathematical abstractions called *point sets*[6]. The user is supplied with the concept that space is composed of infinitely many points and contains spatial objects. Each object consists of an infinite set of points that can be described by finite means. The point set of a spatial object is thus given by the set of points in space occupied by that object. Aspects like dimensionality or shape of an object are not considered.

---

[6] Note that at the designer's level point sets (as we will see in Section 2.3) are a comfortable, simple, and frequently used means to formally define spatial operations, because the fact that point sets are sets enables the use of set operations to define spatial operations.

This concept has been supported by the Probe system [MO86, OM88, Or90], an extensible, object-oriented database system which has been especially designed for geometric applications and which supports a generalization hierarchy. The Probe data model provides a general, dimension- and representation-independent spatial data type called POINT-SET for point sets[7] together with a standard collection of geometric operations. The goal of introducing this special data type is to give the user a concept which is as general as possible, which is applicable to a lot of application areas where different spatial data types are needed, and which forms the basis for more specialized spatial data types. Probe treats the POINT-SET type at the same level as standard types. In contrast to the POINT-SET type, types for lines or polygons are not part of the Probe data model but are application-specific. They are defined and added by users as specializations (subtypes) of the generic POINT-SET type. The detailed description of the inherent characteristics and behaviour of spatial objects is left to the various specialized subtypes of POINT-SET. For example, a polygon is usually described by a list of points in (counter)clockwise order.

Spatial data types modelling spatial objects by using half planes

Another approach of modelling spatial objects is that of using *half planes* [CG82, SV89] where each half plane is defined by a *half plane segment.* A half plane segment uniquely determines a straight line which passes this segment and which forms the one-sided boundary of a half plane. It is described by its two end points and the information at which side of the segment the half plane is situated. For constructing a polygonal region, an appropriately arranged sequence of union, intersection, and difference operations defined on half plane segments and half planes, respectively, is used. Figure 2-4 shows an example of the results of these operations on two intersecting half plane segments.

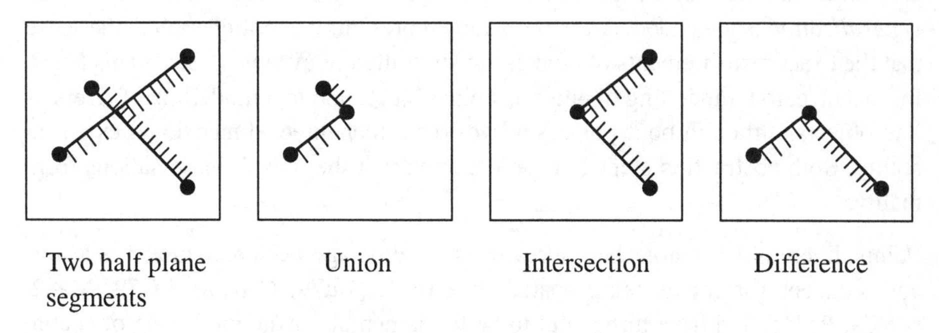

Figure 2-4

[7] Since no statement is made about dimension and shape of objects of the POINT-SET type, this type could also be classed in the category of universal spatial data types.

Data types associated with spatial properties

In [IP87] spatially-referenced objects are modelled as entities where the user receives a high-level view of spatial objects and is relieved from dealing with their internal representation. The geometric description of a spatially-referenced object is associated with a special spatial attribute called *map*. The peculiarity of this approach is that a number of predefined data types for spatial properties like *area*, *perimeter*, and *length* is implicitly available for such an object. While other approaches normally employ operations (functions) to compute geometric properties, this approach models properties as attributes which are treated in the same way as other attributes and can be used in a spatial query language like other attributes. The properties appear to the user as if they were explicitly stored in the database. But in reality the properties are computed when the attributes are required. According to our classification, these special data types describing geometric properties are no spatial data types.

Structure-oriented spatial data types

From the user's perspective the most popular and fundamental abstractions of spatial objects fall into this category. They organize these data in *points*, *lines*, *polygons*[8], and similarly structured entities. Thus, this approach considers the structural shape and spatial extent of spatial objects, i.e., their geometry. Statements about the dimension of spatial objects are made implicitly, since it is known to the user that points have a null-dimensional, lines a one-dimensional, and polygons a two-dimensional extent. Points, lines, and polygons are the results of an abstraction process which maps a geometric structure in three-dimensional space into a suitable geometric structure in two-dimensional space. This mapping process is done by a projection and termed *reduction*. Frequently, a second abstraction step called *generalization* is appended. A generalization represents a simplification in the sense that the exact measurements of objects are simplified or even neglected. This leads, for instance, to a modelling of cities as point objects and to a modelling of rivers as line objects, although both possess a two- (or rather three-)dimensional extent in reality. Both abstraction steps have in common that they result in deviations from reality.

Points, lines, and polygons have already a long time ago been recognized as a suitable concept for representing spatial data (e.g., [Bu79, CAR80, LC79, Ma82, NW79, Pe84]) and have turned out to be fundamental for the modelling of spatial data in *maps* within the framework of geographical information systems [Bu86,

[8] The terminology for the notions *line* and *polygon* differs in the literature. Sometimes for the notion *line* the notions *curve* or *arc* are used. For the notion *polygon* frequently the notions *region* or *area* are applied. We use the notions synonymously.

Da90, Fr91]. Many data models have been proposed which integrate spatial data types for points, lines, and polygons beside traditional types for alphanumerical data. An essential feature of all modelling approaches summarized in this category is a high-level treatment of spatial objects. The structure of spatial objects is *not* modelled by a DBMS data model. At the user's level spatial objects are visible as entities of a spatial data type. In a relational setting, for instance, in addition to the usual standard attributes, spatial data types can be used in the same way as attribute types within a relation scheme (e.g., look at [Ab89, AS90, AS91, BDQV90, Gü88a, Gü88b, RFS88, SH91, SV91]).

The design of a spatial data model, the strength of a spatial database system, and the expressiveness of spatial query languages are to a high degree dependent on the definition of the underlying spatial type system. It is therefore striking that in the literature about spatial data models the description of the structure of spatial data types and the description of the semantics of spatial operations is either neglected or carried out in a non-uniform and informal manner. There is no standardization of spatial data types, and several descriptions are given for spatial data types bearing the same name. A brief example concerning polygons shows how the definition of a spatial data type can influence the modelling power of a data model and that a spatial data type without an understandable and told semantics is unsatisfactory for the user. A geographical situation where a country A lies completely within a country B cannot be modelled by a data model which has no knowledge of polygons with holes (here necessary to model country B). Problems also arise if the task is to model islands belonging to a country. The data model should then contain the facility to model a set of disjoint polygons as a *single* spatial object.

Spatial data types can be characterized by two different criteria, on the one hand by the complexity of the internal structure of a spatial object and on the other hand by the fact whether a spatial object consists of one or more components. Concerning the structure of spatial objects, simple (primitive) and extended structures can be distinguished. Most approaches model spatial objects as simple objects, for instance [Ab89, BDQV90, Bu79, CAR80, Eg89b, Gü88a, Gü88b, Oo90, OSM89, RFS88, RNLE85, SH91]. In the sequel, these approaches are briefly summarized. Let us assume that a type for a point object is called POINT, a type for a line object LINE, and a type for a polygon object POLYGON.

Then the POINT type is the simplest type. Each object represents a *point* in the (Euclidean) plane and is described by a pair (x, y) of coordinates in the form of real numbers. A straight connection between two distinct points is called a *line segment*. The two distinct points themselves are the *end points* of the line segment. A *chain of line segments* is a finite, connected sequence of line segments such that any two adjacent segments share an end point, and no end point belongs to more than two segments. A chain is called *simple* if no other point than an end point is shared by

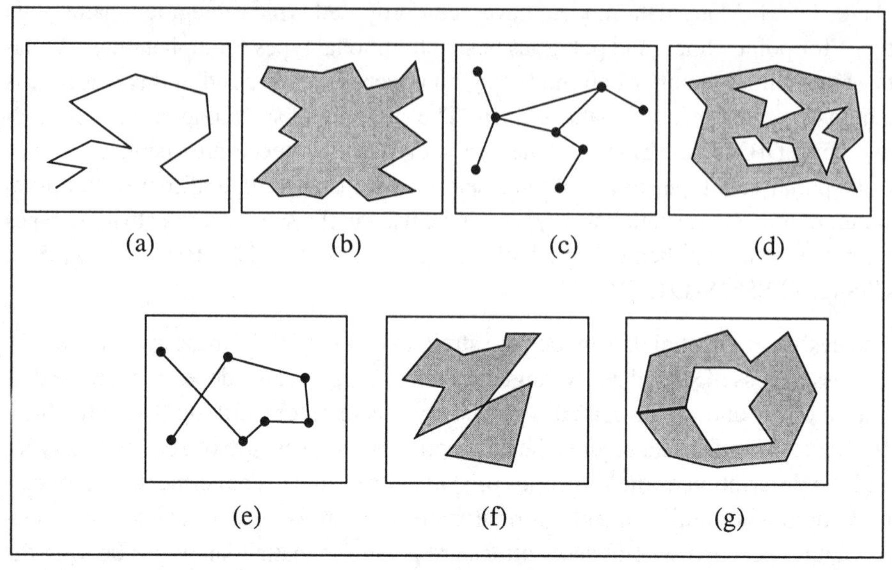

Figure 2-5

any two segments. An object of type LINE is defined as a non-closed, simple chain of line segments (Figure 2-5(a)). Since each inner point of a line segment only belongs to exactly that line segment, self-intersecting line objects are not allowed (Figure 2-5(e)). The reason for excluding self-intersecting lines is that no objects in space with a linear structure exist that can be adequately modelled by such structures.

A simple chain is *closed* and then called *simple polygon* (Figure 2-5(b)) if the first and the last segment of the chain share an end point. This kind of representation is also described as *boundary representation* [KW87]. An object of type POLYGON is defined as a simple polygon. A simple polygon divides the plane into two disjoint regions, the interior and the exterior. The interior, that is, the region enclosed by the closed, simple chain, is considered as belonging to the polygon. Self-intersecting polygon objects are not allowed (Figure 2-5(f)). This is motivated by the objective to use a polygon object to model a connected region enclosed by a boundary line. The situation in Figure 2-5(f) is either interpreted as two different polygons meeting at a common point, or, if a region with a bottleneck is to be modelled, the crosspoint has to be split into two distinct points with a distance greater than zero.

Above we have drawn the reader's attention to the fact that the simple structures limit the modelling power of a data model, because certain situations of reality cannot be modelled in a satisfactory manner. Reasons for modelling spatial objects in such a way lie in the simplicity of the resulting data model and especially in implementation reasons. Representations for storing simple structures and spatial index structures are frequently quite easier to design than those for extended structures.

Many well-known and efficient algorithms from Computational Geometry [PS85] have been developed for geometric operations such as algorithms for the point-in-polygon test or the calculation of the intersection of two line objects. These algorithms are based on simple geometric structures. The treatment of spatial objects with more complex structures like polygons with holes is of course feasible but leads to more complex algorithms.

Data models containing extended spatial data types with a more complex semantics are rare. In [SV91] a line type is proposed which is based on a graph structure. Each line object is a spatially embedded, undirected, connected, and planar graph (Figure 2-5(c)). The nodes of a graph are interpreted as the end points of a line segment, and the edges between nodes as line segments between end points. Thus, for instance, the Nile delta in Egypt can be modelled as a single object in the form of a graph.

For polygons we have the special case of *polygons with holes* [BDQV90, CAR80, ECF94, Gü91, KBS91a, KBS91b, KVW89, Oo90, OSM89, SV91, WB93]. A polygon with holes is a simple polygon where one or more simple polygons are cut out (Figure 2-5(d)). Another modelling approach especially known from geography represents a polygon with holes as a single loop by introducing an additional segment for each hole to join a point on the outer boundary to a point on a hole (Figure 2-5(g)). The polygon resulting from inserting such a "bridge" for each hole is no longer simple. But many graphical and geometric algorithms handle this type of polygon successfully so that this approach has more technical than conceptual reasons. Polygons with holes are necessary, since first they increase the modelling power of a spatial data model, and second they are needed to ensure the validity of closure properties of spatial operations. The approaches of [KVW89, WB93] additionally allow the modelling of a hierarchy of polygons with holes, that is, polygons with holes which contain islands with holes to any finite level.

The second criterion relates to the aspect whether a spatial object is composed of one or more components. Only a few approaches model spatial objects solely as one-component objects [BDQV90, Gü88a, Gü88b], since this leads to simpler data models and simpler geometric algorithms. But one-component spatial data types are not closed under the geometric operations union, intersection, and difference so that these and derived operations cannot be cleanly formulated within such data models. Some approaches support both one-component and many-component forms for lines and polygons [Ab89, SH91]. Purely many-component spatial data types are given in [CZ94, ELNR87, Gü91, Ne88, OSM89, Oo90, RNLE85, SV91]. They all impose the constraint on a spatial object, which consists of a set of points, lines, or simple polygons, that the interiors of all components of such a set have to be disjoint. For a spatial object representing a set of simple polygons with holes, this means, for instance, that any two components may not intersect but they may touch and have common line segments (Figure 2-6).

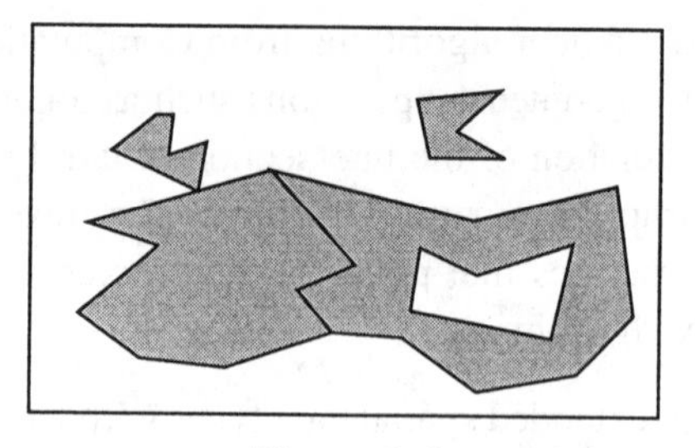

Figure 2-6

So far, we have described the structure and properties of *single* spatial objects. But within the framework of structure modelling (see Section 2.1.1) it is interesting to consider constraints and properties of a *set of spatially-referenced objects* like the logical relationships of objects of such a set. In geographical applications frequently the problem arises to model *partitions*, that is, to model subdivisions of the plane into disjoint polygonal regions, possibly with gaps not covered by regions. Thus, in this case the logical relationship of the concerned objects is a topological one. For instance, the area of a continent may be divided into countries, or a land area may be divided into regions classified with regard to agricultural use (Figure 2-7).

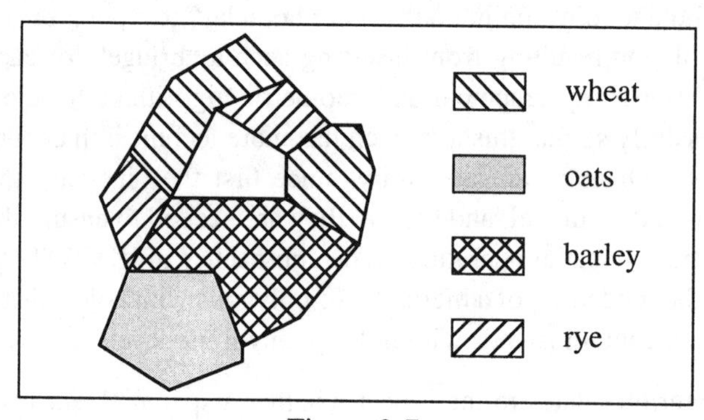

Figure 2-7

In [Gü88a, Gü88b] the approach to solve this problem is based on the attempt to guarantee the disjointedness constraint at the user's level by introducing a spatial data type *area*. Within the framework of an extended relational data model called the geo-relational algebra, the set of polygons occurring in a relation as a column of an attribute of type *area* has to fulfil the constraint that all polygons are disjoint. More precisely, it is required that for any two polygons of type *area* their interiors are disjoint; they may share a boundary segment or a point of a boundary segment. Furthermore, it is not required that the plane is completely covered by regions. Unfortunately, the user is responsible for maintaining this disjointedness constraint; the maintenance of this property is not supported by the data model. In the literature, no other spatial data types or mechanisms have been proposed which express the same or different special properties of a set of spatially-referenced objects.

Points, lines, and polygons are the most important but, of course, not the only data types proposed for geometric applications, since various applications can require different spatial data types. Abel [Ab89], for instance, suggests a type for rectangles which are a special case of simple polygons. Here, the aspect of *extensibility*, one of our design criteria, plays a crucial role. A characteristic feature of most spatial data models offering structure-oriented spatial data types is their extensibility for new data types and operations (see Section 2.2.3) so that the introduction of a new type is (at least) theoretically no problem. Without the facility of an extensible type system, structure-oriented spatial data types would be too restrictive. But experience shows that points, lines, and polygons turn out to be an adequate collection of spatial data types for many spatial applications. It seems the greatest modelling power is achieved if spatial objects of these types are designed as extended structures so that lines as graphs and polygons with holes are possible and if they can be simultaneously modelled as many-component objects.

While points, lines, and polygons are vector-oriented, another category deals with raster-oriented data types like *raster* [ELNR87, RNLE85, SH91] or *image* [Ab89]. These data types do not support the concept of identifiable objects in space and are not considered within this survey.

2.2.2 Spatial Operations

Spatial objects as occurrences of spatial data types are manipulated by spatial operations. A spatial operation is defined as a function with spatial arguments, i.e., it takes spatial objects as operands and returns either spatial objects or scalar values as results. The importance of spatial operations for spatial query languages has been recognized by many authors (e.g., [Fr82, Eg89a, Gü88a, Gü91]). When speaking about spatial operations, we consider them in two-dimensional space and assume a high-level treatment of spatial objects. Many proposals of spatial operations have been made which can be classified into the following categories:

- Spatial predicates returning boolean values
- Spatial operations returning constituents of a spatial object
- Spatial operations returning numbers
- Spatial operations returning spatial objects
- Spatial operations on sets of spatially-referenced objects
- Spatial selection and spatial join

In the following, this classification will be explained in more detail. The most important operations proposed in the literature will illustrate each category. An embedding of these operations into spatial query languages is not shown. Categorizing and explaining spatial operations is, of course, to a large extent dependent on the definition and structure of the corresponding spatial data types. For reasons of simplicity, for our examples we assume the existence of the three very general data types *Points*, *Lines*, and *Regions* defined as a set of points, as a set of line segments, and as a set of simple polygons possibly with holes, respectively.

Many spatial operations are *polymorphic* (or more precisely: *overloaded*) which means that various combinations of data types are permissible as their arguments. The operation *union* (fourth category), for instance, forms the geometric union either of two *Points* objects, two *Lines* objects, or two *Regions* objects.

Spatial predicates returning boolean values

A *spatial relationship* is a relationship between two or more spatial objects. The first category includes *spatial predicates* which compare two spatial objects with respect to some spatial relationship. They conform with traditional binary relationships and return a boolean value. A lot of proposals for spatial predicates have been made, e.g., in [AS90, AS91, CAR80, CG82, CZ94, Eg89a, Eg89b, ELNR87, Fr75, Gü88a, Hu93, MF89, Oo90, PC87, Pe86, Pe88, Pu88, RFS88, RNLE85]. They can be classified into the following subcategories:

- Topological relationships
- Metric relationships
- Spatial order and strict order relationships
- Relationships expressing the motion of spatial objects
- Fuzzy relationships
- Directional relationships

Topological relationships belong to the most formally investigated spatial relationships (e.g., [CCR93, Eg89a, Eg89b, Gü88a]). They use topological properties for their description and include concepts like continuity, adjacency, overlapping, interior, boundary, connectivity, and inclusion. There is no dependence on a distance function, since topological relationships are not distance-preserving. An essential property is that they are preserved under topological transformations such as translation, rotation, and scaling. Examples of topological relationships (some of them are shown in Figure 2-8) are if two spatial objects are *equal*, *unequal*, *disjoint*, *adjacent* (*neighbouring*), if they *intersect* (*overlap*, *cut*), *meet* (*touch*), or if one object is *inside* (*in*), *outside*, *covered_by* or *contains* the other. The necessity of a formal definition of all these predicates for a precise understanding of their semantics should be obvious.

Metric relationships [AS90, AS91] use measurements such as distances and directions. The *in_circle* and *in_window* predicates test if a spatial object lies within the scope of a predefined circle or rectangle, respectively. *Spatial order and strict order relationships* [Fr75] are based on the definition of order and strict order. As a rule,

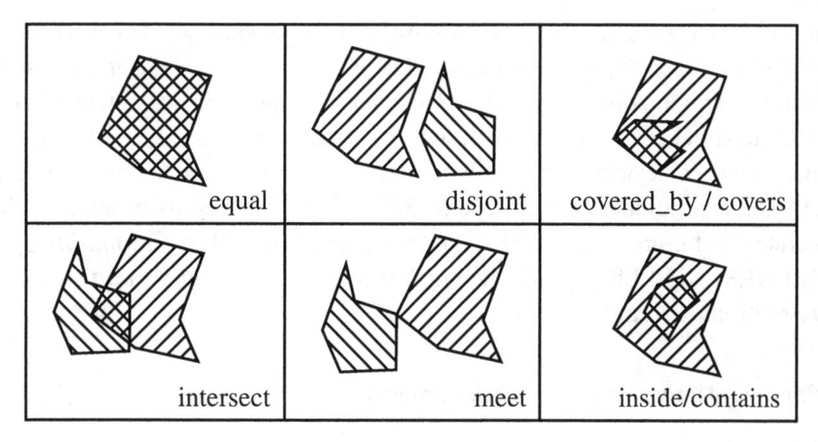

Figure 2-8

each order relation has an inverse relationship. For instance, *behind* is a spatial order relationship based on the order of preference with the inverse relationship *in_front_of*. Other examples are the pairs *above/below* and *over/under*. Also the pair *inside/contains* can be considered as a spatial order relationship. Relationships expressing the *motion* of spatial objects, such as *through* and *into*, combine spatial and temporal aspects. Another subcategory are *fuzzy relationships* [MF89] like *next_to*, *close*, *near*, or *far* which because of their vagueness are very difficult to define. *Directional relationships* [PC87, Pe86, Pe88, PS93, RFS88] like *north/south*, *left/right* compare the relative position between two spatial objects. The problem which makes a definition so difficult is that directional relationships are influenced by the relative size, distance, and the shapes of the two objects, so that these relationships can also be considered as fuzzy concepts.

Spatial operations returning constituents of a spatial object

This category contains unary spatial operations which extract constituents of a spatial object that are conceptually known to the user but not visible at the user's level and only accessible by operations. In [Hu93, SH91] *xc* and *yc* yield the *x*- and *y*-coordinates of a single point as real numbers; *sp* and *ep* compute the start point and the end point of a line modelled as a chain. The operations *nodes* and *segments* produce a set of consecutive points and segments, respectively, representing a line.

Spatial operations returning numbers

This category contains spatial operations which compute a metric property of a spatial object and return a number. The *area* operation returns the area and the *perimeter* operation the perimeter of a *Regions* object. The operation *length* calculates the total length of a *Lines* object. The *diameter* of a spatial object [Gü88a] is defined as the largest distance between any of its components (set elements). Güting [Gü88a] offers three distance operations. The *dist* operation calculates the distance between two single points. The *mindist* and *maxdist* operations calculate the minimal and maximal distances between any two spatial objects. Egenhofer [Eg89a] provides a *direction* operation which computes the angle between any two spatial objects as a real number in the range 0...359°59'59'. The results of all operations enumerated so far are real numbers. The *components* [Bu79] or *cardinality* operation [ELNR87, RNLE85] yields the number of components of a *Points*, *Lines*, or *Regions* object as an integer number.

Spatial operations returning spatial objects

This category consists of operations returning atomic spatial objects as results. Two subcategories can be distinguished:

- Object construction operations which construct new objects from two or more existing objects
- Object transformation operations which transform one or more spatial objects into a new spatial object

The *object construction operations* include geometric *union*, *intersection*, and *difference* operations [Bu79, CG82, CZ94, ELNR87, HSH92, Hu93, RNLE85, SH91] which are applied to two *Points*, *Lines*, or *Regions* objects. Unfortunately, most proposals of such operations do not fulfil closure properties. The intersection of two *Lines* objects can be interpreted in two ways [SV91], either as a set of intersection points and thus as a *Points* object, or as a set of common line segments and thus as a *Lines* object. Usually, the first interpretation is the standard one, and the second interpretation is expressed by another operation. For the intersection of two spatial objects of different types in [Hu93, SH91] a *cut* operation and in [SV91] an *inside* operation are proposed.

In [Gü88a] a *convex_hull* operation is provided which constructs the convex hull of a point set, defined as the smallest convex polygon enclosing all points (Figure 2-9).

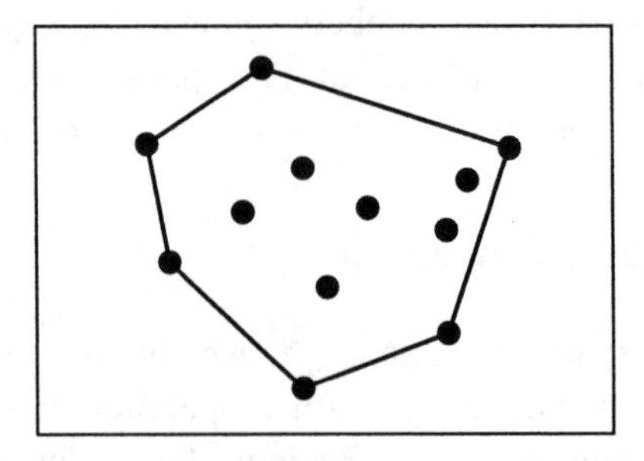

Figure 2-9

The *center* operation [ELNR87, Gü88a, RNLE85] determines the center of a spatial object (Figure 2-10).

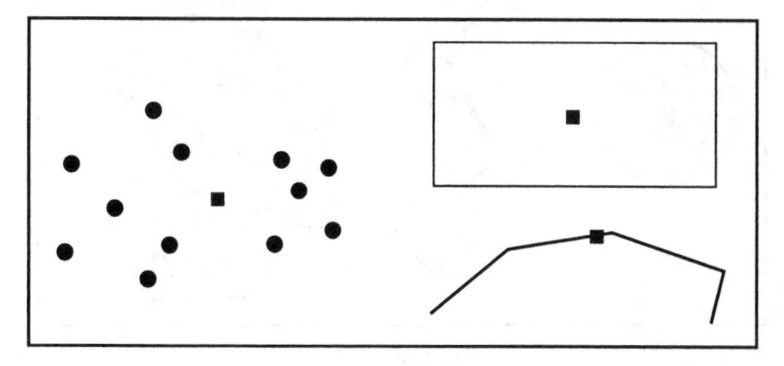

Figure 2-10

Several other operations belonging to the first subcategory have been suggested. The *boundary* operation [Bu79, CG82, Hu93] or *border* operation [ELNR87, RNLE85, SV91] produces the boundary line of a *Regions* object as a *Lines* object, or the end points of a *Lines* object as a *Points* object. The *polygon* operation [Hu93] converts a *Lines* object which is modelled as a chain into a *Regions* object which is a simple polygon. The *box* operation [CG82] or *surrounding* operation [ELNR87, RNLE85] computes the minimal rectangle which bounds a spatial object and is axis-parallel to the selected reference frame. The *components* operation [RNLE85] determines the vertices of a *Lines* object.

The *choose* operation [Bu79] produces an "arbitrary" point on or inside each component of its operand of type *Lines* or *Regions*. This is useful when such an object is to be approximated by a point as a representative. The *interior* operation [Bu79] gets a *Points* object and a *Lines* object as operands and produces a set of smallest regions where each region surrounds a point of the *Points* object and has a boundary that is part of the *Lines* object. That is, if R is a *Regions* object, then the term *interior*(*choose*(R), *boundary*(R)) is equal to R. An application is, for example, if one wishes to define a partition as a set of disjoint *Regions* objects and the set of all boundary lines modelled as a single *Lines* object is given together with a *Points* object. Using this operation, individual and disjoint *Regions* objects may be produced so that common boundaries are consistent.

Object transformation operations as the second subcategory comprise the operations *extend* [Bu79, CG82, CZ94], *rotate* [Hu93, SH91], and *translate* [Hu93, SH91]. The *extend* operation takes a spatial object s and a real number r as operands and creates a polygonal region which is a spatial expansion of s. Parameter r indicates the expansion distance from s (Figure 2-11). If r is negative, a shrinking operation on s is performed. The *extend* operation realizes a well-known operation of GIS and geography called *buffer zoning* [Da90, Va91] or *region expansion* [AS90].

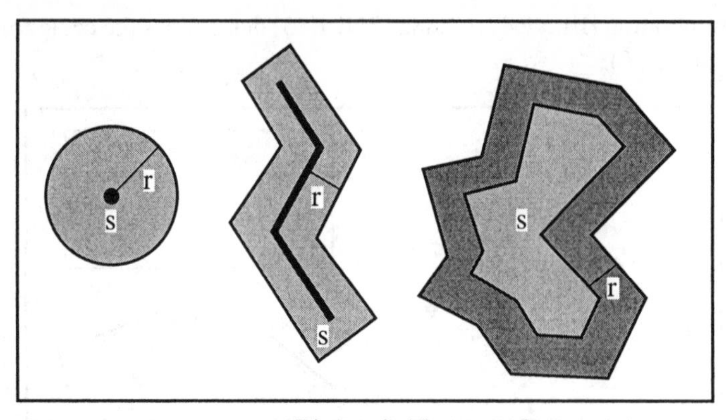

Figure 2-11

The operation *rotate* rotates a spatial object around a point; the operation *translate* moves a spatial object by a defined vector.

Spatial operations on sets of spatially-referenced objects

The operations of this last category support structure modelling. As such they are very complex, since in contrast to the operations of the other categories these operations have to cooperate with a DBMS data model (relational, object-oriented, complex object, etc.) and to manipulate not only spatial objects but also the spatially-referenced objects they are associated with. They take a set of spatially-referenced objects as arguments and produce new sets of spatially-referenced objects as results.

The *overlay* operation [Fr87, Gü88a, HSH92, KBS91a, KBS91b, SV89, Va91], one of the most frequently used spatial operations in a GIS, allows to transparently lay two partitions modelling different topics on top of each other and to combine them into a new partition of disjoint regions. Two different interpretations of the overlay mechanism are given in the literature (Figure 2-12)[9]. In the first interpretation ($overlay_1$) [Gü88a, KBS91a, KBS91b, SV89] the resulting object set contains one spatially-referenced object for each new region obtained as the intersection of a region of the first partition with a region of the second partition. Since the plane need not be completely covered by regions, it is possible that a region of one partition does not intersect any region of the other partition. In this case it will not be part of any new spatially-referenced object. In the second interpretation ($overlay_2$) [HSH92] also those parts of regions are taken into account that do not intersect any region of the other partition.

[9] The dotted boxes in the following figures represent equal reference frames of spatial objects.

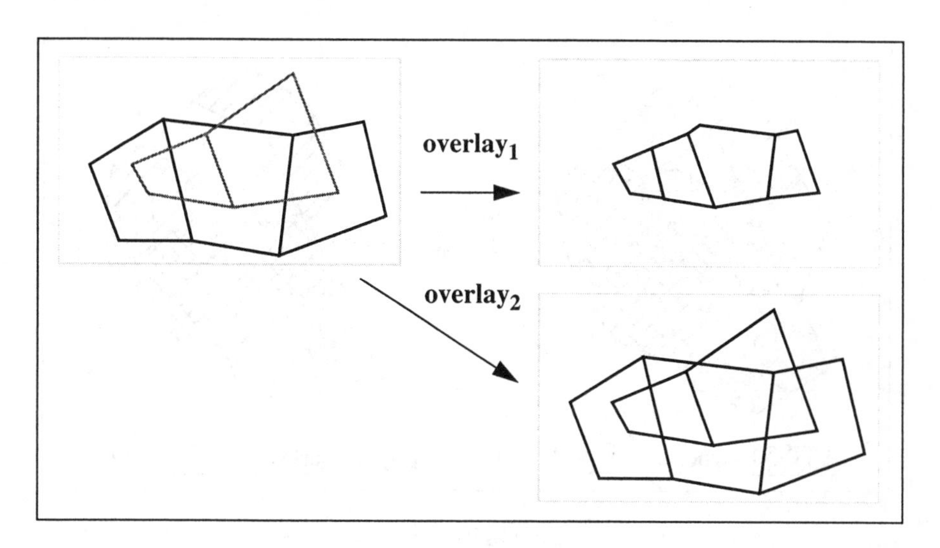

Figure 2-12

The *superimposition* operation [RFS88, SV89] allows to superimpose the spatial objects of a partition on another partition and to cover and erase parts of the other partition (Figure 2-13).

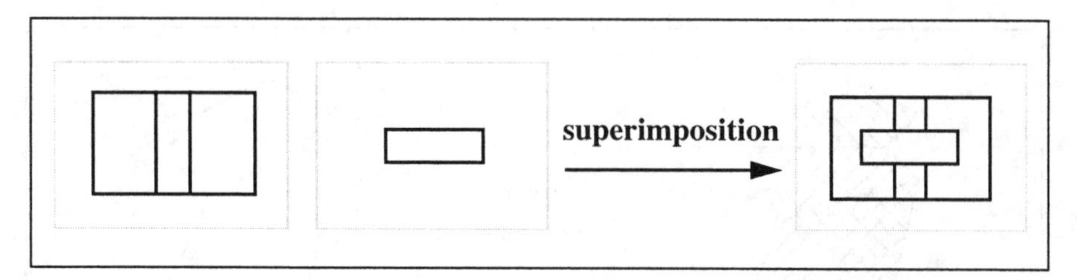

Figure 2-13

The *fusion* operation [CZ94, GNT91, HS93, HSH92, KBS91a, SV89] called *merge* in [KBS91a] and *generalization* in [HSH92] merges the objects of a specified (set of) spatial attribute(s) on the basis of the equality of the objects of another (set of) non-spatial attribute(s). For each group of equal non-spatial objects a (set of) new spatial object(s) is created as the geometric union of a set of spatial objects of the group. Figure 2-14 shows a partition of districts with their land use. The task is to compute the regions with the same land use. Neighbouring districts with the same land use are replaced by a single region, that is, their common boundary line is erased. Each hatched area d_i on the left is part of an object describing a district. On the right after the application of the *fusion* operation all areas belonging to the same group g_i form *a single Regions* object and are hatched in the same way. District boundaries are not distinguished any more. To be feasible, the *fusion* operation requires that *Regions* objects are defined as many-component objects with holes.

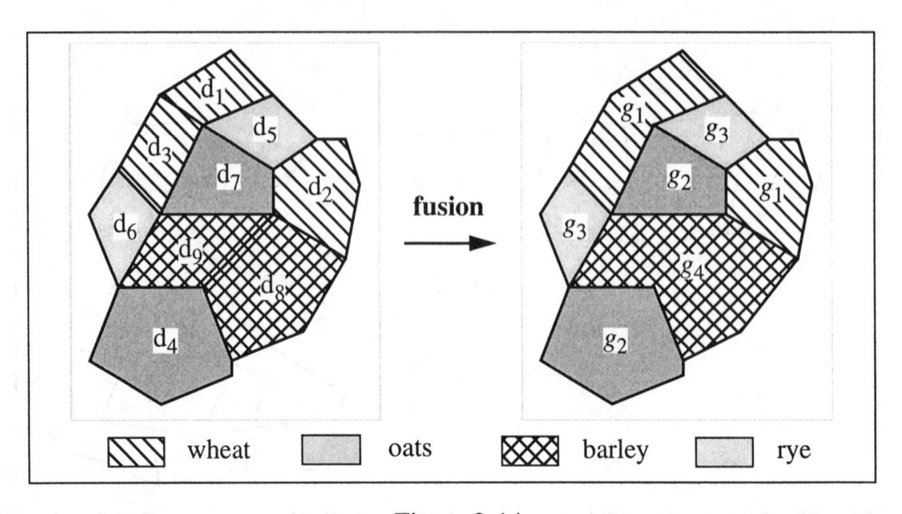

Figure 2-14

The *cover* operation proposed by [SV89] yields a single *Regions* object as the geometric union of all *Regions* objects of a partition (Figure 2-15). Since polygons with holes are not allowed in [SV89], the partition must be complete, i.e., it may not have holes.

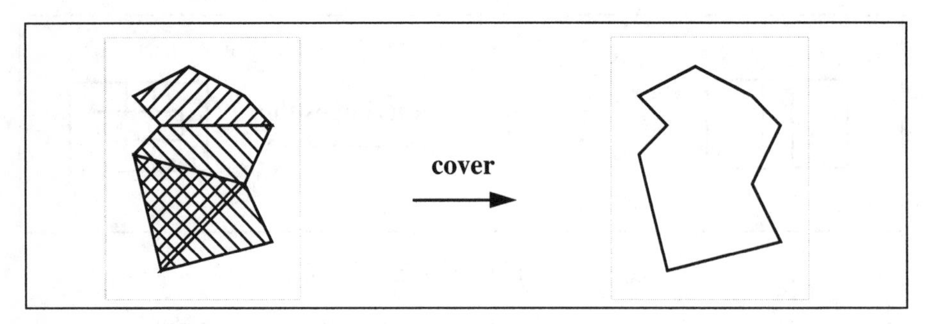

Figure 2-15

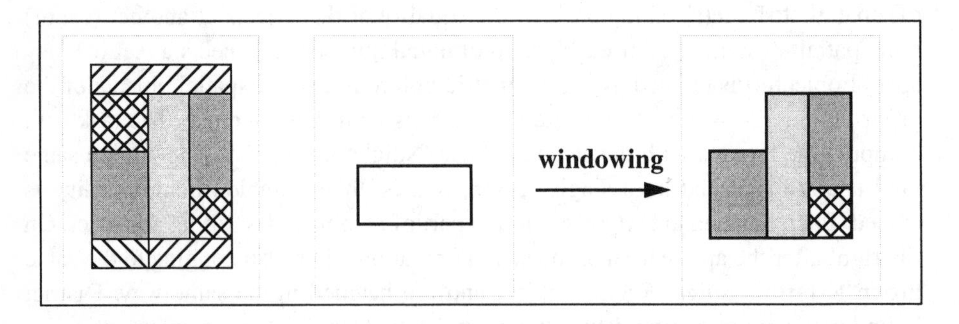

Figure 2-16

The *windowing* operation [SV89] allows to retrieve those regions of a partition whose intersection with a given (rectangular) window is not empty (Figure 2-16). Windowing is also applied in queries where the window is defined as a circle with center p and radius r, for instance, if we ask for all objects whose distance from a given point p is less than r.

The *clipping* operation [RNLE85, SV89] selects those parts of a partition which lie inside a given rectangular window (Figure 2-17).

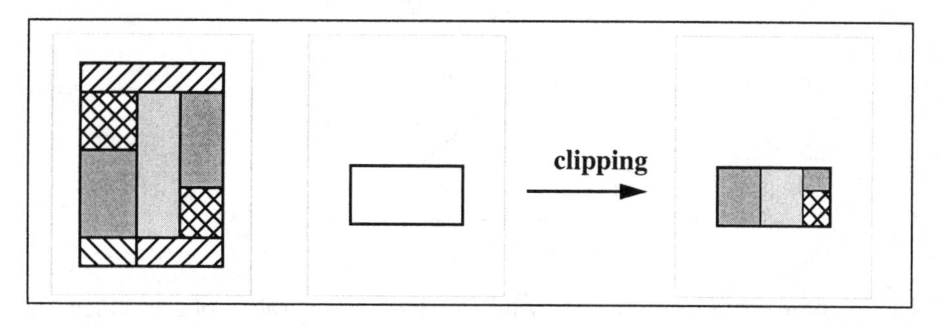

Figure 2-17

The *voronoi* operation [Gü88a, HS93] is based on a well-known structure from computational geometry. For a given set S of points in the plane the *Voronoi* diagram (also called *Thiessen* diagram) associates with each point p from S the region consisting of those points of the plane that are closer to p than to any other point in S (Figure 2-18). Usually, some regions of a Voronoi diagram are infinite, namely those belonging to points on the convex hull of S. But since we assume a rectangular reference frame for our spatial objects, the *voronoi* operation constructs the Voronoi diagram for the points inside this rectangle.

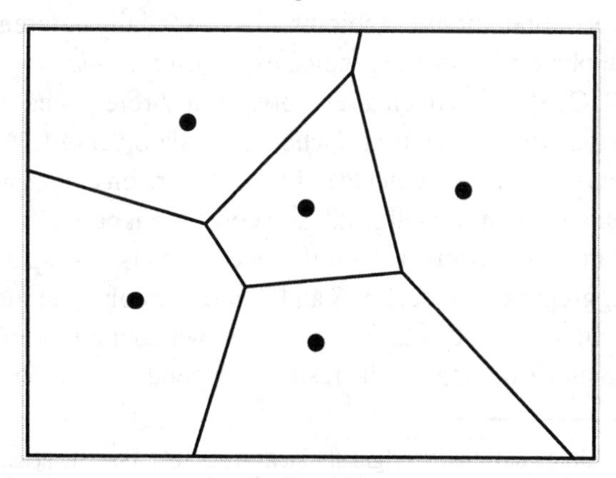

Figure 2-18

The *closest* operation of the geo-relational algebra [Gü88a] yields a (set of) spatially-referenced object(s) whose *Points* object (which is defined as a single point in this algebra) is nearest to a given reference *Points* object (Figure 2-19). As a rule, the resulting object set will consist of exactly one spatially-referenced object.

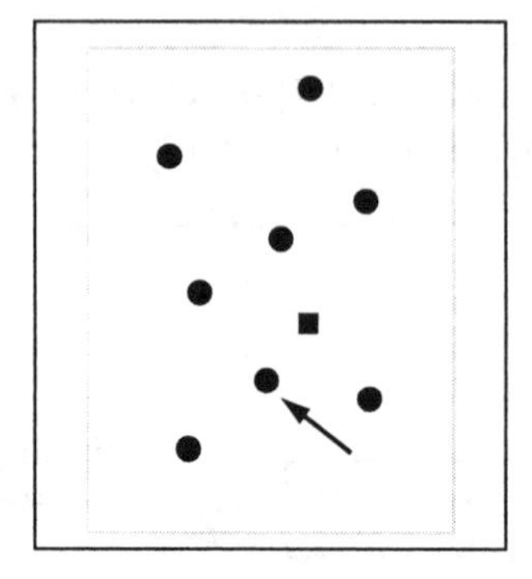

Figure 2-19

The *compose* operation proposed by [GNT91] groups the values of a specified (set of) spatial attribute(s) on the basis of the equality of the values of another (set of) attribute(s). The term "value" is used here, since no spatial objects are meant but segments. Segments can be grouped to sets of segments which then form an object of type *Geometry*. The *decompose* operation disaggregates the values of a specified spatial attribute with respect to a specified (set of) attribute(s), i.e., it multiplies each value according to its number of components (i.e., segments or points) and substitutes it by one of its components.

Spatial selection and spatial join

An important class of operations combining sets of spatially-referenced objects and comparing spatial objects by spatial predicates are *spatial selection* and *spatial join* operations [AS90, Gü88a].[10] Given a set *S* of spatially-referenced objects, a *spatial selection* filters out all those objects of *S* whose spatial object fulfills a selection condition given either by a spatial predicate or by a comparison expression that contains a spatial operation with a numerical result as a constituent part. Given two sets *S* and *T* of spatially-referenced objects, a *spatial join* constructs new spatially-referenced objects that are aggregation objects of *S* and *T* and possibly extended by new computed spatial or non-spatial objects. The decision, whether a pair of spatially-referenced objects of *S* and *T* belongs to the result, is dependent on a boolean expression

[10] The notions "selection" and "join" originally stem from relational database terminology, and indeed, spatial selection and spatial join operations resemble the corresponding relational ones, but in a spatial context. We consider and use these notions regardless of a particular DBMS data model and speak of sets of spatially-referenced objects instead of relations, object classes, etc.

in the join condition which is either a spatial predicate or a comparison expression that contains a spatial operation with a numerical result as a constituent part. Note that within a selection or join condition all spatial predicates (first category) and all spatial operations returning numbers (second category) are allowed.

In the literature, the term "spatial join" is not used in a uniform way. Some authors equate the term with the spatial join intersection operation, i.e., where the spatial predicate in the join condition is restricted to an *intersect* predicate. Other authors do not distinguish between spatial selection and spatial join operations.

2.2.3 Design Criteria for Modelling Spatial Data Types

The last two subsections informally described different proposals for a user's model of spatial data types and spatial operations by intuitively explaining their structure and semantics. This subsection will be devoted to the designer's perspective with respect to this modelling process. It is in particular of interest whether and how the design criteria of Section 1.2 have been taken into account by these proposals. Furthermore, several other aspects are argued like the high-level view of spatial objects, the (in)visibility of the inner structure of spatial objects at the user's level, and the abstract data type mechanism to represent spatial objects.

At the conceptual level the designer is confronted with a multitude of tasks. Three of the most important tasks include

- the design and formal definition of the inner structure and semantics of spatial data types,
- the provision and description of a suitable conceptual view of spatial objects and operations visible at the user's level, and
- the design of cooperation mechanisms for the integration of a spatial type system into a DBMS data model.

Spatial data types and operations are usually embedded in a DBMS data model (see third task). It is beyond the scope of this survey and is subject of object and structure modelling to describe and classify the different approaches for DBMS data models integrating spatial data types. Most DBMS data models emphasize a homogeneous coexistence and treatment of non-spatial and spatial data types at the user's level. This allows the user to have a uniform and integrated view of spatial and non-spatial objects.

What follows is a description whether and how the design criteria are observed by the different proposals in the literature.

Generality and versatility

Most proposals do not take this criterion into account. An exception are the works of [Bu79] and [CAR80] which offer rather a general definition of spatial data types. Both recognize the necessity to treat sets of points, lines, and areas as single values. Moreover, the work of [CAR80] also allows area objects to contain holes, islands within holes, holes within islands within holes and so on. Polygons with holes or even hierarchies of polygons with holes are also available in the data models of [BDQV90, ECF94, KBS91a, KBS91b, KVW89, SV91, WB93]. General regions are defined in [SV89]; an extension to general types for points and lines is presented in [Vo92]. In the literature, line objects (or components of them) are usually modelled as sequences of successive line segments and not as graphs.

Closure properties

The criterion that the domains of spatial data types like *point*, *line*, and *region* should be closed under geometric union, intersection, and difference has been mostly neglected. But this is an important requirement, since the result of such a geometric operation must be a well-defined object, too, and correspond to the definition of spatial data types. For instance, the union of two hole-free simple polygons is, in general, not hole-free. A few approaches (see Section 2.3.4) use regular closed sets and regular set operations which implicitly guarantee closure properties [Ti80, WB93].

Rigorous definition

Several approaches like [Eg89b, FK86, Gü88a, MF89] point to the importance of a formal, clear, and unique definition of spatial data types to clarify their precise meaning and to avoid ambiguities both for the user and the implementor. Abler [Ab87] emphasizes the necessity to develop a coherent, mathematical theory of spatial relationships to overcome shortcomings in almost all geographic applications. Section 2.3 will present a survey of the currently used formal methods for the definition of spatial data types.

Finite resolution, numerical robustness, and topological correctness

In contrast to theory where a number system is commonly infinite (e.g., the algebra of real numbers), number systems in computers are necessarily finite, since a finite sequence of bits can only represent a finite, limited selection of different numbers (e.g., the set of floating point numbers). In Computational Geometry [PS85], which designs geometric algorithms that can be employed for realizing spatial operations, this fact has not been taken into account. The task has been left to the programmer to close the gap between theory and practice which leads inevitably not only to numerical but also to topological errors [EFJ89, EH91, FK86, GY86, KBS91a]. Obviously, a theory for the modelling and representation of spatial data is needed that is compatible with the finiteness of computers. Section 2.4 will present a brief overview of the currently used methods to overcome these problems.

Geometric consistency

Spatial data types to solve the geometric consistency problem have so far not been proposed. There is some weak support in [Gü88a] by introducing an *area* data type (see Section 2.2.1) but it is insufficient.

Efficiency

An efficient execution of spatial operations can be achieved by using the geometric algorithms and methods of Computational Geometry [PS85]. But a central problem is the simultaneous criterion of numerical robustness and topological correctness, because the algorithms of classical computational geometry assume infinite, continuous space and hence conflict with the finite number systems of computers.

Extensibility

The broad spectrum and the diversity of non-standard applications as well as the restrictions of commercial and research database systems led to the development of *extensible database systems*, e.g., DASDBS [PSSWD87], Exodus [CDFGM86], Genesis [BBGST86], Postgres [SR86], PROBE [DMBCG87], and Starburst [SCFLM86]. The concept of extensibility allows the user to adapt the database system to his application needs, for example by integrating new object class representations or index structures, and relates to all levels of the system architecture, from the user interface to storage management. In particular, it includes the facility of extending a spatial data model by new data types and operations.

The question whether a complete, minimal, and application-independent set of data types and operations for spatial applications exists has been doubted by most researchers (e.g., [SH91, SV89, SV91]). Hence, most spatial database systems have been designed as extensible systems (e.g., [AS91, GCKPS89, Gü88b, Gü89, HC91, KGK93, MO86, SH91]). The general strategy is to provide a kernel set of largely application-independent spatial data types which is appropriate to support most geometric applications. For further applications extensibility mechanisms are made available which allow the integration of new spatial data types as well as components of other layers of the system architecture like new spatial index structures. These extensions are usually more application-specific and sometimes restricted with respect to generality.

Data model independence

In [SV89] the observation that the introduction of the complex object model, which is one of the current state-of-the-art DBMS data models, is still insufficient for representing user-defined spatial operations leads to the proposal not to embed the spatial data types into the DBMS data model but to define them regardless of a particular DBMS data model and to associate them to the DBMS data model through simple and general constructs and interfaces.

From a database point of view the cooperation of a spatial type system with a database system can be achieved by (at least) three methods which differ from each other with regard to the degree of openness. Either the DBMS provides mechanisms to integrate predefined and user-defined spatial data types which are then closely connected to an extended database system (e.g., [Gü88a, SV91]), or the DBMS allows to define externally defined types (EDTs) and uses external implementations [SW91, SW93, Wo89, WSSH88], or the DBMS uses an external system as a computation service to perform geometric operations [NSLAB91]. The first alternative is completely under the control of the type system of the DBMS, while the second and third alternatives represent foreign methods. Using the second alternative, the external implementation of a data type is performed in the same system environment as the DBMS either in the form of procedure calls or separate processes. The third alternative abolishes the restriction to the same system environment and allows heterogeneity of computer systems, system software, and programming languages [SW91, SW93].

In summary, no proposal realizes all of the design criteria for modelling spatial data types in one single data model or type system. All of the proposals even do not fulfil most of the design criteria. As a result of the design process the user should receive a suitable conceptual view of spatial objects and operations. A crucial observation is that spatial objects are *complex objects* and should be provided as *high-level objects*, for instance in spatial query languages. The opinion that providing a high-level view of spatial objects is essential has found a large agreement by several authors [Ab89, Eg89b, Eg91b, Eg94, Gü88a, IP87, Oo90, RFS88].

A question which is closely connected to this design principle relates to the *visibility* or *invisibility* of the inner structure of high-level spatial objects at the user's level [Gü88b, Gü89, SH91, SV91]. This design decision has consequences with regard to spatial query languages and application programs. If the inner structure of spatial objects is visible at the user's level, standard operators of the query language can be used to retrieve information about such objects. In the other case, informations about spatial objects are only available through specific operations defined on those objects. This has the advantage that the implementation can be changed without affecting the interfaces of the operations. From the viewpoint of [SV91] it is not clear whether hiding of the internal structure of spatial objects should be totally respected or partially violated. Sometimes it can be favourable to access information about a spatial object directly and not to be forced to call a special operation. The two different ways of handling complex objects have been termed *structural* and *behavioural object orientation*, respectively [Di86].

Most existing spatial database systems support the concept of high-level spatial objects whose structure is invisible and employ some kind of *abstract data type* (ADT) mechanism to represent spatial objects [BDQV90, CW87, Eg89a, GNT91, Gü88a, MO86, Ma82, OH86, RFS88, SRG83, SV91, SR86, St86]. Already Burton [Bu79] has emphasized the necessity of ADTs for points, lines, and polygons together with a set of basic operations. An abstract data type describes a set of objects through a collection of related operations where the type implementation (data structure and operation implementation) is hidden from the user. Thus, the definition of an ADT respects the principle of *encapsulation*, that is, the physical representation of the objects of interest is separated from their logical representation and hidden outwardly. The behaviour of an ADT can only be recognized by observing the results of applying corresponding operations. The advantage for the user is that he is supplied with an appropriate conceptual view of spatial objects and that he does not have to deal with the internal structure of spatial objects. The advantage for the designer is that the internal representation of an ADT is interchangeable and that an exchange does not affect the use of spatial data types, operations and the query language at the user's level, since the interfaces of the operations remain unchanged.

Within the relational setting, the idea of introducing ADTs as new attribute domains into a relational database system has first been mentioned in [SRG83] and later continued in [St86]. An ADT for rectangles ("box") is suggested, and a few operations on rectangles are given. In [RFS88] a spatial query language called PSQL is introduced, and each domain is realized as an ADT. The purpose of ADTs is to relieve the user from dealing with low-level implementation aspects when using spatial operations in queries.

In [BDQV90] two ADT levels are suggested, an outer level which provides user concepts such as points, lines, and polygons and which is solely visible by the user and an inner level which is hidden to the user and deals with internal data. In [MO86] within the Probe data model, user-defined ADTs can be devised. One of the reasons to employ ADTs is that many different ways exist to represent data and that a given representation need not be the best one to support a specific application. Since ADTs hide the internal representations of data types, an exchange of their implementation is possible without great difficulties. Gargano et al. [GNT91] emphasize that from the designer's point of view the use of ADTs offers a formal frame for the correct specification of spatial information.

Güting [Gü88a] advocates that the representation of spatial objects should not be described at the level of the database system's data model but at the programming level. The reasons are conceptual simplicity and efficiency. Standard database query languages do not allow to formulate geometric algorithms like a point-in-polygon test, whereas programming languages offer a sufficient richness to build data structures and to program efficient algorithms. Within the Gral system [Gü88a, Gü88b] at the programming level spatial data types are designed as ADTs and in this context called *opaque* types so that the application programmer can access objects only through corresponding operations.

2.3 Formal Definition Methods

2.3.1 Introduction

As pointed out several times, the description of spatial data types is mostly based on intuitive and informal explanations, and abstract formal models are needed for their definition. It seems that except for a small number of formal approaches designers have frequently defined and currently still define their spatial data types without a very deep understanding of the underlying semantics and that these definitions often do not rest on a formal foundation. This procedure can amount to problems of inconsistency and lack of clarity and entail a weakening of the modelling power of a spatial data model. Consequently, in particular at the designer's level it is advisable to take a formal approach.

An important issue therefore relates to theoretical approaches in the form of complete and consistent mathematical foundations that are appropriate for the modelling of spatial data. Formal methods should fulfil the following requirements:

- At the designer's level a formal definition should and can lead to a better understanding of the complex semantics of spatial objects and the operations performed on these objects.
- Properties of a formal definition of spatial data types should be directly usable for a formal definition of the corresponding spatial operations.
- A formal framework for the correct specification of spatial data types should guarantee clarity and consistency at the user's level.
- Formal methods should take into account the finiteness of computers and the problems of numerical robustness and topological correctness.
- There is no standardization of spatial data types. A mature theoretical foundation could be the first step towards a standard.
- At the implementation level formal specifications of spatial data types should and can serve as a suitable and precise basis for a possible implementation.

Any field of science and engineering dealing with geometric data will benefit from a formalized view of spatial information. The search for suitable formal models which are capable of expressing the essential properties of spatial objects and their behaviour leads, of course, to mathematics. Indeed, some of its branches can to a large degree contribute to an adequate formalization and understanding of spatial data. But the different theoretical approaches have to be assessed by their feasibility in computer practice, which at a closer look restricts the usefulness of some of the formal methods or even makes them useless for practice. The central problem is the assumption of an infinite-precision arithmetic in theoretical approaches and the reality of a finite-precision arithmetic in computers (see Section 2.4). At the implementation level this problem influences the realization of spatial data structures and algorithms, since it causes problems of numerical robustness and topological correctness in geometric computation.

The formal approaches for the definition of spatial data types that are discussed in the literature are dominated by the mathematical concepts of *metric*, *topology*, and *order* which enable spatial queries of various types to be answered. A characteristic feature of almost all proposals is that they are centered around the concept of a *point set*. The idea is that space is composed of infinitely many points and that space contains a set of spatial objects. Hence, each spatial object occupies space and can be regarded as the set of points occupied by that object. All mentioned theoretical approaches formally describe and classify point sets by defining structures on these point sets.

Point set theory employs well-known methods of elementary set theory. Since point sets are sets, set operations (e.g., union, containment) can be applied to point sets. The operations union, intersection, and difference can help to derive new point sets from existing ones and to formally define spatial operations. Analytical geometry is used to represent points, lines, planes, and other geometric structures by numbers and the relations between these structures by equations. It performs the mapping of the two-dimensional space to the space of tuples, each tuple being formed by two real numbers. Metric properties (distances, angles, etc.) and topological properties (adjacency of polygons, inclusion of points in polygons) can be deduced from analytical geometry by numerical computation.

Topology investigates topological structures of point sets which means that it studies the properties of spatial objects that are independent of an underlying distance or coordinate measure (metric). It can thus be regarded as coordinate-free or coordinate-independent geometry. Topology focuses on those characteristics of geometry that are preserved under so-called *continuous topological transformations* which can be intuitively interpreted as elastic transformations or spatial distortions that stretch, twist, or otherwise deform without cutting. Properties that are preserved (invariant) under topological transformations are called *topological invariants*. Examples are translation, rotation, and scaling. Moreover, topology provides a concept of "closeness" of points and point sets. Topology has developed as an independent field of study in mathematics and turns out to be a useful theory both for modelling and for analysis especially of spatial relationships like adjacency, connectivity, and containment.

Two branches of topology are of interest: *point set topology*, also called *general topology*, and *algebraic topology*, also named *combinatorial topology*. Point set topology uses concepts of analysis, deals with continuous functions, and has been strongly influenced by general set theory. Algebraic topology describes the structure of a topological space by associating an algebraic system with it, is not originally based on general set theory, and plays an important role in algebra. Both point set topology and algebraic topology pursue the common aim to determine the nature of (a topological) space with the aid of properties that are invariant under topological transformations.

Order theory allows the comparison of two or more elements of a set and can be used to answer queries of inclusion and containment. Two of the most important and interesting kinds of order are *strict order* and *partial order*.

The following subsections present mathematical concepts of space and an outline of the fundamental concepts of point set theory, point set topology, algebraic topology, and order theory as far as they are needed for a sufficient understanding of the formal modelling approaches discussed in the literature. Furthermore, the subsections show how these concepts are used, which problems they have, and whether they are applicable to practice. In particular, these approaches are confronted with the problem of numerical accuracy. All expositions are related to two-dimensional space. Theorems are stated without proofs.

2.3.2 Mathematical Concepts of Space

The formal treatment of spatial information implies the answer to the following three topics:

- Which views or concepts of *space* exist in mathematics, *and* which of these concepts are appropriate to model spatial reality?
- Which spatial objects and properties can be represented within these views?
- Which spatial operations exist and can be defined for the objects within these views?

Similarly to geometric application areas which combine most ideas and structures with the notion *space*, also mathematics has concepts of space. Mathematics knows numerous variations of the term *space*, but common sense and the question of applicability of a concrete interpretation to reality allow to reduce the consideration to topological, metric, vector, and Euclidean spaces. *Topological spaces*, which have turned out to be a very important class of spaces, on a very abstract and general level comprise concepts of closeness, connectedness, and continuity. *Metric spaces* introduce notions of distances. *Vector spaces* include concepts like direction, dimension, and coordinates. *Euclidean spaces* contain notions of orthogonality, norm, and angle. While in everyday life Euclidean properties seem to be dominant, in geometric computation vectorial, metric, and especially topological properties are more relevant.

The different space structures are not independent from each other. Figure 2-20 (taken from [Bu89]) shows their interrelations. The shape for each kind of space has been chosen according to the structure of the space. Euclidean spaces (symbolized by a rectangle for Euclidean orthogonality) are contained in metric spaces (symbolized by a circle respectively a bowl which can be defined by distances) which are

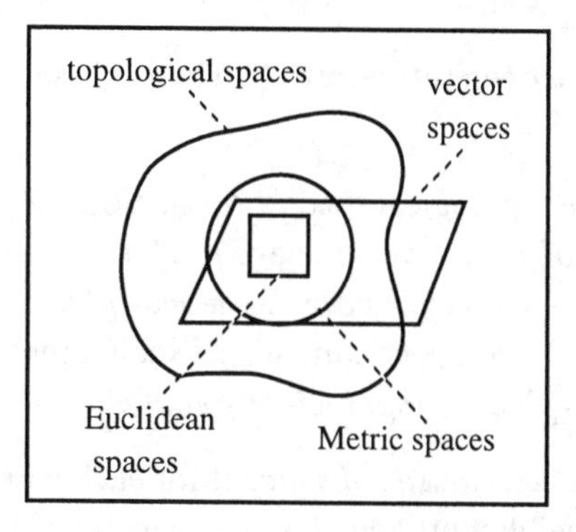

Figure 2-20

themselves included in topological spaces. Euclidean spaces are also vector spaces (symbolized by a parallelogram alluding to the notion of direction) and have all their properties. The containment relation means for instance that any metric space can be provided with a topological structure which is analogously valid for other inclusions. In the following, the different space structures are formalized.

Topological spaces

A *topological space* consists of a set X and a *topology* on X. The notion of a topology in a very general sense establishes a formal and precise framework for the intuitive ideas of closeness and continuity which can be formalized without reference to any kind of metric or distance function. The great generality of the concept of a topological space leads to a wide applicability of topology.

Definition. Let X be a set and $\mathcal{T} \subseteq \mathcal{P}(X)$ a subset of the power set of X. The pair $(X, \mathcal{T})$ is called a *topological space*, if the following three axioms are satisfied:

(T1) $X \in \mathcal{T}, \varnothing \in \mathcal{T}$.
(T2) $U \in \mathcal{T}, V \in \mathcal{T} \Rightarrow U \cap V \in \mathcal{T}$.
(T3) $S \subseteq \mathcal{T} \Rightarrow \bigcup_{A \in S} A \in \mathcal{T}$.

$\mathcal{T}$ is called a *topology* for X. The elements of $\mathcal{T}$ are called *open sets*, their complements in X *closed sets*. The elements of X are called *points*.

Open sets are closed under finite intersections (T2) and under arbitrary unions (T3). Frequently, when no confusion can arise, $\mathcal{T}$ is not mentioned, and X denotes a topological space.

The open sets in a topology allow to formulate a set-theoretic notion of spatial closeness:

Definition. Let U be an open set and $x \in U$. Then U is said to be a *neighbourhood of* x.

Definition. Let X, Y be topological spaces. A function $f: X \rightarrow Y$ is *continuous* if for each open subset U of Y the inverse image $f^{-1}(U) = \{x \in X \mid f(x) \in U\}$ is open in X. A function $f: X \rightarrow Y$ is called a *homeomorphism* if it is bijective[11] and continuous, and $f^{-1}: Y \rightarrow X$ is continuous. If such a function exists, X and Y are said to be *homeomorphic* (or *topologically equivalent*) spaces.

Definition. A space X is a *Hausdorff space* if for each pair $x_1, x_2 \in X$ with $x_1 \neq x_2$ there exist disjoint neighbourhoods U_1 of x_1 and U_2 of x_2.

Metric spaces

A *metric space* is a topological space with special properties and includes an abstract formulation of the Euclidean idea of distance and measurement. With the aid of metric spaces concepts like location, direction, and distance can be realized.

Definition. Let X be a set. A *metric* on X is a function $d: X \times X \rightarrow \mathbf{R}$ which[12] satisfies the following axioms for all $x, y, z \in X$:

(M1) $d(x, y) \geq 0$,
(M2) $d(x, y) = 0 \Leftrightarrow x = y$,
(M3) $d(x, y) = d(y, x)$,
(M4) $d(x, z) \leq d(x, y) + d(y, z)$.

---

[11] A function $f: X \rightarrow Y$ is *bijective* if a function $g: Y \rightarrow X$ exists such that $g(f(X)) = X$ and $f(g(Y)) = Y$.

[12] **R** denotes the domain of real numbers.

The metric d is called *distance* between two points and induces a topology on X, termed the *metric topology defined by d*. The pair (X, d) is called a *metric space*.

The first axiom (M1) states that the distance between two points x and y is non-negative. Axiom (M2) says that the distance between two points x and y is zero if and only if the two points coincide. The third axiom (M3) states that d is symmetric so that the distance between x and y is independent from ordering. That is, distances do not have an orientation and do not underlie the orientation of any coordinate system. Axiom (M4) represents the property of *triangle inequality*. Frequently, metric d is not mentioned, and X itself is referred to as a metric space.

A metric d defined as the distance between two points can be generalized to a metric defined as the distance between two point sets.

Definition. Let (X, d) be a metric space. A metric on $\mathcal{P}(X)$ is defined by

$$\forall A \subseteq X \quad \forall B \subseteq X : d(A, B) := \begin{cases} 0, & \text{if } A \cap B \neq \varnothing \\ \min \{d(a, b) \mid a \in A \wedge b \in B\}, & \text{otherwise} \end{cases}$$

Definition. Let (X, d) be a metric space. Then $B(x, \varepsilon) = \{y \mid y \in X \text{ and } d(x, y) < \varepsilon\}$ is called the *open ball* (ε*-neighbourhood, open disc*) with center $x \in X$ and radius $\varepsilon > 0$. The *closed ball* (*closed disc, cell*) is the set $D(x, \varepsilon) = \{y \mid y \in X \text{ and } d(x, y) \leq \varepsilon\}$. The *sphere* is the set $S(x, \varepsilon) = \{y \mid y \in X \text{ and } d(x, y) = \varepsilon\}$.

Theorem. Any ε*-neighbourhood* $B(x, \varepsilon)$ in a metric space (X, d) is an open subset of X.

Theorem. Every metric space is a Hausdorff space.

Euclidean spaces

Euclidean spaces are important instances of metric spaces, and by adding the algebraic structure of a vector space they constitute the basis for a theory of analytical geometry. As a special case, we consider the n-dimensional Euclidean space $\mathbf{R}^n$ consisting of all n-tuples $x = (x_1, ..., x_n)$ of real numbers. The *Euclidean metric* is described by the Pythagorean distance function which is based upon real-valued coordinates.

Definition. The *n-dimensional Euclidean space* $\mathbf{R}^n$ (n a positive integer) is the metric space $\mathbf{R}^n = \{x = (x_1, ..., x_n) \mid x_i \in \mathbf{R}, 1 \leq i \leq n\}$ with the topology determined by the *Euclidean metric*

$$d(x, y) = \sqrt{\sum_{i=1}^{n} (x_i - y_i)^2}$$

Vector spaces

Vector spaces include concepts like direction, dimension, and coordinates. We give here a very abstract definition of a vector space.

Definition. A *group* is a tuple $(G, @)$ where G is a non-empty set with an operation "@", such that

(G1) $\forall a, b, c \in G : (a @ b) @ c = a @ (b @ c)$ (associativity)
(G2) $\exists e \in G \, \forall a \in G : a @ e = e @ a = a$ (e is called *identity element* of G)
(G3) $\forall a \in G \, \exists a^* \in G : a @ a^* = a^* @ a = e$ (a^* is called *inverse* of a)

A group $(G, @)$ is *commutative* or *abelian* if

(G4) $\forall a, b \in G : a @ b = b @ a$

Definition. A *field* is a triple $(F, +, \cdot)$ where F is a non-empty set with operations "+" and "·", such that

(F1) $(F, +)$ is a commutative group with identity element 0
(F2) $(F \backslash \{0\}, \cdot)$ is a commutative group with identity element 1
(F3) $\forall a \in F : 0 \cdot a = 0$
(F4) $\forall a, b, c \in F : a \cdot (b + c) = (a \cdot b) + (a \cdot c)$ (distributivity)

Definition. A *vector space* over a field $(F, +, \cdot)$ is a set V with two operations, an operation "⊕" under which V forms a commutative group, and a scalar multiplication "⊗" which associates with each $v \in V$ and $a \in F$ a member $a \otimes v \in V$, such that

(V1) $\forall a, b \in F \, \forall v \in V : a \otimes (b \otimes v) = (a \cdot b) \otimes v$ (associativity)
(V2) $\forall a, b \in F \, \forall v \in V : (a + b) \otimes v = (a \otimes v) \oplus (b \otimes v)$ (distributivity)
$\forall a \in F \, \forall v, w \in V : a \otimes (v \oplus w) = (a \otimes v) \oplus (a \otimes w)$ (distributivity)
(V3) $\forall v \in V : 1 \otimes v = v$

The members of a vector space are called *vectors*.

Definition. A set $\{v_1, ..., v_k\}$ of vectors of a vector space V is *linearly dependent* if there exist elements $a_1, ..., a_k$ of the field F such that $a_1 v_1 + ... + a_k v_k = 0$ and not all the a_i are 0. A set of vectors is *linearly independent* if it is not linearly dependent. A set of vectors $\{v_1, ..., v_k\}$ is said to *span* V if each element $v \in V$ can be represented as a *linear combination* $v = b_1 v_1 + ... + b_k v_k$ for some $b_1, ..., b_k$ in F. A *base* or *basis* for V is a linearly independent set of vectors of V which spans V. If V has a finite basis, then V is called *finite dimensional* and the number of elements of all finite bases of V (which is equal) is called the *dimension* of V.

Theorem. Each Euclidean space is a vector space.

Hence, the Euclidean space $\mathbf{R}^n$ is a vector space.

2.3.3 Point Set Theory

A widespread approach for the definition of spatial data types uses concepts of *point set theory*. Just as the other formalization methods, point set theory applies the point set paradigm, i.e., any spatial object is regarded as being composed of a set of points in space (e.g., [GNT91, Gü88a, HSH92, Or90, Pu88, SV89, Ti80]). Such a set is, in general, infinite but it should always be possible to specify it with finite methods. One of the most frequently used, finite concepts for spatial modelling is *boundary representation* [KW87] where a spatial object like a line or a polygon is finitely defined by its surrounding boundary line in the form of a sequence of consecutive edges which themselves are described by their two end points.

Since point sets are sets, elementary operations of general set theory are applicable. Set operations like union, intersection, difference, and containment allow to construct new objects from existing ones and to formally define spatial operations. A polygon with holes, for example, can be easily defined as the set-theoretic difference of its outer polygon and the union of its inner polygonal holes. An essential deficiency of pure point set theory which differentiates it from other formalization methods is that special parts or subsets of a point set, which represents a spatial object, cannot be distinguished. This fact has given rise to criticism by several authors arguing that spatial operations which are based on the distinction of special parts of point sets cannot be (correctly) defined. The support of the design criteria by the formal approaches using point set theory turns out to be rather poor and unsatisfactory. Especially, none of the approaches deals with the problem of numerical robustness and topological correctness.

Within the framework of the geo-relational algebra [Gü88a, Gü88b] Güting models any spatial object by a set of points in the two-dimensional Euclidean space. The representation of an object and the object itself are logically separated by a function *points*. For an object O *points*(O) denotes the set of points occupied by O in space. A *point* is simply a pair $p = (x, y) \in \mathbf{R} \times \mathbf{R}$ in the Euclidean plane with *points*(p) = $\{(x, y)\}$. For two distinct points $p_1, p_2 \in \mathbf{R} \times \mathbf{R}$ the set $\{\alpha p_1 + (1\text{-}\alpha)p_2 \mid \alpha \in \mathbf{R}\}$ defines a line. A restriction of α to values between 0 and 1 leads to the set $\{\alpha p_1 + (1\text{-}\alpha)p_2 \mid \alpha \in \mathbf{R} \wedge 0 \leq \alpha \leq 1\}$ which describes a *line segment* as the part of the line lying between p_1 and p_2. Based on these few definitions, spatial objects like polygons and lines can be easily defined as has been done informally in Section 2.2.1 when discussing structure-oriented data types. A more formal presentation is omitted here.

The following definitions based on point sets have been given for the binary spatial predicates *equal*, *unequal*, *inside*, *outside*, and *intersects* which are applied to two spatial objects x and y, each being of some spatial data type, and which are formulated in terms of the set operations $=$, $\neq$, $\subseteq$, and $\cap$:

$$
\begin{aligned}
x = y &:= points(x) = points(y) \\
x \neq y &:= points(x) \neq points(y) \\
x \; inside \; y &:= points(x) \subseteq points(y) \\
x \; outside \; y &:= points(x) \cap points(y) = \emptyset \\
x \; intersects \; y &:= points(x) \cap points(y) \neq \emptyset
\end{aligned}
$$

The drawback of these definitions is that they do not provide a complete coverage of all possible topological relationships between two spatial objects and that they are not unique. For example, the definitions of *equal* and *inside* are both covered by the definition of *intersects*. Furthermore, the point set model does not allow the definition of those spatial relationships that are based on the distinction of special parts of point sets such as the interior and the boundary. In this way, the relationship *intersects* could be defined to yield the value *true* if both the intersection of the boundaries and the intersection of the interiors of two spatial objects is non-empty. This has to be topologically distinguished from the situation of two spatial objects having common boundary points but no common interior points. A predicate *meets* could characterize this situation.

For the definition of spatial operations a function *components* is needed. Let S be any set of points in the Euclidean space, i.e., $S \subseteq \mathbf{R}^2$. Two points of S are *connected within* S if there exists a curve containing both points and being a subset of S. S is called a *region* if any two points of S are connected within S. If S is not a region, then S can be subdivided into a set of maximally connected regions, that is, $S = r_1 \cup r_2 \cup \ldots$, such that r_i is a region and $r_i \cap r_j = \emptyset$ for $i \neq j$. In spatial applications, of course, S will always be a union of finitely many regions so that $S = r_1 \cup r_2 \cup \ldots \cup r_n$. The function *components* can now be defined to return for a set of points the collection of disjoint regions, that is, $components(S) = \{r_1, \ldots, r_n\}$.

As examples, Güting's definitions of intersection between two *Lines* objects, between two *Regions* objects, and between a *Lines* and a *Regions* object are presented. Note that within this framework a *Points* object represents a single point, a *Lines* object a non-closed simple chain and a *Regions* object a simple polygon. The function *llsect* is applied to two *Lines* objects l_1, l_2 and returns a set of *Points* objects.

$$
\begin{aligned}
llsect(l_1, l_2) := \; & \{x \in components(points(l_1) \cap points(l_2)) \mid x \text{ is a point}\} \cup \\
& \{y \mid \text{a line } s \in components(points(l_1) \cap points(l_2)) \text{ exists and } y \text{ is} \\
& \text{an end point of } s\}
\end{aligned}
$$

The first subset yields points as the usual elements of an intersection between two *Lines* objects. But as a degenerate case two lines may coincide partially. This poses problems, since objects of a mixed type are not allowed. At least three alternatives to deal with the problem are conceivable. Either lines as the result of such an intersection are simply ignored, or two different operators are defined where one returns intersection points and the other one returns common lines, or the end points of common lines are yielded. Here, the third alternative has been selected. The remaining functions can be defined similarly.

$$lrsect(l, r) \quad := \{x \in components(points(l) \cap points(r)) \mid x \text{ is a line}\}$$

$$rrsect(r_1, r_2) := \{x \in components(points(r_1) \cap points(r_2)) \mid x \text{ is a polygon}\}$$

The function *lrsect* for the intersection between a *Lines* object l and a *Regions* object r returns a set of *Lines* objects. Here, the degenerate case is possible that points are the result of intersecting a line with a region. These points are ignored by interpreting set-theoretic intersection in the definitions as "regularized intersection" (see below). The function *rrsect* takes two *Regions* objects as operands and yields a set of *Regions* objects. Again the result of intersection may contain points and lines which are ignored. For the definition of further operations see [Gü88b].

Pullar [Pu88] equates the point set representation of an object with the object itself and specifies relationships between two spatial objects x and y by considering their intersection. Here, too, the missing uniqueness is noticeable.

$$\begin{aligned}
x \; equal \; y &:= x \cap y = x \wedge x \cap y = y \\
x \; contains \; y &:= x \cap y = y \\
x \; contained_by \; y &:= x \cap y = x \\
x \; disjoint \; y &:= x \cap y = \varnothing \\
x \; incident \; y &:= x \cap y \neq \varnothing
\end{aligned}$$

Tilove [Ti80] shows that applying pure conventional set operations to point sets can lead to geometric anomalies (like those mentioned above). These anomalies rest on the fact that set operations on point sets are not dimension-preserving. For instance, the set-theoretic intersection of two point sets representing *Regions* objects may return not only a *Regions* object but also lines and points (see Figure 2-21 taken from [Ti80]). As a solution he introduces so-called *regular sets* and *regular set operations* which eliminate these anomalies. Since his approach combines methods of set theory with methods of point set topology, this approach will be discussed in Section 2.3.4 when dealing with point set topology. Another view of the same problem is

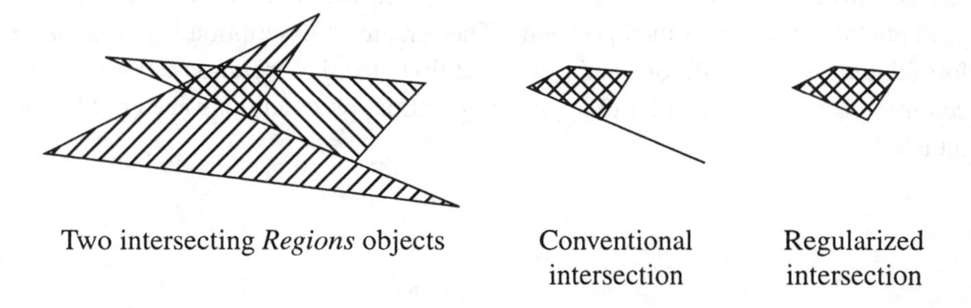

Two intersecting *Regions* objects   Conventional intersection   Regularized intersection

Figure 2-21

that point sets are closed under the *set-theoretic* operations union, intersection, and difference but not necessarily under the corresponding *geometric* operations.

Another proposal by Gargano *et al.* [GNT91] introduces an algebraic structure called the *SHAPES algebra* which uses concepts of set theory and constitutes the basis for the specification of an abstract data type *Geometry* (see Section 2.2.1).

Definition. Let S be a non-empty finite set whose elements are called *atoms*, and let $H_S = \mathcal{P}(\mathcal{P}(S))$ (where $\mathcal{P}(S)$ is the power set of S). The SHAPES *algebra* on S is the 6-tuple $\Psi_S = (H_S, \cup, \cap, \cap^*, geo, compl)$ where $\cup, \cap, \cap^*$, *geo*, and *compl* are operations defined on H_S so that for all $A, B \in H_S$ follows:

$$A \cup B = \{c \in H_S \mid c \in A \vee c \in B\}$$

$$A \cap B = \{c \in H_S \mid c \in A \wedge c \in B\}$$

$$A \cap^* B = \{c \in \mathcal{P}(S) \mid \exists\, a \in A\ \exists\, b \in B : c = a \cap b\}$$

$$geo(A) = \{X\}, \text{ where } X = \{x \in S \mid \exists\, a \in A : x \in a\}$$

$$compl(A, B) = \{Y\}, \text{ where } Y = \{y \in S \mid \forall\, c \in A \cup B : y \notin c\}$$

An element A of H_S is called a *shape*. Hence, H_S is the set of all shapes that can be formed by the atoms of S. Each of the elements of A is called an A-*component*. If A contains exactly one component, it is called *simple*. The $\cup$ and $\cap$ operators form the set-theoretic union and intersection of the components of two shapes. The $\cap^*$ operator applied to shapes A and B returns a shape C whose components are the result of the geometric intersection of every A-component with every B-component. The *geo* operator applied to a shape A with more than one component returns a simple shape whose single component is the union of the atoms of all A-components. For a simple shape A it returns shape A itself. The *compl* operator applied to shapes A and B returns a simple shape whose single component is the difference of the atoms of S and the atoms of the components of A and B. Since Ψ_S is closed under the operations $\cup$, $\cap$, $\cap^*$, *geo*, and *compl*, Ψ_S is an algebra.

The introduction of the SHAPES algebra corresponds to the definition of a parameterized abstract data type (ADT) *Geometry*(S). The syntax description of the ADT presents the signature of the operations. The semantics description is based on the formulation of preconditions and post-conditions for the operations. But since no condition is required in order to apply an operation, preconditions need not be formulated.

adt

Geometry(S : set)

sets

S	set of atoms
G	set of elements of type Geometry(S)
Bool	set of Boolean values consisting of *true* and *false*

syntax

	$\rightarrow$	G	create
G	$\rightarrow$	*Bool*	empty
$G \times G$	$\rightarrow$	G	$\cup, \cap, \cap^*$, *compl*
G	$\rightarrow$	G	*geo*

semantics

range $A, B, C = G$
range $a, b, c, c' = \mathcal{P}(S)$
range $el = S$
range $x = Bool$

post-create(; A) ::= $A = \varnothing$
post-empty(A; x) ::= $(A = \varnothing \wedge x = true) \vee (\neg (A = \varnothing) \wedge x = false)$
post-$\cup$ (A, B; C) ::= $\forall\, c \in C : c \in A \vee c \in B$
post-$\cap$ (A, B; C) ::= $\forall\, c \in C : c \in A \wedge c \in B$
post-$\cap^*$ (A, B; C) ::= $\forall\, c \in C\, \exists\, a \in A\, \exists\, b \in B : c = a \cap b$
post-compl(A, B; C) ::= $C = \{c\} \wedge \forall\, el \in c\, \forall\, a \in A \cup B : el \in c \Leftrightarrow el \notin a$
post-geo(A; B) ::= $B = \{b\} \wedge \forall\, el \in b\, \exists\, a \in A : el \in a \wedge \forall\, a \in A : el \in a \Rightarrow el \in b$

The notation “range a, b, ... = T” means that the variables a, b, ... are elements of the set T. The ADT *Geometry* can be based on any finite set of atomic elements and is therefore parametric, since the nature of the atoms does not influence the definition. For the definition of further operations like *compose*, *decompose*, and *fusion* which are embedded into the relational data model the reader is referred to [GNT91].

Although theoretically correct, from an application’s point of view the practicability and efficiency of the definitions are quite questionable. Gargano *et. al* give an example for the definition of S and specify it as the set of points belonging to a raster decomposition of the plane. The problem is that sets of raster points describing spatial objects can become enormous when stored in a computer system.

2.3.4 Point Set Topology

Point set topology (also named *general topology*) has its origin in the nineteenth century and is most relevant to the mathematical field of analysis. It has gained increasing popularity in spatial modelling for the definition of topological spatial relationships like *inside*, *meet*, and *overlap*, especially since it is not tied to any distance or coordinate measure and is hence coordinate-free (coordinate-independent). Point set topological notions include the concepts of *interior*, *boundary*, *closure*, *exterior*, *separation*, and *connectedness* [Al61, Ar83, Ga64] and are based on the classical point set model. In this subsection, let X be a topological space and $Y \subseteq X$.

Definition. The *interior* of Y, denoted by Y°, is the union of all open sets that are contained in Y. The *closure* of Y, denoted by $\overline{Y}$, is the intersection of all closed sets that contain Y. The *exterior* of Y, denoted by Y^-, is the union of all open sets that are not contained in Y. The *boundary* of Y, denoted by ∂Y, is the intersection of the closure of Y and the closure of the complement of Y, i.e., $\partial Y = \overline{Y} \cap \overline{X - Y}$.

The interior of Y is the largest open set contained in Y. An element $y \in Y$ is in the interior of Y if and only if there exists a neighbourhood of y contained in Y. The interior of X is X itself. If Y is an open set, then $Y^\circ = Y$. If $Z \subseteq Y$, then $Z^\circ \subseteq Y^\circ$. The closure of Y is the smallest closed set containing Y. An element $y \in Y$ is in the closure of Y if and only if every neighbourhood of y intersects Y. The closure of X is X itself. If Y is a closed set, then $\overline{Y} = Y$. If $Z \subseteq Y$, then $\overline{Z} \subseteq \overline{Y}$. The exterior of Y is the largest open set not contained in Y. The boundary of Y is a closed set. An element $y \in Y$ is in the boundary of Y if and only if every neighbourhood of y intersects both Y and its complement. The relationships between the notions of interior, closure, exterior, and boundary are given by the provable statements (1) $Y^\circ \cap \partial Y = \varnothing$, (2) $Y^\circ \cup \partial Y = \overline{Y}$, (3) $Y^- \cap \partial Y = \varnothing$, and (4) $Y^\circ \cap Y^- = \varnothing$. Obviously, $X = \partial Y \cup Y^\circ \cup Y^-$.

For the definition of topological relationships between point sets the concepts of separation and connectedness are needed.

Definition. A *separation* of Y is a pair A, B of subsets of X satisfying the following three conditions: (1) $A \neq \varnothing$ and $B \neq \varnothing$, (2) $A \cup B = Y$, and (3) $\overline{A} \cap B = \varnothing$ and $A \cap \overline{B} = \varnothing$. If a separation of Y exists, then Y is said to be *disconnected*, otherwise Y is said to be *connected*.

If A, B form a separation of Y and if Z is a connected subset of Y, then either $Z \subseteq A$ or $Z \subseteq B$. If $Y^\circ \neq \varnothing$ and $\overline{Y} \neq X$, then Y° and $X - \overline{Y}$ form a separation of $X - \partial Y$, and hence ∂Y separates X.

Tilove's solution [Ti80] to overcome possible geometric anomalies when applying pure set-theoretic operations to point sets (see Section 2.3.3) depends heavily on point set topological notions, in particular on the concept of *regularity*.

Definition. Y is called *regular closed* if $Y = \overline{Y^\circ}$.

Intuitively, regular closed sets model regions containing their boundaries and avoid both isolated or dangling line or point features and missing lines and points in the form of cuts and punctures. Hence, it makes sense to define a *regularization* function r which associates a set Y with a regular closed set, as follows:

$$r(Y) := \overline{Y^\circ}$$

Regular closed sets with the exception of union are not closed under finite conventional set operations so that *regular set operations* must be defined that preserve regularity. Let A, B be regular closed sets with $A, B \subseteq X$.

$$A \cup_r B := r(A \cup B) = A \cup B, \qquad A \cap_r B := r(A \cap B), \qquad A \setminus_r B := r(A \setminus B)$$

Regular closed sets and regular set operations express a natural formalization of the dimension-preserving property taken for granted by many geometric algorithms.

Worboys and Bofakos [WB93] use regular closed sets and regular set operations for the modelling of complex spatial regions that may contain holes and islands within holes to any finite level. The basic objects of their spatial data model are areal objects called *atoms* which are defined as those subsets of $\mathbf{R}^2$ that are topologically equivalent to the closed ball. This model is extended by aggregating atoms to regular closed areal objects termed *base areas* whose structure is described by skeleton graphs. The *skeleton* Σ_A of a finite set A of atoms is a graph whose vertices are labelled by the atoms of A. Given two distinct atoms $a, b \in A$, an edge exists between vertices v_a and v_b if $a \cap b \neq \varnothing$. A finite set of atoms $A = \{a_1, ..., a_n\}$ is called a *base area* if the following conditions are satisfied: (1) For every pair of distinct atoms a_i, a_j in A their intersection is either empty or a singleton set, and (2) Σ_A is acyclic. The *embedding* of A is defined as $emb(A) := \bigcup_{i=1}^{n} a_i$. The embedding of a base area is a regular closed set, because it is a finite union of regular closed sets. Figure 2-22 gives an example of a base area together with its skeleton.

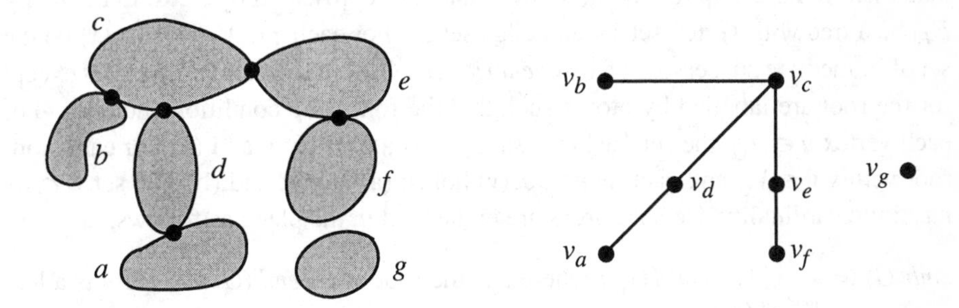

Figure 2-22

Condition (1) in the definition above guarantees that A does not have two atoms whose union is an atom. Additionally, it avoids that two atoms produce a hole between them. Condition (2) ensures that no holes can be formed from more than two atoms of A.

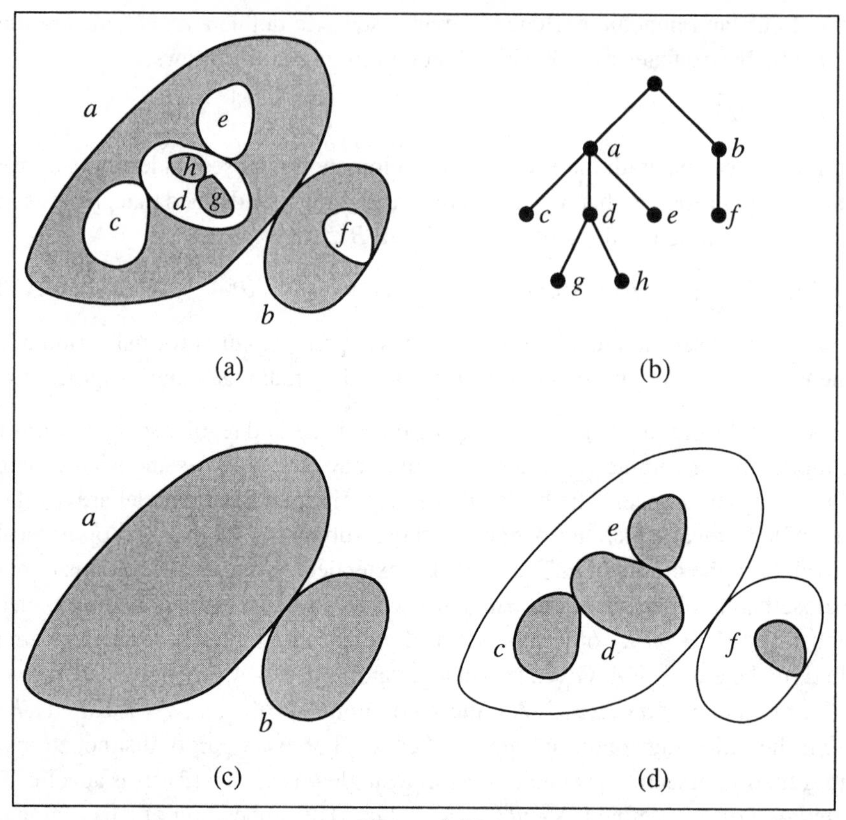

Figure 2-23

Generic areas allow the recursive construction of areal objects with holes and islands within holes, up to any finite level, and are represented by trees. Let $T = (V_T, E_T)$ be a tree with vertex set V_T and edge set E_T. For each $x \in V_T$ let $succ(x)$ be the set of immediate successors of x. A *generic area* G is a tree where all vertices except for the root are labelled by atoms such that the following conditions hold: (1) For each vertex $v \in V_G$, the set $\{w \mid w \in succ(v)\}$ is a base area, and (2) For each non-root vertex $v \in V_G$ and each $w \in succ(v)$ holds: (a) $w \subset v$, and (b) The set $w \cap \partial v$ has finite cardinality. Generic areas are embedded in the plane as follows:

$emb(G) := \bigcup_{w \in succ(r)} emb(G_w)$ where r is the root of G, $emb(G_v) := v$ if v is a leaf vertex of G, and $emb(G_v) := v \setminus \bigcup_{w \in succ(v)} emb(G_w)^\circ$ otherwise.

The embedding of a generic area is a regular closed set. Figure 2-23 gives an example of the embedding of a generic area (a), presents a labelled tree representation of the generic area (b), and shows the base areas of the first two levels ((c), (d)). The unique representation of a generic area is given by the provable statement that for two generic areas G_1 and G_2 holds: $G_1 = G_2 \Leftrightarrow emb(G_1) = emb(G_2)$. The function *emb* is one-one and thus has an inverse.

Operations on areal objects are defined on three levels. Operations on base areas are constructed from primitive operations on atoms, operations on generic areas from operations on base areas. The set of primitive operations on atoms is *equals*, *spatial intersection* ($\cap_s$), *spatial difference* ($\setminus_s$), *spatial union* ($\cup_s$), *boundary* (∂), *adjacent*, *centroid*, *area*, and *perimeter*. The regularized intersection of two atoms has no holes and is thus equal to the embedding of a base area: $a \cap_s b = emb^{-1}(a \cap_r b)$ for two atoms a, b where emb^{-1} is the inverse of the embedding function for base areas. The spatial difference between two atoms is a generic area defined as $a \setminus_s b = emb^{-1}(a \setminus_r b)$ where emb^{-1} is the inverse of the embedding function for generic areas. Four situations have to be distinguished: (1) $a \cap_r b = \varnothing \Rightarrow a \setminus_s b = \{a\}$, (2) $a \subseteq b \Rightarrow a \setminus_s b = \varnothing$, (3) $b \subset a \wedge b \cap \partial a$ is finite $\Rightarrow a \setminus_s b$ is a generic area, and (4) in all other cases $a \setminus_s b$ is a base area. The third case amounts to a hole in the spatial difference of two atoms. The spatial union of two atoms which is defined as $a \cup_s b = emb^{-1}(a \cup_r b)$ may lead to holes and is thus, in general, a generic area. It can be proven that the result of such a spatial union has at most two levels. For a more detailed consideration of the other operations see [WB93].

Operations on base areas and on generic areas cannot be defined by simple mathematical formulas like those above but are defined by partially quite complex construction algorithms. Operations provided are for example *equals*, *spatial intersection* ($\cap_s$), *spatial difference* ($\setminus_s$), *spatial union* ($\cup_s$), *boundary* (∂), *spatial subset* ($\subseteq_s$), *adjacent*, *centroid*, *area*, *perimeter*, *cardinality*, *components*, *connected*, *strongly_connected*, and others. The reader who is interested in the definition of these operations and in their construction algorithms is referred to [WB93].

Two early approaches of Pullar [Pu88] and Wagner [Wa88] use notions from point set topology to describe topological spatial relationships. The *boundary* and the *interior* of a point set are identified as the central components for a description of such relationships. Pullar enriches his point set approach by using these two notions so that the predicates *overlap* and *neighbour* can be defined:

$$
\begin{aligned}
x\ overlap\ y &:= boundary(x) \cap boundary(y) \neq \varnothing \wedge \\
&\quad\ interior(x) \cap interior(y) \neq \varnothing \\
x\ neighbour\ y &:= boundary(x) \cap boundary(y) \neq \varnothing \wedge \\
&\quad\ interior(x) \cap interior(y) = \varnothing
\end{aligned}
$$

Wagner in a more systematic approach views combinations of intersecting and non-intersecting boundaries and interiors of two spatial regions. He distinguishes four relationships: (1) *separation* where neither boundaries nor interiors intersect, (2) *neighbourhood* where boundaries intersect but interiors do not, (3) *strict inclusion* where the boundaries do not intersect but the interiors do, and (4) *intersection* where both boundaries and interiors intersect. This approach uses a formal and coherent framework for the description of topological relationships but it is not carried out in all its consequences. For example, intersection and equality of two spatial regions cannot be distinguished, since for both relationships boundaries and interiors intersect.

The most known, formal model for describing (binary) topological relationships has been developed by Egenhofer *et al.* [Eg89b, Eg91c, EF91, EF95, EH90] and is called *4-intersection model*. Their approach amounts to the definition of a "complete" set of topological relationships which covers any possible constellation between two spatial objects. Given two subsets A and B of a topological space X, this approach is based on the four intersections of the boundaries and interiors of A and B, i.e. $\partial A \cap \partial B$, $\partial A \cap B^\circ$, $A^\circ \cap \partial B$, $A^\circ \cap B^\circ$. Each intersection is tested with regard to the topologically invariant criteria of *emptiness* ($\varnothing$) and *non-emptiness* ($\neq\varnothing$). The pair empty/non-empty is the simplest and most general topological invariant so that

$\partial A \cap \partial B$	$\partial A \cap B^\circ$	$A^\circ \cap \partial B$	$A^\circ \cap B^\circ$	relationship name
$\varnothing$	$\varnothing$	$\varnothing$	$\varnothing$	A and B are *disjoint*
$\varnothing$	$\varnothing$	$\varnothing$	$\neq\varnothing$	
$\varnothing$	$\varnothing$	$\neq\varnothing$	$\varnothing$	
$\varnothing$	$\varnothing$	$\neq\varnothing$	$\neq\varnothing$	*A contains B / B inside A*
$\varnothing$	$\neq\varnothing$	$\varnothing$	$\varnothing$	
$\varnothing$	$\neq\varnothing$	$\varnothing$	$\neq\varnothing$	*A inside B / B contains A*
$\varnothing$	$\neq\varnothing$	$\neq\varnothing$	$\varnothing$	
$\varnothing$	$\neq\varnothing$	$\neq\varnothing$	$\neq\varnothing$	
$\neq\varnothing$	$\varnothing$	$\varnothing$	$\varnothing$	A and B *meet*
$\neq\varnothing$	$\varnothing$	$\varnothing$	$\neq\varnothing$	A and B are *equal*
$\neq\varnothing$	$\varnothing$	$\neq\varnothing$	$\varnothing$	
$\neq\varnothing$	$\varnothing$	$\neq\varnothing$	$\neq\varnothing$	*A covers B / B covered_by A*
$\neq\varnothing$	$\neq\varnothing$	$\varnothing$	$\varnothing$	
$\neq\varnothing$	$\neq\varnothing$	$\varnothing$	$\neq\varnothing$	*A covered_by B / B covers A*
$\neq\varnothing$	$\neq\varnothing$	$\neq\varnothing$	$\varnothing$	
$\neq\varnothing$	$\neq\varnothing$	$\neq\varnothing$	$\neq\varnothing$	A and B *overlap*

Figure 2-24

any other invariant like the dimension of a set or the number of connected components may be considered as more restrictive. Each quadruple of intersections describes one possible topological relationship so that $2^4 = 16$ mutually exclusive specifications of topological relationships can be enumerated for any pair of sets A and B in X (Figure 2-24).

The geometric interpretation of these quadruples depends both on the topological space in which A and B lie and on the properties of A and B themselves. Given a connected topological space X, a *spatial region* is defined as a non-empty proper subset A of X which is regular closed and whose interior is connected. Restricting X to $\mathbf{R}^2$ and arbitrary subsets of X to spatial regions in X, it is proven in [EF91, EH90] that eight of the combinations of Figure 2-24 are not valid and that two of them have symmetric counterparts so that six different relationships result which are called *disjoint*, *inside*, *meet*, *equal*, *covers*, and *overlap*. A graphic representation of these relationships has been given in Figure 2-8 of Section 2.2.2.

Egenhofer and Franzosa [EF91, EF95] show that more detailed topological relationships can be deduced by considering further topological invariants like the *dimension* of the boundary intersections of spatial regions in addition to the emptiness/non-emptiness criterion. For example, two spatial regions can meet in a null-dimensional object (i.e. a point) or in a one-dimensional object (i.e. a common line), which leads to the more special relationships 0-*meet* and 1-*meet*. Similarly, the relationships 0-*covers*, 1-*covers*, 0-*overlap*, and 1-*overlap* can be defined. The other relationships *disjoint* and *inside/contains* are excluded from this consideration because they have empty boundary intersections.

Egenhofer's fundamental model of 4-intersection has been extended in various ways. Pullar and Egenhofer [PE88] before examined topological relationships between one-dimensional, closed, and connected spatial objects (i.e. one-dimensional intervals) embedded in a one-dimensional space. The results deviate only slightly from those about spatial regions in $\mathbf{R}^2$. Egenhofer [Eg91c, Eg91d] has extended the original 4-intersection model to a *9-intersection model* by also taking into account the intersections with the exterior A^- of a set A. For spatial regions, A^- is equal to the complement of A. The 9-intersection method does not only consider the relationships between the object parts but also their relationships to the embedding space. In [Eg91c, SP92] the *composition* of topological relationships is studied which enables one to derive new spatial information and which leads to a complete set of composed binary topological relationships. For example, if region A *meets* region B, and region B *contains* region C, then A and C must be *disjoint*.

Egenhofer's standard approach has also been extended to point and line features [EH92] which results in six major groups of binary relationships: region/region, line/region, point/region, line/line, point/line, and point/point. A problem of this approach in practice is the large number of different relationships whose names are difficult to remember for the user. Another drawback is that some intuitively different situations cannot be distinguished. For example, two regions that share exactly one point (i.e. touch) and two regions that have a complete line in common (i.e., are adjacent) are represented by the same relationship *meet*, since the intersection of the boundaries is non-empty and the other three intersections are all empty. Here, Clementini *et al.* [CFO93] propose a *dimension-extended method* which also takes into account the dimension of an intersection as a second topological invariant in addition to the emptiness/non-emptiness criterion. Thus, in two-dimensional space an intersection can be empty, 0D (point), 1D (line), or 2D (region). In principle this results in $4^4 = 256$ combinations, but many of them are not valid so that in total 52 relationships between points, lines, and regions remain.

Because the number of relationships is too large, five basic and overloaded relationship names are introduced: *touch*, *in*, *cross*, *overlap*, and *disjoint*. Their definition is based on the formal framework of the dimension-extended method and general in the sense that they are applicable (with exceptions) to point, line, and region features. Let λ_1 and λ_2 be either features or boundaries of features, and let *dim* be a function which returns the dimension of a point set. Then

- $(\lambda_1, touch, \lambda_2) \Leftrightarrow (\lambda_1^\circ \cap \lambda_2^\circ = \varnothing) \wedge (\lambda_1 \cap \lambda_2 \neq \varnothing)$
- $(\lambda_1, in, \lambda_2) \Leftrightarrow (\lambda_1 \cap \lambda_2 = \lambda_1) \wedge (\lambda_1^\circ \cap \lambda_2^\circ \neq \varnothing)$
- $(\lambda_1, cross, \lambda_2) \Leftrightarrow (dim(\lambda_1^\circ \cap \lambda_2^\circ) = max(dim(\lambda_1^\circ), dim(\lambda_2^\circ)) - 1) \wedge$
  $(\lambda_1 \cap \lambda_2 \neq \lambda_1) \wedge (\lambda_1 \cap \lambda_2 \neq \lambda_2)$
- $(\lambda_1, overlap, \lambda_2) \Leftrightarrow (dim(\lambda_1^\circ) = dim(\lambda_2^\circ) = dim(\lambda_1^\circ \cap \lambda_2^\circ)) \wedge (\lambda_1 \cap \lambda_2 \neq \lambda_1) \wedge$
  $(\lambda_1 \cap \lambda_2 \neq \lambda_2)$
- $(\lambda_1, disjoint, \lambda_2) \Leftrightarrow \lambda_1 \cap \lambda_2 = \varnothing$

The relationships *in* and *disjoint* are applicable to all six region/region, line/region, point/region, line/line, point/line, point/point situations. The *touch* relationship is not applicable to the point/point situation. The *cross* relationship can only be applied to line/line and line/region situations and the *overlap* relationship only to region/region and line/line situations.

Additionally to the above relationships, three operators are provided to isolate the boundaries of line and region features: Two operators *f* (from) and *t* (to) return the end points of a line, and an operator *b* yields the boundary line ∂A of a region *A*. A proof is given in [CFO93] that (1) the five relationships are mutually exclusive, i.e., no two different relationships can be simultaneously valid for any two features, that (2) the relationships are complete in a sense that any relationship between two features must be one of the five, and that (3) all situations described by the dimension-extended method can be defined using the five relationships and the three boundary operators.

The approaches discussed so far assume spatial objects whose interiors, boundaries, and exteriors are connected which in the two-dimensional case leads to simple regions. In [ECF94] Egenhofer *et al.* view topological relationships between spatial objects that have a more complex structure, namely regions that may contain *holes*. Whereas their boundaries and exteriors are allowed to be disconnected, their closures must be connected. Objects with disconnected closures would form separations, i.e., they would consist of disjoint components, a case which is excluded from their investigations.

The exterior of a region with holes may be separated into one *outer exterior*, denoted by A_0^- and $n > 0$ *inner exteriors*, denoted by $A_1^-, ..., A_n^-$. The union of outer exterior and inner exteriors forms the entire exterior: $A^- = \bigcup_{i=0}^{n} A_i^-$. A *region A with embedded holes* is a non-empty subset of $\mathbf{R}^2$ with a connected interior such that (1) the closure of any two different exteriors are disjoint and (2) A is regular closed:

$$(1)\ \forall\, i, j \in \{0, ..., n\}, i \neq j : \overline{A_i^-} \cap \overline{A_j^-} = \varnothing \text{ and } (2)\ A = \overline{A^\circ}.$$

A *hole* of A is defined as the closure of an inner exterior (i.e., as a simple region), i.e., $H_i^A = \overline{A_i^-}$ for $i > 0$. A hole is a connected set that is strictly contained in A and disjoint from all other holes in A. The *generalized region* A^* corresponds to the union of A and all holes contained in A:

$$A^* = A \cup \bigcup_{i=1}^{n} H_i^A$$

A region with embedded holes can be mapped into a group of simple regions A^*, $H_1^A, ..., H_n^A$, each without any holes. By considering the holes as separate objects, the modelling of topological relationships between regions with embedded holes can be expressed in terms of topological relationships between simple regions. The topology of a set of n spatial objects is fully specified by n^2 topological relationships. For two regions A and B with n and m holes, respectively, we consider the set $S = \{A^*, H_1^A, ..., H_n^A, B^*, H_1^B, ..., H_m^B\}$ which consists of $n + m + 2$ simple regions so that $\mu = (n + m + 2)^2$ topological relationships can be distinguished between the regions and their holes. Egenhofer *et al.* [ECF94] show that because of redundancy the number of relationships can be reduced to $\mu' = nm + n + m + 1$.

The different approaches surveyed in this subsection satisfy some of the design criteria and demonstrate the expressive power of type systems based on point set topology. Generality is supported by (hierarchies of) regions with holes [ECF94, WB93]. Modelling approaches based on regular closed sets and regular set operations fulfil closure properties [Ti80, WB93]. Furthermore, point set topology allows to design a "complete" set of topological relationships between two spatial objects [CFO93, Eg89b, Eg91c, EF91, EH90]. The problems of numerical robustness and topological correctness are not treated by these approaches.

2.3.5 Algebraic Topology

Algebraic topology (also called *combinatorial topology*) [Ar83, Cr78, Gi77, He90] is older than point set topology and has its origin in the mid-17th century. Its methods play an important role in the mathematical field of algebra, and it has even developed to a large, independent field of mathematics. Spatial modelling based on algebraic topology is not impeded by the finiteness and imprecision of the numbering systems of computers and not by the possibly arising topological incorrectness of algorithms, since the inherent problems of implementations of Euclidean geometry are overcome by recording topological properties explicitly.

Algebraic topology investigates topological structures for classifying and formally describing point sets by using algebraic means. Its techniques rest on problem translation and algebraic manipulation of symbols that represent spatial configurations and their relationships. The basic method consists of three steps: (1) conversion of the problem from a spatial environment to an algebraic environment, (2) solving the algebraic form of the problem, and (3) conversion of the algebraic solution back to the spatial environment. The intersection of two lines, for example, becomes the search for common nodes. The coincidence of two line objects, the neighbourhood of two region objects, and the neighbourhood of a line and a region object result in a search for common edges or common nodes (depending on the definition of neighbourhood) of the lines and the boundaries of regions.

Modelling approaches based on algebraic topology aim at decomposing space into a collection of irregular geometric shapes. Such a collection allows to model the complete underlying geometry of an application and forms a geometric framework from which meaningful spatial objects can be built. Depending on the structure of the shapes, two main branches of algebraic topology are distinguished called *simplicial topology* and *cellular topology*. They lead to the two basic decomposition techniques of *simplicial decomposition* and *cellular decomposition*.

Simplicial decomposition is the algebraically and logically simpler decomposition technique and is based on the structure of a *simplex*.

Definition. Given $k+1$ points $v_0, ..., v_k \in \mathbf{R}^n$ where the k vectors $\{v_1 - v_0, v_2 - v_0, ..., v_k - v_0\}$ are linearly independent. Then the set $\{v_0, ..., v_k\}$ is called *geometrically independent*, and the point set

$$\sigma = \sigma_k = \{x \in \mathbf{R}^n \mid x = \sum_{i=0}^{k} \lambda_i v_i \text{ with } \sum_{i=0}^{k} \lambda_i = 1, \lambda_i \in \mathbf{R}, \lambda_0, ..., \lambda_k \geq 0\} \subset \mathbf{R}^n$$

is called the (closed) *simplex* of dimension k (or the *k-simplex*) with the vertices $v_0, ..., v_k$, or the *k-simplex spanned* by $\{v_0, ..., v_k\}$.

A k-simplex is homeomorphic to the convex hull of $k+1$ geometrically independent points and is embedded in a Euclidean space of dimension k or greater. For a given dimension k, a k-simplex is the minimal and elementary spatial object, i.e., a building block from which all more complex spatial objects of this dimension can be constructed. In three-dimensional space, a 0-simplex is a single point or a node, a 1-simplex is a straight line or an edge between two distinct points including the end points, a 2-simplex is a filled triangle connecting three non-collinear points, and a 3-simplex is a solid tetrahedron connecting four non-coplanar points (Figure 2-25).

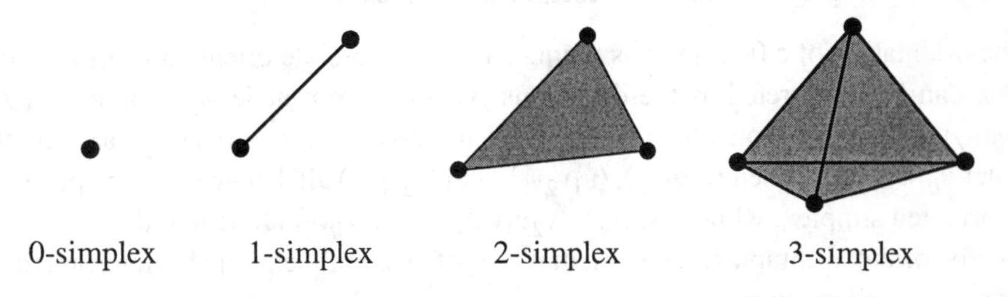

Figure 2-25

Any k-simplex is composed of $k + 1$ geometrically independent simplexes of dimension $k - 1$. For example, a triangle, a 2-simplex, is composed of three 1-simplexes which are geometrically independent if no two edges are parallel and no edge has length 0. A *face* of a simplex is any simplex that contributes to the composition of the simplex. For example, a node of a bounding edge of a triangle and a bounding edge itself are faces.

Definition. A (*simplicial*) *complex* C is a finite set of simplexes so that each face of a simplex in C is also in C and the intersection of two simplexes in C is either empty or a face of both simplexes. The *dimension* of C is the largest dimension of the simplexes in C.

Figure 2-26 shows two examples of a 1-complex (a) and a 2-complex (b) and three configurations which are not simplicial complexes, because the intersection of some of their simplexes is either not a face ((c), (d)) or not a simplex (e).

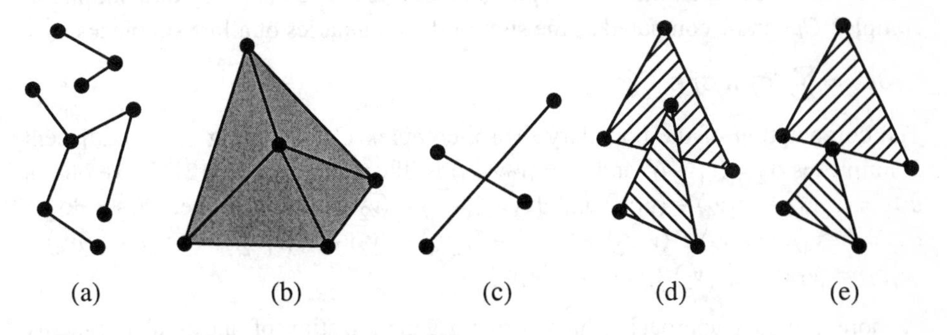

Figure 2-26

Definition. An *oriented k-simplex* is obtained from a k-simplex σ with vertices v_0, ..., v_k by choosing an order for the vertices. We write an oriented k-simplex as an ordered sequence $\sigma = \langle v_0 v_1 v_2 ... v_k \rangle$. The equivalence class of even permutations of the chosen order determines the positively oriented simplex, while the equivalence class of odd permutations determines the negatively oriented simplex. An *oriented simplicial complex* is obtained from a simplicial complex by assigning an orientation to each of its simplexes.

The orientation of a 0-simplex is unique. The two possible orientations of a 1-simplex can be interpreted as the directions (vectors) from node v_0 to node v_1 ($\sigma = \langle v_0 v_1 \rangle$) and from v_1 to v_0 ($\sigma = \langle v_1 v_0 \rangle$). Given a 2-simplex $\sigma = \langle v_0 v_1 v_2 \rangle$, assume the order $v_0 < v_1 < v_2$. Then $\langle v_0 v_1 v_2 \rangle$, $\langle v_1 v_2 v_0 \rangle$, and $\langle v_2 v_0 v_1 \rangle$ all denote the same positively oriented simplex, while $\langle v_0 v_2 v_1 \rangle$, $\langle v_2 v_1 v_0 \rangle$, and $\langle v_1 v_0 v_2 \rangle$ all denote the same negatively oriented simplex. The orientations of a 2-simplex can be interpreted as clockwise and counterclockwise.

Definition. The *boundary* of a k-simplex σ_k, denoted by $\partial \sigma_k$, is the union of all $k + 1$ $(k - 1)$- simplexes. The *boundary* of a k-complex C, denoted by ∂C, is the smallest complex that contains the symmetric difference of the boundaries of its constituent k-simplexes. The *interior* of a k-complex C, denoted by C°, is the union of all $(k-1)$-simplexes which are not part of the boundary of C.

For example, the boundary of a 2-simplex $\sigma = \langle v_0 v_1 v_2 \rangle$ is $\partial \sigma = \{\langle v_0 v_1 \rangle, \langle v_1 v_2 \rangle, \langle v_2 v_0 \rangle\}$. The boundary of a 2-complex $C = \{\langle x_1 y_1 z_1 \rangle, ..., \langle x_n y_n z_n \rangle\}$ is $\partial C = \Delta(\partial\langle x_1 y_1 z_1 \rangle, ..., \partial\langle x_n y_n z_n \rangle)$ where Δ denotes the symmetric difference operation. The common edges of the 2-dimensional simplexes of C form the interior of C. If $\sigma_k = \langle v_0 ... v_k \rangle$ is an ordered k-simplex, the boundary of σ_k can be computed algebraically [Sc68] by

$$\partial \sigma_k = \sum_{i=0}^{k} (-1)^i \langle v_0 ... \hat{v}_i ... v_k \rangle$$

where $\hat{v}_i$ indicates that the vertex v_i is to be omitted. The boundary of a simplicial complex C_k can be computed as the sum of the boundaries of all its simplexes σ_k:

$$\partial C_k = \sum \partial \sigma_k \text{ if } \sigma_k \in C_k.$$

The computation of the boundary of a 2-complex C_2 consisting of two adjacent 2-simplexes $\sigma_2 = \langle v_1 v_2 v_4 \rangle$ and $\tau_2 = \langle v_2 v_3 v_4 \rangle$ is illustrated in Figure 2-27. We obtain $\partial \sigma_2 = \langle v_2 v_4 \rangle - \langle v_1 v_4 \rangle + \langle v_1 v_2 \rangle$ and $\partial \tau_2 = \langle v_3 v_4 \rangle - \langle v_2 v_4 \rangle + \langle v_2 v_3 \rangle$. Then $\partial C_2 = \partial \sigma_2 + \partial \tau_2 = \langle v_2 v_4 \rangle - \langle v_1 v_4 \rangle + \langle v_1 v_2 \rangle + \langle v_3 v_4 \rangle - \langle v_2 v_4 \rangle + \langle v_2 v_3 \rangle = \langle v_1 v_2 \rangle + \langle v_2 v_3 \rangle + \langle v_3 v_4 \rangle - \langle v_1 v_4 \rangle = \langle v_1 v_2 \rangle + \langle v_2 v_3 \rangle + \langle v_3 v_4 \rangle + \langle v_4 v_1 \rangle$.

A more complex approach which forms a generalization of and is algebraically nearly equivalent to simplicial decomposition is *cellular decomposition* of space

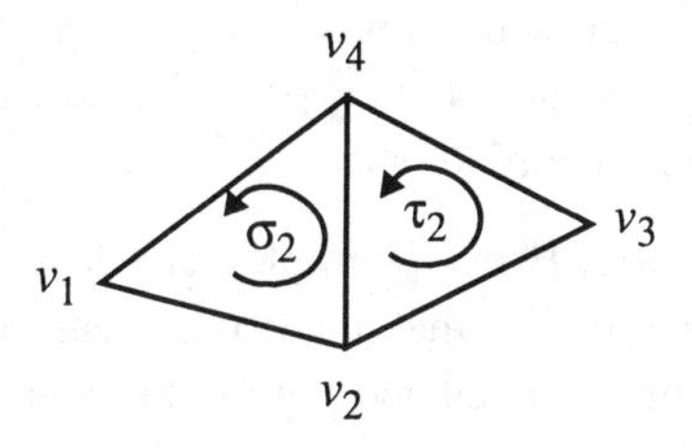

Figure 2-27

which is only briefly mentioned here. It is based on *k-cells* which are point sets homeomorphic to a k-simplex.

Definition. A *k-cell* is a point set which is homeomorphic to a closed k-disc (this includes a k-simplex). The *boundary* of a k-cell is that portion of the k-cell mapped onto by the $(k - 1)$-sphere by any homeomorphism. A *k-dimensional cellular complex* (*k-cell complex*) is a collection of k-cells such that portions of their boundary (homeomorphic to $(k - 1)$-cells) have been identified to one another.

A k-complex of special interest is the punctured cell consisting of a k-cell from which a smaller k-cell has been cut out. In two-dimensional space this is homeomorphic to an annulus, in three-dimensional space to a cylinder. In contrast to simplexes which are convex hulls and therefore straightly bounded, cells allow to model non-straight lines as well as non-convex regions. Furthermore, cells make it possible to model objects with holes. One main difference between simplexes and cells is that of storage. Normally for the same spatial object a cellular realization needs fewer cells than a simplicial realization needs simplexes.

Most proposed spatial data models using methods of algebraic topology rest on simplicial topology, since it is the simpler theory and since data structures and algorithms for simplicial complexes are easier to design and implement. A simplicial decomposition forms a complete partition of space and determines a geometric framework as a closed world from which meaningful spatial objects can be built. Spatial objects are all located in this world and constructed by aggregating simplexes and complexes of the partition. The use of simplexes and complexes implies that spatial objects are connected, not self-intersecting, and have no holes.

Two *completeness principles* [EFJ89, Eg89b, FK86] guarantee that the space contains each simplex only once and that no isolated simplex exists:

- *Completeness of incidence*: The intersection of two k-simplexes is either empty or a face of both simplexes. This rule implies that no two distinct spatial objects exist that occupy the same space and that in the two-dimensional case the intersection of lines at points which are neither start nor end points of lines is excluded. The rule enforces geometric consistency of related spatial objects. For example, an edge may represent both a part of a state boundary and a part of a river but the edge is only recorded once.

- *Completeness of inclusion*: Every k-simplex is a face (boundary simplex) of a (k+1)-simplex. This rule implies that in a two-dimensional space every point is a start or end point of a line (no isolated points exist) and that every line is part of the boundary of a triangle.

Frank and Kuhn [FK86] use algebraic topology to solve the conflict between the infinite precision real numbers of Euclidean geometry and the finite precision number systems of computers. They propose an approach in which the topological structures and relationships of point sets underlying spatial objects are separately recorded and independent of metric positions. Their topological data model (later continued in [EFJ89]) is based on simplicial complexes (cell complexes) and forms a discrete geometric basis both for modelling and for implementation. Essentially they offer an irregular triangular network partition of the plane as a geometric domain over which spatial objects can be defined. A many-sorted algebra for cell complexes is introduced which provides a variety of operations like creating an initial cell complex, adding a point to a cell complex, connecting two points in a cell complex with a line, deleting a point, or deleting a line.

Egenhofer *et al.* [EFJ89] show the simplicity of an implementation of simplicial structures. They present a simplicial algebra with only a small set of operations which are closed within the simplicial decomposition, i.e., an operation manipulating a simplicial complex can produce only a simplicial complex. Update operations must be consistent, and the completeness principles must be ensured after each modification. Algorithms proposed relate to the insertion of a node, a line, and a polygon into a simplicial decomposition.

Egenhofer [Eg89b, Eg91a] uses simplicial topology for the definition of topological relationships which requires an extension of the concepts of *boundary* and *interior*. Applied to a k-simplex, the *boundary* operation yields only faces of dimension $k-1$ so that only these faces can be taken into account by relationships. The treatment of relationships which are based on faces of dimension $k-2$ or less is impossible. For example, the coincidence of the boundaries of two 2-simplexes (regions) in a 1-simplex (line) can be treated by the boundary operation; the coincidence in a 0-simplex (point) cannot be treated, since the intersection of the two boundaries does not identify any common parts. Hence, the operations *boundary* and *interior* are generalized to the operations *boundingFaces* and *interiorFaces*, respectively, which consider all boundary and interior faces down to dimension 0. Their algebraic definition is based on the notion of *r-skeleton*. The r-skeleton of a complex C_k, denoted by $C_k^{(r)}$, is defined as the union of all simplexes of dimension at most r:

$$C_k^{(r)} = \bigcup_{i=0}^{r} \{\sigma_i \in C_k\}.$$

The *boundingFaces* of a k-complex C_k, denoted by $\partial^f C_k$, is defined as the $(k-1)$-skeleton of the boundary of C_k:

$$\partial^f C_k = \bigcup_{r=0}^{k-1} C_k^{(r)} \in \partial C_k.$$

The *interiorFaces* of a k-complex C_k, denoted by $C_k^{\circ f}$, is the set of all faces of the k-skeleton of C_k which are not part of the *boundingFaces*:

$$C_k^{\circ f} = C_k^{(k)} \setminus \partial^f C_k.$$

The dimension of the *boundingFaces* of a k-complex C_k is defined to be the largest dimension of all faces in $\partial^f C_k$, i.e. k–1. The dimension of the *interiorFaces* $C_k^{\circ f}$ is k.

Topological relationships are now defined by intersecting all combinations of bounding and interior faces of two complexes ($\partial^f \cap \partial^f$, $\partial^f \cap {}^{\circ f}$, ${}^{\circ f} \cap \partial^f$, ${}^{\circ f} \cap {}^{\circ f}$) and by examining the emptiness and non-emptiness of the intersection results. This procedure corresponds to Egenhofer's 4-intersection model (see Section 2.3.4) and leads to the same six possible relationships (see Figure 2-24). Adding dimension as a second topological invariant allows to define refined topological relationships like 0-*meet* and 1-*meet*.

Approaches based on algebraic topology support some of the design criteria we are interested in. Generality is supported by cell complexes which allow to model regions with holes. Problems of numerical robustness and topological correctness are solved by focussing only on topological properties of spatial objects and by excluding the handling of metric properties. Connections to the underlying finite arithmetic are missing so that approaches based on algebraic topology are independent of the limitations of number systems in computers and thus well suited for the implementation in a computer. Predicates describe topological relationships among spatial objects and allow user queries about neighbourhood and inclusion to be processed without performing numerical calculations.

2.3.6 Order Theory

So far, little attention has been paid to *order theory* as a complementary or even alternative mathematical concept for the modelling of spatial information and in particular of spatial relationships. Order theory provides a different way of looking at spatial relationships in comparison with topological and metric approaches and can be applied to describe spatial situations of inclusion and containment. Two important kinds of order have to be distinguished: *strict order* and *partial order*.

A strict order relation establishes a hierarchy of the elements of a set. A subdivision of space into regions, for instance, is commonly regarded as a hierarchy in which a region belongs to exactly one larger region. For example, political subdivisions form a hierarchy: countries are divided into states, states into districts, and districts into municipal areas. Hence, a municipal area belongs to exactly one district, and a district can contain several municipal areas. Another application example of strict order are perspectives such as left/right or in front of/behind.

Definition. A *strict order* on a set S is a binary relation $<$ on S, such that

$$\begin{array}{llll} \text{(S1)} & \forall\, x, y \in S: & x < y & \Rightarrow \quad \neg(y < x) \text{ (asymmetry)} \\ \text{(S2)} & \forall\, x, y, z \in S: & x < y \wedge y < z & \Rightarrow \quad x < z \text{ (transitivity)} \end{array}$$

A set S with an asymmetric and transitive relation $<$ is called a *strictly ordered set*. If for two elements $x, y \in S$ with $x \neq y$ there exists a relation $x < y$, x and y are called *comparable*. If within a strictly ordered set S any two elements are comparable with each other, i.e., $\forall\, x, y \in S, x \neq y : x < y \vee y < x$, we obtain a *total order* or *linear order* and S is called a *totally (linearly) ordered set* or a *sequence*. As an example, consider all railway stations on a selected railway line from Hamburg to Berlin.

Hierarchical structures are too restrictive to model all spatial relationships that are based on the order of spatial objects. The attempt to combine several hierarchies of subdivisions of space, for instance, inevitably results in overlapping regions, a situation which cannot be expressed by hierarchical structures. One way of overcoming the limitations of a hierarchy is the use of a partial order as a more general order leading to the structures of *partially ordered sets* and *lattices* [Bi67, DP90, Gr78]. For example, the relationship between school districts and voting districts is a partial order. While some school districts may be completely included within a voting district, other school districts may either overlap or do not have any area in common with voting districts. A hierarchy is a special type of a partially ordered set with exactly one path from any element to the top.

Definition. A *partial order* on a set P is a binary relation $\leq$ on P, such that

$$\begin{array}{llll} \text{(P1)} & \forall\, x \in P: & x \leq x \text{ (reflexivity)} & \\ \text{(P2)} & \forall\, x, y \in P: & x \leq y \wedge y \leq x & \Rightarrow \quad x = y \text{ (antisymmetry)} \\ \text{(P3)} & \forall\, x, y, z \in P: & x \leq y \wedge y \leq z & \Rightarrow \quad x \leq z \text{ (transitivity)} \end{array}$$

A set P with a reflexive, antisymmetric, and transitive relation (*order relation*) $\leq$ is called a *partially ordered set* or *poset*. The relation $x \leq y$ is read as "less than or equal" or "is contained in". If for two elements $x, y \in P$ there exists a relation $x \leq y$, x and y are called *comparable*. The inverse relation $\geq$, read as "greater than or equal" or "contains", defines a poset, too, the *dual* of P. An example of a poset is the power set of a set with its set-theoretic inclusion $\subseteq$.

Each (finite) poset can be represented graphically by an order diagram, also called the *Hasse diagram*. The construction of such a diagram is based on the concept of a covering relation which only allows for the direct neighbours of an order relation and which omits the drawing of the transitive relations of a poset. For two elements $A, B \in P$, A *covers* B (or B *is covered by* A) means that $B \leq A$ and there exists no $x \in P$ such that $B < x < A$. A Hasse diagram of a poset P consists of cycles representing the elements of P and connecting straight lines indicating the covering relation. If A covers B, the circle of element A is drawn immediately above the circle of element B, and both are connected with a straight line.

As an example, let us take the containment of regions that can be regarded as an order relation in the sense of posets. In Figure 2-28(a) five regions over the same space are drawn where regions B and C are contained in region A and regions D and E are contained in both B and C. B and C overlap. Figure 2-28(b) shows the Hasse diagram of the corresponding poset. Another view of the Hasse diagram of a poset is its interpretation as a directed acyclic graph from the greatest to the least element (or vice versa) so that data structures and algorithms of graph theory can be used to store posets and to perform operations on them.

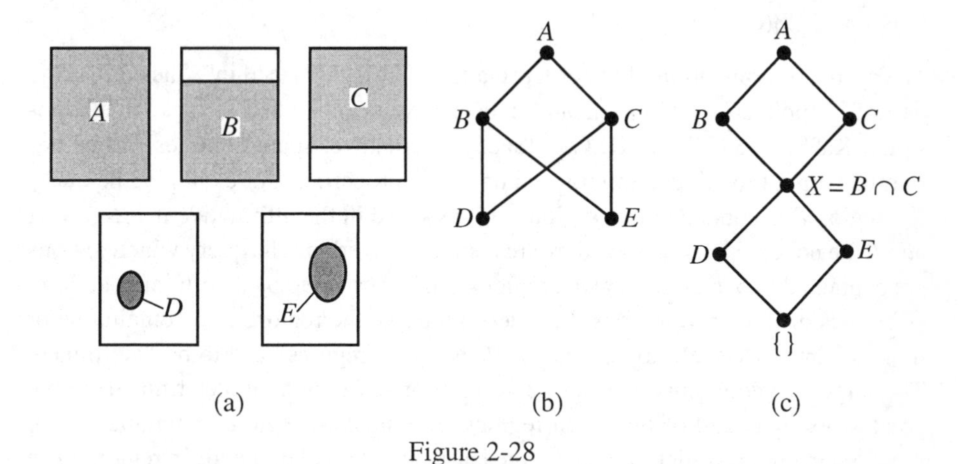

Figure 2-28

For posets and subsets of ordered sets it is often desirable to consider elements that are greater or less than all other elements. Let P be a poset and $S \subseteq P$. An element $a \in S$ is the *greatest* (*least*) *element* of S if $a \geq x$ ($a \leq x$) for every $x \in S$. An element $x \in P$ is an *upper bound* (*lower bound*) of S if $s \leq x$ ($s \geq x$) for all $s \in S$. The set of all upper bounds of S is denoted by S^* (read as "S upper"), and the set of all lower bounds is denoted by S_* (read as "S lower"). If S^* has a least element, it is called the *least upper bound* of S. If S_* has a greatest element, it is called the *greatest lower bound* of S. A least upper bound or a greatest lower bound is always unique.

By using posets not all queries regarding the order of spatial objects can be answered. An example is the question to look for the largest region that is contained in a set of regions, or in terms of order theory for the greatest lower bound. Unfortunately, for a given set of poset elements there need not always exist a greatest lower bound (and a least upper bound). Hence, a more specific order structure is required.

Definition. A *lattice* L is a poset in which every pair of elements has a least upper bound and a greatest lower bound. A lattice is called *complete* if a least upper bound and a greatest lower bound exist for every subset of the poset.

The poset in Figure 2-28(b) is no lattice, since D and E have no greatest lower bound. An important mathematical theorem of lattice theory states that every finite lattice is complete so that only complete lattices occur in spatial applications. Another theorem states that every poset P can be embedded into a complete lattice L by suitably adding elements to P so that inclusion and all greatest lower bounds and least upper bounds in P are preserved. The process of the *normal completion* of P [Ma37] specifies how to find a smallest set of elements that has to be added to P in order to create a lattice, i.e., to construct the minimal containing lattice for a poset. The normal completion of the poset in Figure 2-28(b) to a complete lattice is shown in Figure 2-28(c).

The spatial interpretation of the order relation $\leq$ is "is contained in", "lies in", or "is part of". Applications of posets and lattices have been studied for spatial subdivisions [Ka88, Ka89, Ka90, KEG93, Sa85]. In a lattice, spatial queries can be answered without looking at topological or metric properties. For example, the query if a region B is contained in a region A is answered in the affirmative if $B \leq A$. If A and B are not comparable, then they are disjoint or overlap. The query which regions are contained in one or more given regions can be answered by determining all lower bounds of the given regions. In order to retrieve the regions that contain one or more given regions, all upper bounds of the given regions have to be determined. The largest region contained in one or more given regions is determined by the greatest lower bound of the given regions. The smallest region containing one or more given regions is determined by the least upper bound of the given regions. The queries show the importance of greatest lower bounds and least upper bounds and the necessity of complete lattices and normal completion in order to find an answer in all possible cases.

New elements added to the poset to form a lattice can be interpreted as the intersection of poset elements. In the lattice of Figure 2-28(c) two new elements have been created, one of them being the empty set. The element X can be interpreted as the spatial intersection of B and C. The least upper bound X of D and E is not their spatial union but the least element that contains both D and E, namely the intersection of B and C. It is obvious that any element "somewhere between" the intersection of B and C and the union of D and E could have been taken as element X. In [Ka88] Kainz shows how to use lattice theory for merging several spatial subdivisions by simultaneously keeping regions ordered under inclusion, and how to use it for creating new intersections and for preserving lattice structures. In [KEG93] lattices are used as a high-level model for spatial regions and their relationships. This model uses simplicial complexes for the topological representation of regions.

Comparing topological and order relations, the differences between them can be illustrated by the example of the two regions Europe and Germany. According to order theory Germany is contained in Europe (is part of Europe); from a topological point of view, however, the polygon Germany is an island in the polygon Europe (topologically speaking, it does not belong to Europe). An overview in [Ka90] enumerates the correspondences between topological and order relations and reveals that by topological means more specific spatial relationships can be described, whereas order relations are more general (Figure 2-29).

Topological Relationships	Corresponding Order Relations
A and B are disjoint	A and B are not comparable
A and B meet	A and B are not comparable
A and B overlap	A and B are not comparable
A is equal to B	$A = B$, i.e. $A \leq B \wedge B \leq A$
A is inside of B	$A \leq B$
A is covered by B	$A \leq B$
A contains B	$B \leq A$
A covers B	$B \leq A$

Figure 2-29

2.4 Numerical Robustness and Topological Correctness of Geo-Algorithms

The design criterion of finite resolution, numerical robustness, and topological correctness relates to the underlying geometry of a spatial algebra. Spatial operations are, in general, realized by methods, principles, and algorithms of Computational Geometry. Theoretical computational geometry [Fo87, LP84, Me84, PS85, Ya92] has provided a large number of useful and efficient geometric algorithms which rest on coordinate-based Euclidean geometry as their underlying model and are designed with continuous parameters and infinite-precision arithmetic in mind. However, it is not easy to create efficient and reliable programs from these algorithms, because number systems of computers are necessarily finite and consequently the programs suffer from rounding errors. The central problem is the assumption of an infinite-precision arithmetic in geometric algorithms and the reality of a finite-precision arithmetic in computer systems [EFJ89, EH91, FK86, GY86, KBS91a].

In the programmer's practice this conflict is mostly neglected, and computer arithmetic is used as if it were infinitely precise[13]. Precision, robustness, and correctness of geometric primitives (like segment intersection) within geometric algorithms are simply taken for granted. But geometric programs are quite sensitive regarding this procedure which leads not only to numerical errors but also to topological inconsistencies and degeneracies, since topological information must often be inferred from coordinate-based geometric data. The implementor of a geometric algorithm has to take into account that computers do not always yield exact and correct results, even if the input data are assumed not to be affected with inaccuracy, and that the result of a computation might crucially depend on the order of the single computation steps. Forrest [Fo85] demonstrates the discrepancy between theoretical and applied computational geometry by discussing the problems of finding numerically robust algorithms for the point-in-polygon test and segment intersection. [BKK84] shows that degeneracies can even occur in simple point-in-polygon tests.

In a computer system providing floating-point arithmetic, a coordinate-based representation of spatial objects can lead to rounding errors, since it is only feasible with limited precision, i.e., over a finite, discrete, and heterogeneous grid. A new calculated intersection point, for instance, is moved by rounding it to the nearest grid point. This procedure cannot only cause numerical errors, which, in general, are relatively small and could be perhaps accepted in some situations, but unfortunately topological inconsistencies and degeneracies. Due to the rounding process the intersection point can walk to the wrong side of a line. Topological errors cannot be accepted, since they are visible in the results of spatial operations integrated into spatial query languages at the user interface of a system. The negative influence of rounding errors is intensified by performing iterations of calculations reusing already computed imprecise results. Not only each single calculation but a series of calculations as well must be robust and topologically correct. Many other topological errors are conceivable; some examples are:

- Spurious intersections can be produced if the intersection of a line *a* with a line *b*, when rounded, creates an intersection of line *a* with a line *c* that was not originally present [GY86].

---

[13] The other extreme is that symbolic computation is used so that numerical problems in geometric computations disappear and all computations lead to exact results. But this approach suffers from high inefficiency, and furthermore, Hoffmann [Ho89] shows that a purely symbolic approach raises logical existence problems of certain geometric configurations.

- The scaling of coordinates may change topology by moving points, initially close to a line, from one side to another [Fr84].
- A series of intersections of a line a with lines $b_1, b_2, \ldots$ by applying rounding each time can cause line a to drift away from its original position [GY86].
- The intersection of two lines does not necessarily lie on both lines [NS88].
- Applying two inverse geometric operations may generate a geometry which differs from the original geometry [DS88, Ho89].
- By rounding, the order of lines can be changed and lead to topological inversion [GY86].
- The important map overlay operation known from GIS introduces so-called gaps and slivers which have to be removed with computationally expensive and conceptually dubious methods producing further errors in the spatial data [EFJ89, Fr87].

Concepts have to be devised which guarantee topological immunity to slight numerical changes of coordinate data and invariance under basic object transformations like rotation, scaling, and translation. A set of fundamental *geometric primitives* [CG82, FK86] has to be provided that use finite-precision arithmetic and lead to the construction of more complex geometric operations which are numerically *stable*, i.e., *correct*, *robust*, and *efficient*. Correctness means that the implementation should always yield the right topological result compared to reality. Robustness implies that all cases occurring in a geometric operation should be handled correctly. Efficiency emphasizes the requirement that the possible advantages of an elaborate finite-precision algorithm should not be neutralized by time-consuming numerical computations. The conclusion is that a *spatial information theory* is needed pursuing two main goals, namely the modelling and representation of spatial objects which is compatible with the finiteness of computers, and the design of stable geometric algorithms that can be implemented by using geometric primitives based on finite-precision arithmetic.

The problem of numerical robustness and topological correctness in geometric computations has been studied for several years in computational geometry. But so far no general agreement has been achieved what strategy might succeed in solving this problem. The problem is treated from three different points of view: either it is considered as a matter of achieving sufficient numerical precision, or as a fundamental difficulty in dealing with interacting numerical and topological data, or as a problem of avoiding degenerate configurations. Proposals for solving the problem can be distinguished into three categories: the *ad hoc* approach, *perturbation-free* approaches, and *perturbation* approaches. *Perturbation* relates to slightly changing input data or computed values in a suitable way when they are assigned to variables.

The principle of perturbation is that the approximate result computed by a program can be viewed as the exact result on slightly changed input data. In the following, a brief survey of attempts to solve the problem of numerical robustness and topological correctness is given. More details can be found in [DS88, DS90, Ho89].

The most frequently applied strategy is the *ad hoc* approach which tries to handle each occurrence of a rounding error individually. This usually results in arbitrarily increased precision or arbitrarily selected tolerances in the form of epsilon values. It should be clear that this approach solves problems only temporarily and that it is far from being robust and consistent.

Perturbation-free approaches aim at performing exact geometric computations with such sufficiently high precision that correct and robust numerical results must be obtained. Provided that the input data are exact, the task is to determine how many digits of precision are required by numerical computations so that the algorithm produces correct results and takes desired accuracy into account. One approach proposes modified floating-point systems. Specific techniques are suggested for implementing a sequence of floating-point computations in a way that minimizes rounding errors. Kulisch and Miranker [KM81] introduce an exact *scalar product* function that computes the scalar product of two vectors by rounding only at the end instead of after each individual addition or multiplication. Using this scalar product and assuming that the given input is exact, i.e., the coordinate values are machine numbers in the floating-point system of the given machine, Ottmann, Thiemt, and Ulrich [OTU87] show how to implement numerically "stable" geometric primitives like a robust segment-intersection algorithm. Matula and Kornerup [MK80, KM83, MK85] try to avoid floating-point computation and use finite-precision *rational arithmetic*. Their approach is based on number theoretic foundations like continued fractions and Farey series. The problem of rational arithmetic is that it can be slow and unwieldy for repeated calculations and that the numerators and denominators grow rapidly and can become very large.

Another approach uses *interval arithmetic* which represents a real number as an interval bounded by its two nearest floating point numbers. Arithmetic operations are defined on intervals, and result intervals are extended if necessary. Mudur and Koparkar [MK84] apply interval arithmetic to the processing of spatial objects and try to restrict computed intervals returned by interval-based geometric functions. Geometric algorithms use these functions for tasks like curve drawing and intersection detection. The drawback of interval arithmetic is that the intervals extend rapidly as the computations advance. While spatial objects, in general, become smaller during iterated intersections, the intervals become larger.

Fortune [Fo89] introduces approximate ε-*arithmetic* with corresponding operations and explains correctness, robustness, and stability of a geometric algorithm implemented in ε-arithmetic. The approach which is demonstrated for geometric predicates, the calculation of convex hulls, and point set triangulations in the plane has the drawback that the precision required by operations increases with the number of inputs. Sugihara and Iri [SI88] show how to compute Voronoi diagrams in a numerically robust and topologically consistent way using finite-precision arithmetic.

Perturbation approaches allow to slightly change input data or computed results. Because in many applications the input data are approximate from the beginning, such slight alterations seem to be acceptable. Using floating-point arithmetic Dobkin and Silver [DS88, DS90] illustrate the difficulty of precisely performing iterations of geometric computations. They show that even the numerical output of very simple geometric operations can be rather inaccurate. Their proposal is a statistical approach which computes an accuracy estimate for each geometric calculation. The input data for the calculations are systematically perturbed and the effect of the perturbation on the output is measured, especially for iterations of calculations. In contrast to their encouraging results, experiments of Hoffmann [Ho89] testing their approach are less promising.

Milenkovic [Mi88] suggests two methods for verifiable correct finite-precision implementations of geometric algorithms. The first method is *data normalization* which changes the structure of a spatial object by *vertex shifting* and *edge cracking* to maintain a distance of at least ε between vertices and edges of an object. Neither two vertices nor a vertex and an edge may be closer to each other than some predefined tolerance ε. Vertex shifting merges two vertices that are closer than ε into a single vertex. Edge cracking splits an edge passing within distance ε of a vertex into two edges both having the vertex as an end point. The second method is the *hidden variable method* which models a geometric line by an approximation deviating at most ε from the line it represents. In a similar approach Hoffmann *et al.* [HHK88] compute an exact result for perturbed input data and assume a *minimum feature separation* of spatial objects and their constituent parts like vertices and edges. Segal and Sequin [SS85] introduce the methods of *minimum feature size* and *face thickness* to spatial objects. These methods either merge or pull apart those objects lying within some separation distance from each other so that it is possible to distinguish precisely between incident and non-incident objects. The technique is similar to Milenkovic's data normalization method but differs in that the coordinates are rounded to grid points.

Greene and Yao [GY86] show a discrete version of the segment-intersection problem by transforming spatial objects from the continuous domain to the discrete domain (which is a uniform grid) and performing all computations in the discrete domain. End points of line segments and all segment intersections lie on grid points. For this purpose, true intersection points are rounded to the nearest grid points, and the line segments participating in an intersection are suitably replaced by their *redrawings*. A redrawing of a line segment is a polygonal line which is the shortest path traversing the rounded intersection point and the end points of the segment within the *envelope* of this segment. The envelope of a segment is roughly the set of grid points surrounding and lying on the segment. This procedure which assumes precise computations and allows slight perturbations of the original line segments guarantees topological correctness and consistency. Similar to [GY86] Franklin *et al.* [FCKSA88] achieve numerical robustness when performing intersections on a uniform grid.

Edelsbrunner and Mücke [EM88] deal with the problem of handling degeneracies which is related to the problem of robustness. They describe a programming technique, called *Simulation of Simplicity*, which handles these situations without having to specify all the single special cases explicitly and which removes existing degeneracies without creating new ones. As an example, they consider points and remove collinearity of three and coplanarity of four or more points. Yap [Ya88] also studies the problem of geometric degeneracies, such as the possibility of three collinear points. The method he suggests uses perturbations on the input data. Schorn [Sc94] investigates the problem of degeneracy in geometric algorithms and the suitability of the perturbation approach for solving the problem. Distinguishing between problem-dependent and algorithm-dependent degeneracies he comes to the conclusion that only algorithm-dependent degeneracies can be reliably eliminated when following the principle of perturbation.

In [Mi89] Milenkovic proposes a topology-preserving strategy called *double precision geometry* for performing numerical computations on lines and line segments by using rounded arithmetic and by eliminating expensive computations. The technique is demonstrated for the intersection of line segments. Guibas *et al.* [GSS89] introduce *epsilon geometry* which allows to build robust primitives and algorithms using imprecise computations. The algorithms compute an exact solution for suitably perturbed input data together with a bound on the size of this perturbation. Nagy *et al.* [NME90] apply two techniques for dealing with numerical inaccuracy. The first technique rests on explicit constraints about the intended coincidence of two or more vertices, the collinearity of three or more vertices, and the parallelism of two or more line segments. The second technique is a sophisticated form of rounding. A best integer approximation depending on given constraints is introduced for the rational coordinates of an intersection of two line segments with integer coordinates.

As a conclusion, implementing numerically stable algorithms in practice is a complex task and an open problem. For special situations successful approaches have been developed, but so far no universally valid and applicable solution has been found for the problem of numerical robustness and topological correctness of geometric algorithms implemented on a computer system.

2.5 Finite Precision Computational Geometry

The knowledge about the problems of numerical robustness and topological correctness in geometric algorithms, application areas like computer graphics and VLSI design, and the requirement to develop robust, efficient, consistent, reliable, and accurate graphics and geometry systems have raised growing interest in solving problems of computational geometry in a *discrete* space on a uniform integer grid. That is, points, end points of line segments, vertices of polygons etc. have integer coordinates instead of arbitrary floating-point coordinates. This has led to a small subfield of computational geometry called *finite precision computational geometry* [Ya92] or *finite resolution computational geometry* [GY86] dealing with geometry performed on a discrete domain. On a grid in a number of situations problems can be solved more efficiently and often simpler than in arbitrary Euclidean space. The main reason is that techniques like table look-up or perfect hashing can be used which frequently replace searching or the need for rebalancing structures. On the other hand, problems like how to handle the intersection point of two integer line segments are more complicated.

Finite precision computational geometry has so far only been studied by a few researchers (overviews can be found in [KK81, Ov88b, Ov88c]). Problems considered for example are the nearest neighbour searching problem [KM85], range searching on a grid [Ov88a, Ov88b], the point location problem [Mü85], the computation of rectangle intersections and maximal elements by divide-and-conquer [KO88b], computing the convex hull of a set of points, reporting all intersections of a set of arbitrarily oriented line segments, and the calculation of rectangle intersections and maximal elements by using the plane sweep technique [KO88a, Ov88b]. A main problem is the treatment of the intersection point of two integer line segments which usually does not fall on a grid point with integer coordinates but has rational coordinates. Greene and Yao [GY86] and Milenkovic [Mi89] have proposed solutions for this problem which have been sketched in the preceding subsection, allow perturbations of the original segments, and maintain topological correctness.

Chapter 3

Realms: A Foundation for Spatial Data Types in Database Systems

In this chapter we introduce a new concept which represents a formal foundation for the definition of spatial data types in database systems. This concept is called *realm* [GS93], a structure which is composed of points and non-intersecting line segments defined over a discrete domain, i.e., a grid. The first section illustrates the basic ideas of the realm concept and describes its main purposes. The second section together with an appendix introduces integer arithmetic and robust geometric primitives which form the basis for higher structures. One of the most important and difficult problems over a discrete domain is the treatment of the segment intersection problem. The third section presents a topology-preserving solution of this problem called *redrawing*. After an introduction into the problem, the main concepts, properties, and number-theoretical foundations of the redrawing process are explained. For a restricted case a redrawing algorithm is given which is afterwards generalized. The fourth section gives a formal realm definition which emphasizes its nearness to *spatially embedded planar graphs*. The fifth section presents operations on realms which are summarized in the *realm interface*. Algorithms are given for the three most important realm operations, namely for inserting points and segments and for splitting a realm segment by a point. The sixth section introduces realm-based structures (e.g., cycles, faces) and primitives on these structures (e.g., inside, meet, intersect) that can be discovered within a realm and that are useful for the definition of spatial data types. The seventh and last section compares the realm concept with related work.

3.1 The Realm Concept

For spatial modelling usually Euclidean space is used or implicitly assumed which represents points in the plane by pairs of infinitely precise, real numbers. Unfortunately, number systems in computers are finite and offer only limited approximations like floating point numbers. Hence, system and user data types are always based on some underlying "finite domains of possible values", and several data types may depend on the same domain. The fact that numerical data types are sim-

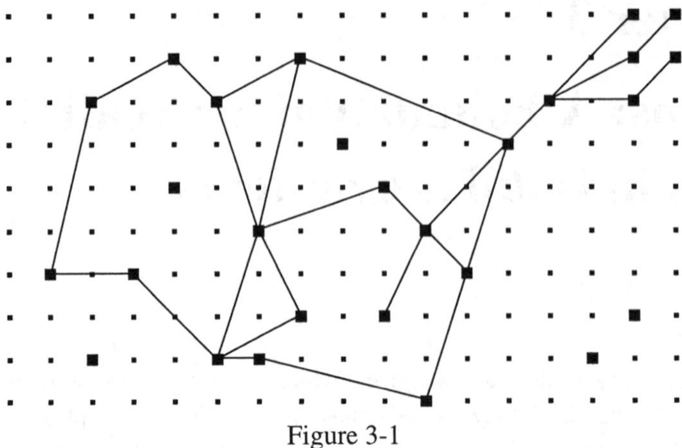

Figure 3-1

ply regarded as arbitrarily precise leads to many problems in geometric computation (see Section 2.4). In order to avoid confusion with other terms, the notion of a *realm* is introduced. In general, a realm is a finite, user-definable structure with certain properties that is used as a basis for the definition of one or more user data types. A realm is somewhat similar to an enumeration type in programming languages. Within the framework of this book we are in particular interested in realms that allow us to model the complete underlying geometry of a spatial application by providing a discrete geometric basis.

A *realm* [GS93] used as a basis for the definition of spatial data types is a finite set of points and *non-intersecting* line segments over a discrete domain (Figure 3-1) and can from a graph-theoretical point of view be considered as a *spatially embedded planar graph* over a finite resolution grid. Intuitively, it describes the complete underlying geometry for a collection of spatially-referenced objects of an application. All spatial objects like points, lines, and regions can be defined in terms of realm objects, that is, in terms of points and line segments present in the realm. In fact, spatial objects are never created directly but only by suitably selecting some realm objects and composing them to spatial objects. They are never updated directly. Instead, updates are performed on the realm and from there propagated to the dependent spatial objects. Hence, all spatial objects occurring in a database are *realm-based*. The construction and modification of a realm is feasible by using a special interface (see Section 3.5) or interactively by employing a graphical user interface (see Section 6.3.3).

The underlying grid of a realm arises from the fact that numbers have a finite representation in computer memory. In practice, these representations will be of fixed length and correspond to INTEGER or REAL data types available in programming languages or to special, higher precision implementations of number systems. Of

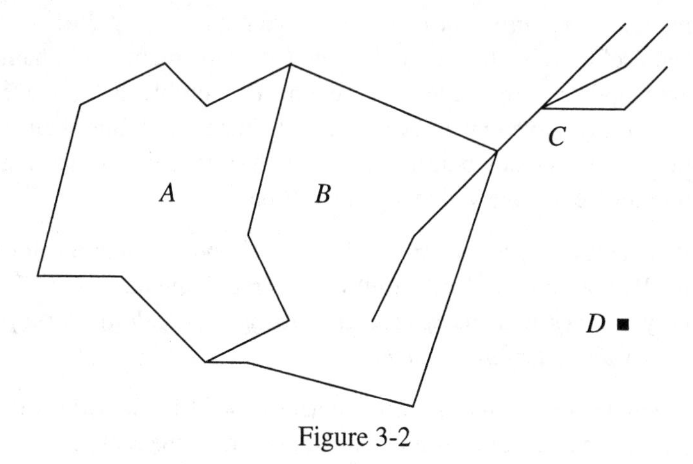

Figure 3-2

course, the grid resolution selected for a concrete application will be much finer than could be shown in Figure 3-1.

Figure 3-2 shows some spatial objects definable over the realm of Figure 3-1. The *realm-based spatial data types* are called points, lines, and regions and represent the geometric sorts (types) of the ROSE algebra. Hence, *A* and *B* represent regions objects, *C* is a lines object, and *D* a points object. One can imagine *A* and *B* to belong to two adjacent countries, *C* to represent a river, and *D* a city.

The realm concept as a foundation for the definition of spatial data types supports the design criteria established in Section 1.2 and integrates several apparently distinct problems of spatial modelling in one coherent approach. It serves the following main purposes:

- It allows to base several distinct spatial data types on a common underlying domain which is manipulable by the user. These data types can be used as component types (attribute types) in spatially-referenced objects. In operations of query languages they are used in the same way as standard data types.

- It guarantees nice *closure properties* for the computation with spatial data types above the realm. The algebraic operations for the spatial data types are defined in a way to construct only spatial objects that are realm-based as well. For example, the intersection of region *B* with line *C* (the part of river *C* lying within country *B*) is also a realm-based lines object. Hence, the spatial algebra is closed with respect to a given realm.

- It shields geometric computation in query processing from problems of *numerical robustness* and *topological correctness*. Problems arise from the computation of intersection points of line segments which normally do not lie on the grid. With realm-based spatial data types, there are *never any new intersection points to be computed* in query processing. Instead, the numerical problems are treated *within and below* the realm level, namely, whenever updates are made to a realm. The problems of correctness are essentially solved by an explicit management of all line segment intersection points as realm objects together with slight geometric changes of those line segments involved in an intersection.

- It provides the programmer with a precise specification on all levels of the model that directly lends itself to a correct implementation. This particularly implies that the *spatial algebra obeys algebraic laws precisely in theory as well as in practice.*
- It enables the design of very efficient geometric algorithms if spatial objects as their arguments are defined over the same realm.
- At the level of the data model, it solves the problem to keep spatial information consistent that is simultaneously contained in several objects by enforcing *geometric consistency.* Regions may now consistently share common borders, a country boundary may be precisely defined by the course of a river, or a power supply line may go along a road. The common part of the borders of countries *A* and *B* in Figure 3-2 is exactly the same for both objects. This also means that distinct spatial objects sharing common parts can be kept consistent when performing operations on them.
- A data structure representing a realm can be used as a spatial index structure into the database, either as a secondary or a primary (clustering) index. Our implementation concept assumes that each point and segment in a realm has an associated list of logical pointers to the spatial attribute values defined over it in the database. This also allows one to support a common spatial clustering and indexing of spatially-referenced objects of *distinct* classes. Since realms represent not only geometric but also topological information, they can be used to efficiently support operations like neighbourhood testing.

3.2 Integer Arithmetic and Robust Geometric Primitives

The formal definition of realm-based spatial data types is organized as a series of layers (see Appendix B) that are described bottom-up. Each layer defines its own structures and primitives and uses the notions of the layers below. To be able to distinguish operations of the various layers we use the following typographical conventions:

- Layer 1 - integer arithmetic: italic (e.g., *div*)
- Layer 2 - robust geometric primitives: underscore (e.g., intersect)
- Layer 3 - realms and realm-based primitives: underscore italic (e.g., *area-disjoint*)[14]
- Layer 4 - spatial algebra primitives: bold italic (e.g., ***area-disjoint***)
- Layer 5 - ROSE operations: bold (e.g., **inside**)

Integer arithmetic forms the bottom layer of the formal development and is the basis for the definition of all higher structures and operations. It can be equal to the built-in integer arithmetic of a computer, but usually the domain of such an arithmetic is too small to allow the definition of a discrete grid space with a very high resolution. We assume that the employed integer arithmetic is error-free with respect to overflow and that it provides the following standard arithmetic primitives:

$Integer \times Integer$	$\rightarrow Integer$	$+, -, *, div, mod$
$Integer \times Integer$	$\rightarrow Bool$	$=, \neq, <, \leq, \geq, >$

The next higher definition layer introduces a finite discrete space $N \times N$ with $N = \{0, ..., n-1\} \subseteq \mathbf{N}$, points and line segments with coordinates in N as objects over this space, and some *robust geometric primitives* on them. An *N-point* is a pair $(x, y) \in N \times N$. An *N-segment* is a pair of distinct N-points (p, q). Intuitively, it describes the straight, closed real line joining p and q. P_N denotes the set of all N-points and S_N the set of all N-segments. Robust geometric primitives are simple predicates and operations defined on N-points and N-segments. Figure 3-3 shows the primitives defined between two N-segments and between an N-point and an N-segment.

In Appendix A formal definitions of these primitives are given that are based on integer arithmetic which allows direct and robust implementation free of numerical errors. We explain the primitives informally here: Two N-segments are equal (v) if their end points coincide. Two N-segments meet (iii) if they have exactly one end point in common. They overlap (vi) if they are collinear and share a (partial) N-segment. If they are collinear and do not share a (partial) N-segment, we call them aligned (viii). If they have the same slope, they are parallel (iv). The on (x) primitive tests if an N-point lies on an N-segment; the in (ix) primitive does nearly the same but the N-point must not coincide with one of the end points of the N-segment. An N-segment touches (vii) another N-segment if both N-segments do not overlap and exactly one of the end points of the first N-segment lies in the second N-segment. Two N-segments touch each other if the first N-segment touches the second one, or if the reverse relation holds. If two N-segments have exactly one common point but

[14] Unfortunately, there is a collision between the typographical conventions for realm-based primitives and for data types (both underscore italic). It is not avoided in order to remain consistent with [GS93], [GS95], and [Gü93] (the last reference will be used later as a formal framework for defining signatures).

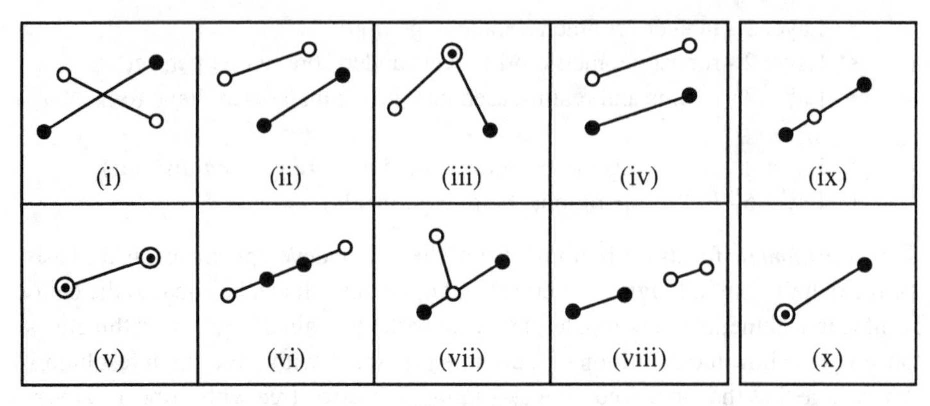

Figure 3-3

do neither meet nor touch, they intersect (i). They are disjoint (ii) if they are neither equal nor meet nor touch nor intersect nor overlap. The intersection primitive as the only non-predicate calculates the intersection point of two N-segments and rounds it to the nearest N-point.

3.3 Redrawing as a Solution of the Discrete Segment Intersection Problem

As mentioned before, numerical robustness and topological correctness problems are treated within and below the realm level. One of the main problems over a finite resolution grid is the *discrete segment intersection problem.* The first subsection outlines the underlying problem and shows that straightforward approaches can lead to topological incorrectness. The second subsection introduces a solution approach called *redrawing* and presents its underlying concepts and properties. The third subsection sketches the number-theoretical foundations of the redrawing process. How this mathematical background is applied for computing a redrawing is presented in the fourth subsection. Furthermore, a redrawing algorithm is developed for a special case. Its generalization is performed in the fifth subsection.

3.3.1 The Segment Intersection Problem Defined Over a Discrete Space

The fundamental problem when intersecting segments defined over a grid is that intersection points usually do not lie on the grid but must be moved to a grid point (here especially in order to ensure the realm properties) and that a solution strategy for this problem must be topology-preserving (see Section 2.4). The treatment of this problem is necessary, because geometric data coming from an application are, in general, not intersection-free, as required for a realm. Application data can at the lowest level of abstraction be viewed as a set of points and intersecting line segments. Hence, there is the task to transform an application's set of *intersecting* line segments into a realm's set of *non-intersecting* line segments.

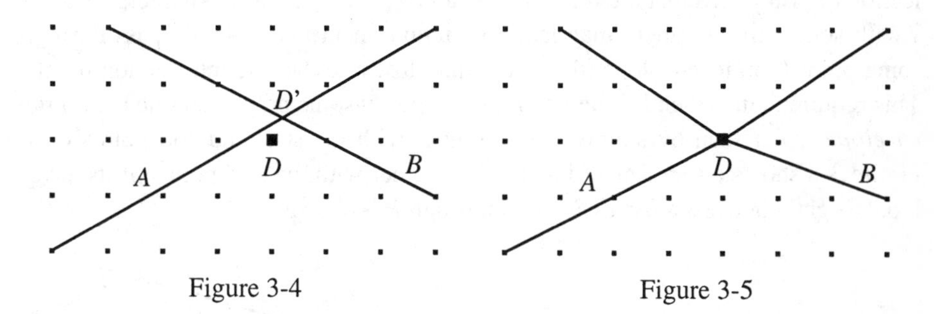

Figure 3-4 Figure 3-5

In Figure 3-4, the intersection point *D*' of line segments *A* and *B* will be moved to the closest grid point *D*. This leads, for example, to the following topological errors: (1) A test whether *D* lies *on A* or *B* fails. (2) A test whether *D* lies properly within some area defined below *A* and *B* will incorrectly yield *true*. (3) If there is another segment *C* between the true intersection point and *D*, *D* will be reported to lie on the wrong side of *C*. The basic idea to avoid these errors is to slightly change segments *A* and *B* by transforming them into chains of segments going through *D*, as shown in Figure 3-5. However, this does not suffice, since it allows a segment to drift (through a series of intersections) by an arbitrary distance from its original position. For example, a further intersection of *A* with some segment *C* (Figure 3-6) is resolved as shown in Figure 3-7, where intersection point *E* has already a considerable distance from the true intersection point of *A* and *C*. Note in particular that segment *A* has in Figure 3-7 been moved to the other side of a grid point (indicated by the arrow) which may later be reported to lie on the wrong side of *A*.

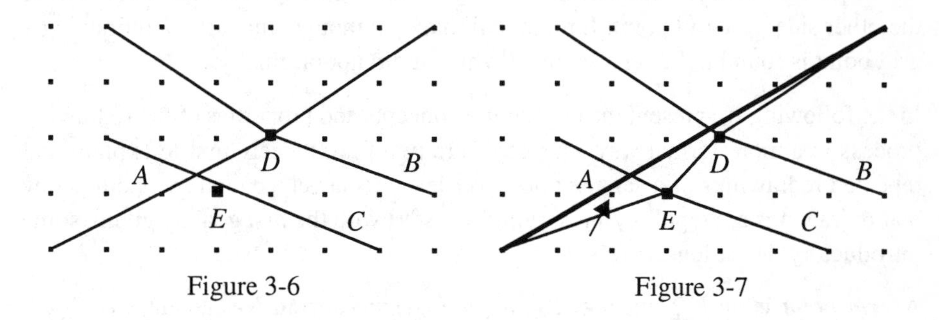

Figure 3-6 Figure 3-7

3.3.2 Concepts and Properties of the Redrawing Process

A refined and topology-preserving solution was proposed by Greene and Yao [GY86]. The idea is to define for a segment s an *envelope* $E(s)$ roughly as the collection of grid points that are immediately above, below, or on s. An intersection of s with some other segment may lead to a requirement that s should pass through some point P on its envelope (the grid point closest to the true intersection point). This requirement is then fulfilled by *redrawing* s by some polygonal line *within the envelope* rather than by simply connecting P with the start and end points of s. Figure 3-8 shows a segment s (drawn fat) together with the grid points of its envelope. Slightly above s a *redrawing* of s through P is shown.

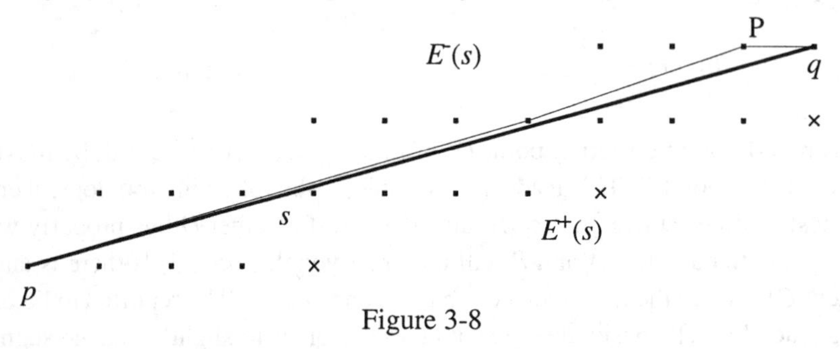

Figure 3-8

Intuitively, the process of redrawing can be understood as follows: Think of segment s as a rubber band and the points of the envelope as nails on a board. Now grip s at the true intersection point and pull it around the closest grid point P. The resulting polygonal path is the redrawing. The number of segments of this path is in the worst case logarithmic in the size of the grid, but it seems that in most cases only very few segments are created. This approach guarantees that the polygonal line describing a segment always remains within the envelope of the original segment. We adopt this technique for realms. It then means that by redrawing a segment can never drift to the other side of a grid point. It might still happen, though, that after a redrawing a grid point is found to lie *on* a segment which it did not originally.

In the following we present the underlying concepts and properties of the redrawing process in a more formal way. Our goals are two-fold, namely first to explain that repeated redrawings of a segment do never leave its envelope, and second to show that redrawings are topology-preserving. We start with the first goal by giving some introductory definitions.

A *grid point* is an N-point $p \in P_N$; a *(line) segment* is an N-segment $s \in S_N$. A *polygonal path* $T = p_1p_2...p_n = < p_1, p_2, ..., p_n >$ is a sequence of line segments (p_i, p_{i+1}), $i \in \{1, ..., n-1\}$. T is a polygon if $p_1 = p_n$. An *open line segment* corresponds to a line segment without its end points. A *unit segment* is defined as a horizontal or vertical line segment of unit length. For a grid point $p = (x, y)$ we define the square $SQ(p) = [x - 0.5, x + 0.5) \times [y - 0.5, y + 0.5)$. The set $\{SQ(p) \mid p \in P_N\}$ forms a tiling of the discrete space $N \times N$, and p is said to be the *nearest* grid point to a point q with rational or real coordinates if $q \in SQ(p)$. The length of a line segment s is denoted by $|s|$. The notation s_∞ relates to the infinite line passing through s.

For a segment s the set of *bounding points* $B(s)$ is defined as $B(s) := \{p \in P_N \mid p \underline{\text{on}}\ s \vee p$ is end point of an open unit segment intersecting $s\}$. The set of *auxiliary points* $A(s)$ of s is defined as $A(s) := \{p = (x, y) \in P_N \mid p \notin B(s) \wedge (x - 1, y) \in B(s) \wedge (x, y + 1) \in B(s)\}$. The *envelope* $E(s)$ of s is the union $E(s) := A(s) \cup B(s)$. Its elements are called *envelope points*. If $s = (p, q)$, then $\bar{E}(s) := E(s) \setminus \{p, q\}$ is called the *proper envelope* of s. If we consider $s = (p, q)$ as a directed segment from p to q with $p < q$, we can define the *left* (*right*) *envelope* of s to be the subset of $E(s)$ that lies in the left (right) closed halfplane of s_∞, and we name it $E^-(s)$ ($E^+(s)$).

Figure 3-8 illustrates some of the terms just defined. The dotted points are the bounding points of the example segment $s = (p, q)$; the crosses mark the auxiliary points. Auxiliary points are included in the envelope only for technical reasons in order to be able to handle 45°-segments. If a 45°-segment intersects another segment at the center q of a square, the convention is to map q to the nearest grid point bottom-right, i.e., to an auxiliary point. In all other cases the intersection point is rounded to a bounding point. The points of $E(s)$ define a *border polygon* formed by the points of $E^-(s)$ along the direction of s from p to q and by the points of $E^+(s)$ in the opposite direction. We will identify this border polygon with $E(s)$. The following lemma demonstrates some "closeness" properties of an envelope.

Lemma 3-1 For an envelope $E(s)$ of a segment s the following statements hold: (i) There are no grid points in the interior of $E(s)$. (ii) The nearest grid point to any point $p \in s$ is in $E(s)$. (iii) Any grid point on $E(s)$ has a Euclidean distance which is less than $\sqrt{2}$ from s.

For horizontal, vertical, and diagonal segments the three statements are obviously fulfilled. Let for statement (i) s have a slope less than 45°. Consider groups of grid points with the same x-coordinate which lie parallel to the y-axis and belong to $E(s)$. According to the definition of envelope points each group consists of either one or two grid points (auxiliary points play no role here and are not counted) lying on or within the border polygon $E(s)$. If a group contains one grid point, this grid point lies on s and hence on $E(s)$. If both grid points of a group lie on or within the border polygon $E(s)$, s necessarily intersects the vertical open unit segment between them. Hence, both grid points lie on $E(s)$, one grid point immediately above and one grid point immediately below s, and no grid point lies between them in the interior of $E(s)$. If the slope of s is greater than 45°, the argumentation is analogous and relates to grid points with the same y-coordinate. The two cases with a negative slope of s are treated in the same way as the corresponding two cases with a positive slope.

The argumentation for statement (ii) is as follows: If $p \in s$ is a grid point, then $p \in E(s)$. If p is the intersection point of s and an open unit segment whose end points are envelope points, then one of these two envelope points is the nearest grid point. Otherwise p lies in a unit square traversed by s. This unit square consists of exactly four (or three for an end point of s) corner grid points being all elements of $E(s)$ and contains no other grid points as shown in statement (i). Hence, one of these four (three) envelope points is the nearest grid point to p.

Statement (iii) can be proved as follows: If $p \in E(s)$ lies on s, the distance of p to s is equal to 0 and thus less than $\sqrt{2}$. Otherwise, p is a corner grid point of one or two unit squares traversed by s. The longest distance within each such unit square between one of its corner grid points and p is the length of the diagonal line which is equal to $\sqrt{2}$. Since s traverses each unit square, the shortest distance between p and s is less than $\sqrt{2}$. ❑

The rounding of an intersection point p to the nearest envelope point q is marked by a vector called *hook* from p to q. A *hook* h on a segment s is a vector $\langle p, q\rangle$ where the tail $p \in s$ and the head $q \in E(s)$ is the nearest grid point to p. A segment s together with a set of hooks $h_1, ..., h_n$ is called a *hooked segment* and denoted by $L = (s; h_1, ..., h_n)$; s is called the *base segment* of L.

The next step is to define the redrawing process of a hooked segment in a way that is topology-preserving. We must not simply connect the heads of the hooks, as we have seen in the previous subsection. Instead, the redrawing scheme is as follows: The *redrawing* $\hat{L}$ of a hooked segment $L = (s; h_1, ..., h_n)$ where $s = (a, b)$ and $h_i = \langle p_i, q_i\rangle$, $i \in \{1, ..., n\}$, is the shortest path whose segments lie in or on the border polygon $E(s)$ and which traverses the set of points $\{a, b, q_1, ..., q_n\}$ (see Figure 3-8).

The following, very important lemma shows the stability of the redrawing process, namely that repeated redrawings of a segment do not leave the segment's envelope. We generalize the definition of an envelope. For a polygonal path $T = p_1p_2...p_n$, $E(T)$ shall denote the union of all envelopes of the individual segments (p_i, p_{i+1}), and for a hooked segment L, $E(L)$ shall denote the envelope of its base segment.

Lemma 3-2 $E(\hat{L}) \subseteq E(L)$.

The redrawing $\hat{L}$ of a hooked segment L forms a polygonal path which contains two kinds of segments. Segments of the first kind lie (except for their end points) *within* the border polygon $E(L)$. Since there are no grid points in the interior of $E(L)$, a segment of $\hat{L}$ can never drift to the other side of a grid point. Hence, for a segment of the first kind there cannot be an envelope point which is in $E(\hat{L})$ but not in $E(L)$. If an envelope point $p \in E(L)$ which is not traversed by the base segment of L is now traversed by the redrawing of L, it is possible that envelope points of L below or above p are no envelope points of $\hat{L}$ any more. Hence, for the first kind of segments

$E(\hat{L})$ is a subset of $E(L)$. Segments of the second kind lie *on* the border polygon $E(L)$ and are horizontal and vertical segments of unit length as well as diagonal segments of a unit square. The proper envelope of a diagonal 45°-segment consists of exactly one bounding point to the bottom-right of the unit square of the segment and is an element of $E(L)$. The proper envelope of a diagonal segment of -45°, of a horizontal unit segment, and of a vertical unit segment contains no envelope points so that $E(\hat{L})$ is reduced in comparison with $E(L)$. Hence, also for the second kind of segments $E(\hat{L})$ is a subset of $E(L)$. ❑

If an intersection of two segments has been detected and marked by a hook, we might immediately construct the redrawings of the two hooked segments, then compute intersections of these redrawings with other segments, create the corresponding hooks, redraw the intersecting segments again, and so on. The following equivalent strategy avoids the alternating procedure of creating hooks and repeated redrawings. In a first step, all necessary hooks are determined for a collection of line segments. Finally in a second step, all redrawings are computed for the hooked segments in one process. This can be done for each hooked segment individually and does not depend on any of the other segments.

Let $H = \{L_1, ..., L_n\}$ be a collection of hooked segments with the corresponding base segments $\{s_1, ..., s_n\}$, and apply the following rules in arbitrary order whenever applicable: (i) If s_i and s_j intersect at p, then add the hook $\langle p, q\rangle$ to both L_i and L_j, where q is the closest grid point to p. (ii) If s_i intersects a hook $h = \langle p, q\rangle$ of L_j at p', then add the hook $\langle p', q\rangle$ to L_i. If an application of neither of the two rules is possible any more, the resulting set H' is called the set of *completely hooked segments*. H' is said to be derived from H, and it is obvious that H' is unique with respect to H. Furthermore, it is recommendable for algorithmic reasons to derive H' by repeated applications of rule (i), followed by repeated applications of rule (ii).

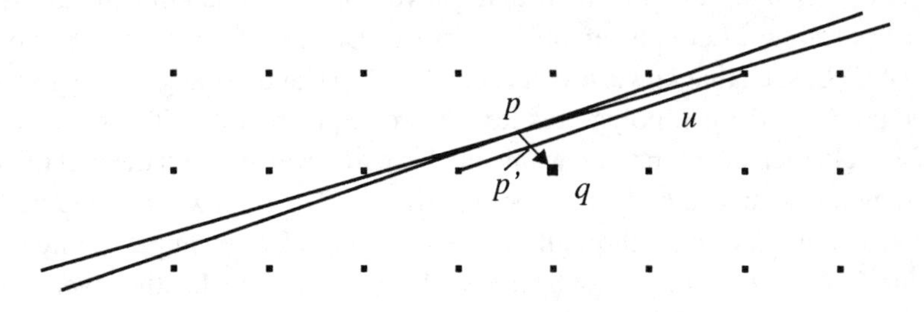

Figure 3-9

Figure 3-9 illustrates the two rules. Hook $\langle p, q\rangle$ arises from intersecting two segments according to rule (i) and is called *original hook*. Furthermore, the figure shows the importance of rule (ii) to prevent topological incorrectness. For each hook created one generally needs to check whether any segments are intersected by it. Otherwise, after a redrawing of two intersecting segments their common rounded intersection point on the grid can lie on the wrong side of a third segment. In the example an intersection of hook $\langle p, q\rangle$ with segment u is discovered and a corresponding hook $\langle p', q\rangle$ called *induced hook* is added to segment u (where p' is the intersection point of the hook and u).

The following definitions and lemmata lead to the second goal and show that the redrawings of a set of completely hooked segments are topology-preserving, i.e., preserve the topology of the original base segments. The possibly self-intersecting polygon formed by segment s and $\hat{L}$ is called the *gap polygon* between $\hat{L}$ and s, and is denoted by $G(\hat{L}, s)$. Let $s = (a, b)$ be a line segment and $T = p_1p_2...p_n$ a polygonal path. T is called *monotone* with respect to s if the scalar product of each vector $\langle p_i, p_{i+1}\rangle$, $i \in \{1, ..., n-1\}$, with vector $\langle a, b\rangle$ is non-negative, i.e., if $|\varphi_i| \leq 90°$ for all i where φ_i is the smaller angle between a pair of vectors $\langle p_i, p_{i+1}\rangle$ and $\langle a, b\rangle$. A vertex p_i of a polygon is said to be *inflex* if the angle $\angle\, p_{i-1}\, p_i\, p_{i+1}$ is less than or equal to 180° when viewed from the inside of the polygon.

Lemma 3-3 Let L be a hooked segment with base line $s = (a, b)$ and hooks $\langle p_1, q_1\rangle$, ..., $\langle p_n, q_n\rangle$. Then the following statements hold: (i) $\hat{L}$ is monotone with respect to s. (ii) Each inflex vertex of $G(\hat{L}, s)$ is an element of the set $\{a, b, q_1, ..., q_n\}$.

For horizontal, vertical, and diagonal hooked segments the two statements are obviously fulfilled. For the other cases the argumentation for statement (i) is as follows: Let w.l.o.g. $a.y < b.y$, and let first s have positive slope. For each segment (p_i, p_{i+1}) of the redrawing $\hat{L}$ of s holds that $(p_i.x < p_{i+1}.x \wedge p_i.y \leq p_{i+1}.y) \vee (p_i.x = p_{i+1}.x \wedge p_i.y < p_{i+1}.y)$. This property shows that the redrawing approximately "follows" s from a to b and that segments of the redrawing with maximal slope are horizontal or vertical. The scalar product of these vectors with vector $\langle a, b\rangle$ is non-negative. The argumentation for a segment s with negative slope is similar. Statement (ii) can be shown as follows: Let q_i be an inflex vertex of $G(\hat{L}, s)$. Then there are two vertices $k, l \in E(L)$ such that kq_il is a sequence of $G(\hat{L}, s)$, and the angle $\angle\, kq_il$ viewed from the inside of the gap polygon is less than or equal to 180°. If q_i were not the head of a hook, then the redrawing would not have to traverse q_i and the part of $G(\hat{L}$, $s)$ including the sequence kq_il would be superfluous. The redrawing would pass either through k and l, or only through k, or only through l, or even pass more closer to s within $G(\hat{L}, s)$. Hence, q_i must be the head of a hook. Note that the reverse statement that each element of the set $\{a, b, q_1, ..., q_n\}$ is inflex does not hold. ❑

Let for a segment $s = (a, b)$ a *bent line* from a to b be any monotone polygonal path from a to b which lies in $E(s)$ and whose vertices are grid points on $E(s)$. Then a redrawing is an example of a bent line. A bent line z from a to b can be extended to infinity at both ends. We obtain $E^+(s_\infty)$ and $E^-(s_\infty)$ and call the resulting infinite paths z^+ and z^-, respectively. The closed region to the left of z^+ is denoted by z^L, and the closed region to the right of z by z^R. A set A lies to the left (right) of z, if $A \subseteq z^L$ ($A \subseteq z^R$).

For two points a and b on a polygon P let $P[a, b]$ denote the section of P from a to b. Let for a pair (s_i, s_j) of segments z_i and z_j be two corresponding bent lines. (z_i, z_j) is called *faithful* to (s_i, s_j) if (i) in the case that s_i and s_j are disjoint, w.l.o.g. say $s_i \subseteq s_j^L$, we have $z_i \subseteq z_j^L$, and (ii) in the case that s_i and s_j intersect at c, w.l.o.g. say $s_i[a, c] \subseteq s_j^L$ and $s_i[c, b] \subseteq s_j^R$ (where $s_i = (a, b)$), we obtain $q \in z_i \cap z_j$, $z_i[a, q] \subseteq z_j^L$ and $z_i[q, b] \subseteq z_j^R$ where q is the nearest grid point to c. Hence, the term *faithful* expresses the property of preserving topology. Let s_k be a third segment and z_k be a corresponding bent line. (z_i, z_j, z_k) is faithful to (s_i, s_j, s_k) if, in the case that s_i and s_j intersect, their intersection point c lies to the same side of s_k as q does to z_k, where q is the nearest grid point to c.

Lemma 3-4 Let $\{L_1, ..., L_n\}$ be a set of completely hooked segments derived from $\{s_1, ..., s_n\}$. Then (i) $(\hat{L}_i, \hat{L}_j)$ is faithful to (s_i, s_j) for every i, j. (ii) $(\hat{L}_i, \hat{L}_j, \hat{L}_k)$ is faithful to (s_i, s_j, s_k) for every i, j, k.

We only show the lemma for the case that s_i and s_j intersect, since the disjoint case is simpler. Suppose both segments intersect at c. We obtain two hooked segments, each containing a hook with tail c. Suppose further that $s_i[a, c] \subseteq s_j^L$. We show that $\hat{L}_i[a, q] \subseteq \hat{L}_j^L$ where q is the nearest grid point to c. If this statement is false, a simple polygon $\gamma = \alpha\beta$ with non-trivial interior must exist such that α is a part of $\hat{L}_i$, β is a part of $\hat{L}_j$, and α lies to the right of $\hat{L}_j$. Polygon γ has at least three inflex vertices, and there is one inflex vertex t apart from where α and β are joined. According to the preceding lemma, t must be the head of a hook of either L_i or L_j but not of both. But this entails that either $\hat{L}_i$ or $\hat{L}_j$ would have traversed the grid point t which is impossible. This proves statement (i). The proof of (ii) is immediate. ❑

3.3.3 Number-Theoretical Foundations for the Redrawing of a Hooked Segment

When computing the redrawing of a hooked segment, it is important that the segments of the redrawing do not drift too much away from the original base segment but pass as close as possible so that no grid points lie between the base segment and its redrawing. Hence, we are interested in computing best approximations of a base segment on a grid. The foundations for computing such approximations and finally the redrawing of a hooked segment are based on the mathematical field of number theory. The three most important concepts needed are *diophantine approximation*, *Farey series*, and *continued fractions*.

Base segments and redrawing segments are N-segments and have thus integer-valued coordinates. We interpret segments here as vectors and identify the slope of a vector $\langle b, a\rangle$ with the rational number $\frac{a}{b}$. The problem of determining grid-based approximations of a vector can then be transformed into the problem of computing rational approximations of a rational (or more general: real) number. That is, given a real number $\alpha > 1$ we search for a rational number $\frac{a}{b}$, $b > 0$, $\gcd(a, b) = 1$, such that from $b' \leq b$, $\frac{a}{b} \neq \frac{a'}{b'}$ necessarily follows

$$\left|\alpha - \frac{a'}{b'}\right| > \left|\alpha - \frac{a}{b}\right|$$

A slightly stronger definition leads to the *best approximation* of a real number by a rational number. The rational number $\frac{a}{b}$ with $b > 0$, $\gcd(a, b) = 1$ is called *diophantine approximation* (*best approximation*) of a real number α iff $|b'\alpha - a'| > |b\alpha - a|$ for all a', b' with $\frac{a}{b} \neq \frac{a'}{b'}$ and $b' \leq b$,. If $\frac{a}{b}$ is a diophantine approximation of α, we can conclude for $b' \leq b$ and $\frac{a}{b} \neq \frac{a'}{b'}$ that

$$|b\alpha - a| < |b'\alpha - a'| \Rightarrow \left|\alpha - \frac{a}{b}\right| < \frac{1}{b}|b'\alpha - a'| \leq \frac{1}{b'}|b'\alpha - a'| = \left|\alpha - \frac{a'}{b'}\right|$$

Before studying how to find diophantine approximations, it is useful to dwell a moment on how these rational numbers with limited denominator are related to each other. For each denominator bound m, we obtain an infinite discrete set of rational numbers which when arranged in increasing order of magnitude is called the *Farey series* or *Farey sequence of order m* and denoted by $\mathcal{F}_m$. Two rational numbers $\frac{a}{b}$ and $\frac{c}{d}$ are said to be *adjacent in* $\mathcal{F}_m$ if they are successive elements of $\mathcal{F}_m$. Fragments of the sequences for the first few values of m look as follows:

$\mathcal{F}_1$: $\ldots, \frac{-1}{1}, \frac{0}{1}, \frac{1}{1}, \frac{2}{1}, \ldots$

$\mathcal{F}_2$: $\ldots, \frac{-1}{2}, \frac{0}{1}, \frac{1}{2}, \frac{1}{1}, \frac{3}{2}, \ldots$

$\mathcal{F}_3$: $\ldots, \frac{-1}{3}, \frac{0}{1}, \frac{1}{3}, \frac{1}{2}, \frac{2}{3}, \frac{1}{1}, \frac{4}{3}, \ldots$

$\mathcal{F}_4$: $\ldots, \frac{-1}{4}, \frac{0}{1}, \frac{1}{4}, \frac{1}{3}, \frac{1}{2}, \frac{2}{3}, \frac{3}{4}, \frac{1}{1}, \frac{5}{4}, \ldots$

$\mathcal{F}_5$: $\ldots, \frac{-1}{5}, \frac{0}{1}, \frac{1}{5}, \frac{1}{4}, \frac{1}{3}, \frac{2}{5}, \frac{1}{2}, \frac{3}{5}, \frac{2}{3}, \frac{3}{4}, \frac{4}{5}, \frac{1}{1}, \frac{6}{5}, \ldots$

Lemma 3-5 (i) If $\frac{a}{b}$ and $\frac{c}{d}$ are adjacent in $\mathcal{F}_m$ for some m, then $|ad - bc| = 1$.

(ii) If $|ad - bc| = 1$, $\frac{a}{b}$ and $\frac{c}{d}$ are adjacent in $\mathcal{F}_m$ for $\max(b, d) \leq m < b + d$, and they are separated by the single element $\frac{a + c}{b + d}$ in $\mathcal{F}_{b+d}$.

Proofs are given in [Le77]. This lemma on the one hand gives necessary and sufficient conditions that $\frac{a}{b}$ and $\frac{c}{d}$ are adjacent in $\mathcal{F}_m$. On the other hand it points to a method for creating the new elements that are added when going from $\mathcal{F}_m$ to $\mathcal{F}_{m+1}$. The number $\frac{a + c}{b + d}$ is called the *mediant* of $\frac{a}{b}$ and $\frac{c}{d}$. ❑

The question of computing the diophantine approximation $\frac{a}{b}$ (assuming that $b \le m$) of a real number α is equivalent to the question of determining the nearest rational number in $\mathcal{F}_m$ for α. To achieve this goal we do not need to generate the entire $\mathcal{F}_m$. We can accelerate the computation by using the method of *continued fractions*.

Let $\alpha > 1$ be a real number. The determination of the numbers $a_n = \lfloor \alpha_n \rfloor$ through

$$\alpha_n = a_n + \frac{1}{\alpha_{n+1}}, \; a_n \in \mathbf{Z}, \alpha_{n+1} > 1.$$

for $n = 0, 1, 2, \ldots$ (with $\alpha_0 = \alpha$) is called *(regular) continued fraction expansion*. The a_n are called the *partial quotients* of α. We write $\alpha = [a_0, a_1, \ldots, a_n, \alpha_{n+1}]$. The reduced fractions $[a_0, a_1, \ldots, a_n]$ for $n = 0, 1, 2, \ldots$ are the n-th approximate fractions of α and called *convergents* of α. We say that the continued fraction expansion of α terminates if $\alpha_n = a_n$ for some n.

Lemma 3-6 The continued fraction expansion of α terminates if and only if α is rational.

The direction "$\Rightarrow$" is obvious. The direction "$\Leftarrow$" can be shown as follows: Let α be rational, $\alpha = \frac{a}{b} > 1$ with $\gcd(a, b) = 1$. According to Euclid's algorithm we obtain

$$\begin{array}{lcll} a & = & x_0 b + y_0 & (0 < y_0 < b) \\ b & = & x_1 y_0 + y_1 & (0 < y_1 < y_0) \\ & & \ldots & \\ y_{n-1} & = & x_{n+1} y_n + y_{n+1} & (y_{n+1} = 0) \end{array}$$

The algorithm terminates since the y_i form a strictly decreasing sequence of natural numbers. A comparison of Euclid's algorithm with the continued fraction expansion shows their relationship:

$$\begin{array}{lll} \frac{a}{b} = x_0 + \frac{y_0}{b} & a_0 = x_0 & \alpha_1 = \frac{b}{y_0} \\ \frac{b}{y_0} = x_1 + \frac{y_1}{y_0} & a_1 = x_1 & \alpha_2 = \frac{y_0}{y_1} \\ & \ldots & \\ \frac{y_{n-1}}{y_n} = x_{n+1} + \frac{0}{y_n} & a_{n+1} = x_{n+1} = \alpha_{n+1} & \end{array}$$

❑

Lemma 3-7 The computation of the continued fraction expansion for a rational number $\alpha = \frac{a}{b}$ needs time $O(\log N)$ for $0 \le a, b < N$.

The worst time complexity of Euclid's algorithm is known. Knuth [Kn81] shows that on the assumption that $0 \le a, b < N$ the number of division steps required by Euclid's algorithm applied to a and b is at most $\lceil \log_\Phi(\sqrt{5}N) \rceil - 2$. Φ is called the *golden ratio*, and its value is $(1 + \sqrt{5}) / 2$ which is about 1.618. Since $\log_\Phi(\sqrt{5}N)$

is approximately $2.078 \log N + 1.672$, Euclid's algorithm needs time $O(\log N)$. Consequently in the case that $a < b$, Euclid's algorithm needs time $O(\log b)$. ❑

The following algorithm called the *continued fraction algorithm* computes the partial quotients of a rational number α by using only integer arithmetic:

```
algorithm continued_fraction_algorithm
input: Two Integer numbers u and v representing the rational number α = u/v
output: α = [a_0, a_1, ..., a_n]
begin
    u_0 := u; v_0 := v; a_0 := u_0 div v_0;
    i := 0;
    while u_i ≠ v_i a_i do
        u_{i+1} = v_i;
        v_{i+1} = u_i mod v_i; (* α_{i+1} = u_{i+1} / v_{i+1} = v_i / (u_i mod v_i) *)
        a_{i+1} = u_{i+1} div v_{i+1}; (* a_{i+1} = ⌊α_{i+1}⌋ *)
        i := i + 1;
    end-while;
end continued_fraction_algorithm.
```

In order to find out a relationship between the convergents of a real number α, the continued fraction expansion is presented in a simpler way by introducing two-dimensional vector or matrix notation. The equivalence of two column vectors is defined as follows:

$$\begin{bmatrix} a \\ b \end{bmatrix} \sim \begin{bmatrix} c \\ d \end{bmatrix} :\Leftrightarrow \frac{a}{b} = \frac{c}{d}$$

We can now write down the continued fraction expansion of α in the following way:

$$\alpha_n = a_n + \frac{1}{\alpha_{n+1}} = \frac{a_n \alpha_{n+1} + 1}{\alpha_{n+1}}$$

$$\Rightarrow \quad \begin{bmatrix} \alpha_n \\ 1 \end{bmatrix} \sim \begin{bmatrix} a_n \alpha_{n+1} + 1 \\ \alpha_{n+1} \end{bmatrix} \sim A_n \begin{bmatrix} \alpha_{n+1} \\ 1 \end{bmatrix} \text{ with } A_n = \begin{bmatrix} a_n & 1 \\ 1 & 0 \end{bmatrix}$$

We obtain

$$\begin{bmatrix} \alpha_0 \\ 1 \end{bmatrix} \sim A_0 \begin{bmatrix} \alpha_1 \\ 1 \end{bmatrix}, \begin{bmatrix} \alpha_1 \\ 1 \end{bmatrix} \sim A_1 \begin{bmatrix} \alpha_2 \\ 1 \end{bmatrix}, \ldots, \text{ and thus } \begin{bmatrix} \alpha \\ 1 \end{bmatrix} \sim A_0 A_1 \ldots A_n \begin{bmatrix} \alpha_{n+1} \\ 1 \end{bmatrix}$$

and finally with $P_n = A_0 \ldots A_n$

$$\begin{bmatrix} \alpha \\ 1 \end{bmatrix} \sim P_n \begin{bmatrix} \alpha_{n+1} \\ 1 \end{bmatrix}$$

Furthermore, we can compute $\det(A_i) = -1$ and hence conclude $\det(P_n) = (-1)^{n+1}$.

Lemma 3-8 The matrices P_n ($n \geq 0$) have the form

$$P_n = \begin{bmatrix} p_n & p_{n-1} \\ q_n & q_{n-1} \end{bmatrix}$$

where p_n and q_n are given by the recursive formulas $p_n = a_n p_{n-1} + p_{n-2}$ and $q_n = a_n q_{n-1} + q_{n-2}$ with the initial values $p_{-1} = 1$, $p_{-2} = 0$, $q_{-1} = 0$, $q_{-2} = 1$. Furthermore, $\gcd(p_n, q_n) = 1$ for $n = 0, 1, 2, \ldots$, and $p_0 < p_1 < \ldots$ as well as $q_0 < q_1 < \ldots$ holds.

The first assertion can be shown by the induction principle. Since $\det(P_n) = p_n q_{n-1} - q_n p_{n-1} = (-1)^{n+1}$, $\gcd(p_n, q_n) = 1$. The last assertion is obvious because of the recursive formulas for p_n and q_n. ❑

The relationship between the results of the last lemma and the convergents $[a_0, a_1, \ldots, a_n]$ ($n = 0, 1, 2, \ldots$) of α is given by the following lemma:

Lemma 3-9 For a real number α, let $\alpha = [a_0, a_1, \ldots, a_n, \alpha_{n+1}]$. Then $[a_0, a_1, \ldots, a_n] = p_n/q_n$ for $n = 0, 1, 2, \ldots$

Let $\beta_n = [a_0, a_1, \ldots, a_n]$. For $n = 0$, $\beta_0 = [a_0] = a_0 = a_0/1$. On the other hand, we obtain $p_0 = a_0 p_{-1} + p_{-2} = a_0$ and $q_0 = a_0 q_{-1} + q_{-2} = 1$. For $n > 0$, $a_n \geq 1$ holds. Using vector notation we can write:

$$\begin{bmatrix} \beta_n \\ 1 \end{bmatrix} \sim P_{n-1} \begin{bmatrix} a_n \\ 1 \end{bmatrix} \quad \Rightarrow \quad \beta_n = \frac{p_{n-1} a_n + p_{n-2}}{q_{n-1} a_n + q_{n-2}} = \frac{p_n}{q_n}$$

Note that for $a_n = 1$ the equation $[a_0, a_1, \ldots, a_{n-1}, 1] = [a_0, a_1, \ldots, a_{n-1} + 1]$ holds. ❑

The next lemma shows convergence criteria and inequalities regarding the convergents p_n/q_n.

Lemma 3-10 Let for $n = 0, 1, 2, \ldots$ p_n/q_n be the convergents of a real number $\alpha = [a_0, a_1, \ldots, a_n, \alpha_{n+1}]$. Then the following inequalities hold:

(i) $\left|\alpha - \dfrac{p_n}{q_n}\right| < \dfrac{1}{q_n^2}$

(ii) $a_0 = \frac{p_0}{q_0} < \frac{p_2}{q_2} < \frac{p_4}{q_4} < \ldots < \alpha$

(iii) $\alpha < \ldots < \frac{p_5}{q_5} < \frac{p_3}{q_3} < \frac{p_1}{q_1}$

For the first assertion we obtain with $\det(P_n) = (-1)^{n+1}$:

$$\frac{p_n}{q_n} - \alpha = \frac{p_n}{q_n} - \frac{\alpha_{n+1}p_n + p_{n-1}}{\alpha_{n+1}q_n + q_{n-1}} = \frac{(-1)^{n+1}}{q_n(\alpha_{n+1}q_n + q_{n-1})}$$

$$\Rightarrow \left|\alpha - \frac{p_n}{q_n}\right| < \frac{1}{q_n^2} \quad \text{since } \alpha_{n+1}q_n + q_{n-1} > q_n.$$

The inequalities of (ii) and (iii) can be derived from the recursive formulas for p_n and q_n. If the equation $p_n = a_n p_{n-1} + p_{n-2}$ is multiplied by the factor q_{n-2} and the equation $q_n = a_n q_{n-1} + q_{n-2}$ is multiplied by the factor p_{n-2}, the difference of both equations is $p_n q_{n-2} - q_n p_{n-2} = a_n \det(P_{n-1}) = a_n (-1)^n$. A transformation leads to

$$\frac{p_n}{q_n} - \frac{p_{n-2}}{q_{n-2}} = \frac{a_n(-1)^n}{q_n q_{n-2}}$$

For $n = 2k$ the difference is greater than 0 so that assertion (ii) holds. For $n = 2k - 1$ analogously assertion (iii) holds. Hence, the sequence p_n/q_n converges in an oscillating way against α. ❑

In the last step we associate continued fractions with diophantine approximations.

Lemma 3-11 Each diophantine approximation of a real number $\alpha > 1$ is equal to a convergent of α.

Let $\frac{a}{b}$ be a diophantine approximation of $\alpha = [a_0, a_1, \ldots]$. We consider its location relatively to the convergents of α. Four cases have to be distinguished:

(i) If $\frac{a}{b} < \frac{p_0}{q_0} = a_0$, we would obtain: $|1 \cdot \alpha - a_0| < \left|\alpha - \frac{a}{b}\right| \leq |b\alpha - a|$, $1 \leq b$,

that is, $\frac{a}{b}$ would not be a diophantine approximation.

(ii) If $\frac{a}{b} > \frac{p_1}{q_1}$, then $\left|\alpha - \frac{a}{b}\right| > \left|\frac{p_1}{q_1} - \frac{a}{b}\right| \geq \frac{1}{bq_1}$.

Hence $|b\alpha - a| > \frac{1}{q_1} = \frac{1}{a_1} \geq \frac{1}{\alpha_1} = |\alpha - a_0|$, $1 \leq b$.

This contradicts the definition of a diophantine approximation.

Thus $\frac{a}{b}$ is either a convergent or located between two convergents.

(iii) Supposing that $\left|\frac{p_{n+1}}{q_{n+1}}\right| < \frac{a}{b} < \left|\frac{p_{n-1}}{q_{n-1}}\right|$ for a certain even n.

On the one hand $\frac{1}{q_n q_{n-1}} = \left|\frac{p_n}{q_n} - \frac{p_{n-1}}{q_{n-1}}\right| > \left|\frac{a}{b} - \frac{p_{n-1}}{q_{n-1}}\right| \geq \frac{1}{bq_{n-1}}$, i.e., $b \geq q_n$.

On the other hand $\left|\alpha - \frac{a}{b}\right| \geq \left|\frac{p_{n+1}}{q_{n+1}} - \frac{a}{b}\right| \geq \frac{1}{bq_{n+1}}$, i.e., $|b\alpha - a| \geq \frac{1}{q_{n+1}}$.

Furthermore $\frac{1}{q_{n+1}} \geq |\alpha q_n - p_n|$,

since $\left|\alpha - \frac{p_n}{q_n}\right| = \frac{1}{q_n(\alpha_{n+1}q_n + q_{n-1})} \leq \frac{1}{q_n q_{n+1}}$.

Hence, $|\alpha q_n - p_n| \leq |b\alpha - a|$ and $q_n \leq b$, i.e., $\frac{a}{b}$ would not be a diophantine approximation.

(iv) Correspondingly, one concludes for odd n that $\left|\frac{p_{n-1}}{q_{n-1}}\right| < \frac{a}{b} < \left|\frac{p_{n+1}}{q_{n+1}}\right|$ is impossible.

Consequently, there must be a certain n so that $\frac{a}{b} = \frac{p_n}{q_n}$. ❑

Lemma 3-12 Each convergent p_n/q_n, $n \geq 1$, of a real number $\alpha > 1$ is a diophantine approximation of α.

The proof is given in [KP76]. ❑

3.3.4 The Restricted Redrawing Algorithm

Redrawing a hooked segment means computing the shortest polygonal path which passes through the envelope of the segment and traverses both the end points of the hooked segment and the heads of its hooks. In this subsection on a rather abstract level a redrawing algorithm for a restricted case is presented which follows the approach of Greene and Yao [GY86] and is based on the number-theoretical concepts introduced in Section 3.3.3.

The restriction of the algorithm relates to the assumption that without loss of generality (as we will see in the next subsection) the base segment $s = (a, b)$ of a hooked segment has the end points $a = (0, 0)$ and $b = (x, y)$ where $x, y > 0$ and $0 < y/x < 1$, i.e., we only consider segments with slope greater than 0° and less than 45°. In the sequel, segments are regarded as vectors, and the slope of the vector $\langle x, y\rangle$ is identified with the rational number y/x. Both notations will be used synonymously.

We distinguish three different situations during the redrawing of a hooked segment: (i) the treatment of a single hook by means of a traversal from end point a through the head of the hook to end point b, (ii) the construction of the redrawing between two consecutive hooks on the same side of the base segment, and (iii) the computation of the redrawing between two consecutive hooks lying on opposite sides of the base segment. As we will see later, a redrawing of a segment with k hooks needs $O(k \log |s|)$ time and is obtained by composing the different redrawing parts between each pair of consecutive hooks one after the other. Each redrawing part solves one of the three situations just described.

We begin with the first situation of a hooked segment $L = (s; h)$ with a single hook $h = \langle p, q\rangle$ and assume w.l.o.g. that $q = (u, z) \in E^+(s)$, i.e., q lies *below* s. The redrawing will be computed in two parts. In time $O(\log |s|)$ we will at first compute $\hat{L}[a, q]$, the redrawing between a and q, and afterwards the redrawing $\hat{L}[q, b]$. We start with searching for the *maximal best vector from below* which is (a multiple of) the best approximation of y/x with denominator bound u, the x-coordinate of q. More precisely, for a rational number y/x and an integer $u > 0$, the vector $\langle d, e\rangle$ is called the *maximal best vector from below with respect to* $(y/x; u)$ if one of the following two conditions holds: (1) If there is a vector $\langle d', e'\rangle$ with $d' \leq u$, $e' \in \mathbf{N}$, $\gcd(d', e') = 1$, and $e'/d' = y/x$ (i.e., the vector lies on s), select $d := k \cdot d' \leq u$, $e := k \cdot e'$ with $k \geq 1$ such that $e/d = y/x$ and $(k + 1) \cdot d' > u$. (2) Otherwise, there must be a vector $\langle d', e'\rangle$ with $d' \leq u$, $e' \in \mathbf{N}$, $\gcd(d', e') = 1$ such that $e''/d'' \leq e'/d' < y/x$ for all $d'' \leq u$ with $e'' \in \mathbf{N}$. Select $d := k \cdot d' \leq u$, $e := k \cdot e'$ with $k \geq 1$ such that $e'/d' = e/d < y/x$ and $(k + 1) \cdot d' > u$. The vector $\langle d', e'\rangle$ is called the *minimal best vector from below with respect to* $(y/x; u)$. Hence, the maximal best vector results from scalar multiplication of the minimal best vector with an integer $k > 0$.

Figure 3-10 shows the scenario of a redrawing $\hat{L}[a, q] = q_0 q_1 q_2 \ldots q_m$ between $q_0 = a$ and $q_m = q$. The vector $q_1 = \langle d_1, e_1\rangle$ corresponds to the maximal vector from below with respect to $(y/x; u)$. Iteratively, the vector $q_2 - q_1$ corresponds to the maximal vector from below with respect to $(e_1/d_1; u - d_1)$. However, since no grid points lie between s and $\hat{L}$, the vector $q_2 - q_1$ also corresponds to the maximal vector from below with respect to $(y/x; u - d_1)$. Hence, it is unnecessary to update the fraction y/x which is to be approximated but only the denominator bound. This all is stated in the following lemma.

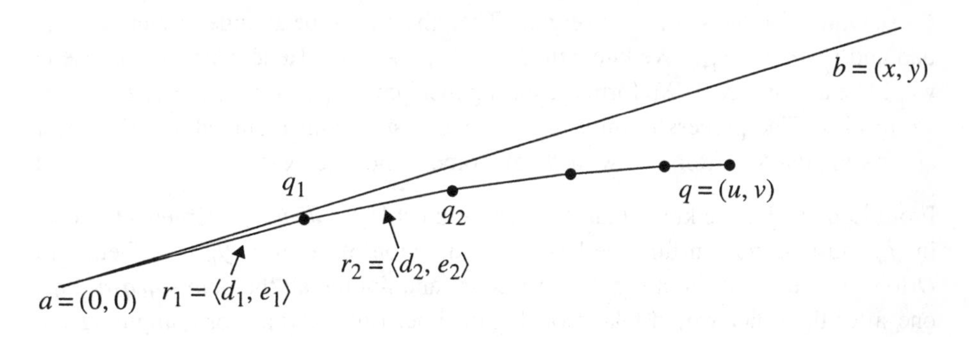

Figure 3-10

Lemma 3-13 Suppose that $L = (s; h)$ is a hooked segment with $s = (a, b)$ and a single hook $h = \langle p, q \rangle$ and that $Q = q_0 q_1 ... q_m$ with $q_0 = a$ and $q_m = q$ is the polygonal path obtained like described above. Let r_i be the *difference vector* $q_i - q_{i-1}$ for $1 \le i \le m$. Then $r_i = \langle d_i, e_i \rangle$ is the maximal vector from below with respect to $(y/x; u_i)$ where $u_i = u - \sum_{k=1}^{k=i-1} d_k$, and $Q = \hat{L}[a, q]$.

The polygonal path Q has the following properties: (1) Each q_i is a grid point of $E^+(s)$. (2) Q only turns right at each q_i. (3) There are no grid points between s and Q. Hence, Q must be the shortest path from a to q within $E(s)$, i.e., $Q = \hat{L}[a, q]$. ❑

From Section 3.3.3 we know that in the general case the best approximation with respect to $(y/x; u)$ is equivalent to the diophantine approximation of y/x with denominator bound u and hence equivalent to the nearest rational number of y/x in the Farey series $\mathcal{F}_u$. Each diophantine approximation of y/x can be calculated as a convergent of y/x by a continued fraction expansion. Let $v = \langle x, y \rangle$ have the continued fraction expansion $v = [a_0, a_1, ..., a_n]$, and let v_j for $j \in \{0, ..., n\}$ be the convergents of v. Since according to our assumption $y/x < 1$, $a_0 = 0$. The convergents v_j satisfy the recursive formula $v_j = a_j v_{j-1} + v_{j-2}$ for $j \ge 2$ and the inequality $v_{2j} \le v \le v_{2j+1}$ for $j \ge 0$.

Our assumption that the hook lies *below* s leads to the constraint that we must search for the best approximation of y/x *from below* in $\mathcal{F}_u$ which leads to the minimal best vector from below with respect to $(y/x; u)$. Since convergents oscillate around y/x, we also have to take mediants into account which are not convergents of y/x. Any mediant which is not a convergent is called a *quasi-convergent* and is a linear combination of two adjacent convergents.

Lemma 3-14 Let $w = \langle p, q \rangle$ be a vector with $\gcd(p, q) = 1$ which is the best approximation to $v = \langle x, y \rangle$ from below in $\mathcal{F}_u$ for some u. Then $w = kv_j + v_{j-1}$ where j is odd and k is in the range $0 \le k < a_{j+1}$.

Suppose that w is a convergent. Since only convergents with even indexes approximate v from below, there must be a largest index j such that j is odd, $w = v_{j-1}$, and

$k = 0$. Otherwise, w is not a convergent. Then there must be an index j such that j is odd and $v_{j-1} < w < v_{j+1}$. We know that $v_{j+1} = v_{j-1} + a_{j+1}v_j$. Hence, $k < a_{j+1}$ or else w would be a convergent. We form increasing mediants $v_{j-1} + kv_j$ with $k = 1, 2, ...$ until we reach w. The process terminates for some k, since w is assumed to be the best approximation to v from below in $\mathcal{F}_u$ for some u and since $k < a_{j+1}$. ❑

From Lemma 3-7 we know that there are at most $\log_\Phi(\sqrt{5}\,u)$ convergents to $\langle x, y\rangle$ in $\mathcal{F}_u$ and that we can find the best approximation of $\langle x, y\rangle$ in $\mathcal{F}_u$ from below in $O(\log u)$ time. Thus, iterating this procedure and finding all the points $q_0q_1q_2...q_m$ one after the other would take $O(m \log u)$ time. But the time for computing the points $q_0q_1q_2...q_m$ can be reduced to a total cost of $O(\log u)$ by a single backward scan of the convergents. To show this, it is necessary to examine the difference vectors r_i more carefully.

Lemma 3-15 Each difference vector r_i can be written uniquely either in the form $r_i = kv_j$ where j is even and $k > 0$ or in the form $r_i = kv_j + v_{j-1}$ where j is odd and $0 < k < a_{j+1}$.

According to Lemma 3-13 each difference vector r_i is the maximal best vector from below with respect to $(y/x; u_i)$. We consider the corresponding unique minimal best vector w which must be either a convergent or a quasi-convergent. According to Lemma 3-14, if w is a convergent, then $w = v_j$ for some even j, and hence r_i is a multiple of v_j. Otherwise, if w is a quasi-convergent, then $w = kv_j + v_{j-1}$ where j is odd and $0 < k < a_{j+1}$. Vector r_i must be equal to w, i.e., r_i must be a reduced rational number, since if it is not, then $2 \cdot (kv_j + v_{j-1})$ does not exceed the denominator bound u_i. This would lead to the contradiction that $(k + 1)v_j + v_{j-1}$ also does not exceed the denominator bound u_i and is a better approximation to $\langle x, y\rangle$ than w. ❑

Theorem 3-16 The redrawing of a segment s with a single hook $h = \langle p, q\rangle$ can be computed in time $O(\log |s|)$.

At first all convergents of $s = (a, b) = ((0, 0), (x, y))$ are generated which needs time $O(\log x)$ and thus $O(\log |s|)$. Then two angular sweeps are used to compute the redrawing of s, one bending away from s and extending from end point a to q and one moving towards s from q to end point b.

The *first angular sweep* starts parallel to s from a and sweeps through shorter difference vectors of decreasing slope towards q. For this purpose, the convergents are processed from largest to smallest in one pass. At each step a difference vector r_i is computed according to Lemma 3-15, and $j(i)$ shall denote the index j for difference vector r_i. If a $k > 0$ is found, the value of k is maximal, and the remaining denominator bound u_{i+1} will be less than the denominator of $v_{j(i)}$. Therefore $j(i)$ is a strictly decreasing function of i, and the computation of the difference vectors r_i takes time $O(\log u)$ and thus $O(\log |s|)$.

The *second angular sweep* starts with the upright vector $\langle 0, 1\rangle$ from q and sweeps through longer difference vectors of decreasing slope towards b until it is parallel to s. Because the same segments are derived forward in the same way as backward, we need only consider difference vectors r_i either of the form $r_i = kv_j$ where j is odd and $k > 0$ or of the form $r_i = kv_j + v_{j-1}$ where j is even and $0 < k < a_{j+1}$. The terms odd and even have been swapped due to symmetry. The function $j(i)$ is now strictly increasing. The only algorithmic difference to the first angular sweep is that we must prevent r_i from crossing s and from exceeding the denominator bound, i.e., there are two tests instead of one. The redrawing process as a whole runs in time $O(\log |s|)$. ❑

The second situation which can occur during the redrawing process is to construct the redrawing between two consecutive hooks lying both w.l.o.g. *below* on the same side of the base segment. The strategy is to begin at the left hook, to move *towards* the base segment according to the second angular sweep of Theorem 3-16, and then *away* from the base segment to the right hook according to the first angular sweep of Theorem 3-16.

Lemma 3-17 The redrawing of a segment s between two consecutive hooks lying on the same side of s can be computed in time $O(\log |s|)$ in one left-to-right pass.

At first the denominator bound is initialized to the difference of the x-coordinates of the heads of the right and the left hook. Then the second angular sweep is used until some difference vector r_i exceeds the denominator bound. This difference vector r_i, a line parallel to the base segment s through q_{i-1} (the left end point of r_i), and a vertical line through the right hook form a triangle which must be free of grid points. Otherwise, there would exist a grid point q_i within the triangle whose distance to s is smaller than the distance of q_{i-1} to s, and we would have found a new difference vector $r_i = q_i - q_{i-1}$ on our way towards s within the denominator bound which is a contradiction to our assumption that r_i exceeds the denominator bound. Hence, we can begin an angular sweep through shorter difference vectors of lesser slope than s (first angular sweep). The function $j(i)$ strictly increases up to the point where the direction of motion changes, and then strictly decreases. It is obvious that the redrawing of a segment s between two consecutive hooks needs time $O(\log |s|)$. ❑

The third situation which can occur during the redrawing process is to compute the redrawing between two consecutive hooks lying on opposite sides of the base segment. Unlike the difference vectors r_i computed so far, it is here difficult to determine the correct difference vector when crossing the base segment, since it is not always a nearest rational number to $\langle x, y\rangle$ in some $\mathcal{F}_u$. Fortunately, only $O(1)$ close rational numbers need to be examined. We call this special difference vector *crossing difference vector*. W.l.o.g. we assume that the *left hook* lies *below* the base segment and that the *right hook* lies *above* the base segment.

At first we define crossing criteria for an index j' of a convergent. An index j' of a convergent satisfies the crossing criteria if (i) j' is odd and the vector $(k + 1)\ v_{j'}$ exceeds the denominator bound where k is such that $r_i = kv_{j'}$ if $j' = j(i)$ for some i or $k = 0$ otherwise, or if (ii) j' is even, $j' = j(i)$ for some i, and $2r_i$ exceeds the denominator bound. Thus, the crossing criteria express that extending the found difference vector r_i would exceed the denominator bound. All indexes that violate the crossing criteria extend to grid points above the base segment that are not right of the next hook. The strategy is to produce difference vectors according to the second angular sweep of Theorem 3-16 until the index j' is found which fulfills the crossing criteria, then to compute the crossing difference vector according to the following lemma, and finally to approach the next hook above the base segment by using the first angular sweep of Theorem 3-16.

We have to show that the crossing difference vector must be part of the redrawing, that is, that it remains within the envelope of the base segment and that it is part of the shortest path between the two hooks. Furthermore, the turns of the shortest path between the two hooks must be to the right at points on the bottom of the envelope and to the left at points on the top of the envelope.

Lemma 3-18 Let j' be the first index found by the second angular sweep of Theorem 3-16 and satisfying the crossing criteria. Then it is possible to select a single crossing difference vector in one of three directions which is computable in time $O(1)$ from $v_{j'}$, $v_{j'-1}$, and k so that the redrawing can be completed up to the next hook by the first angular sweep of Theorem 3-16.

We confine ourselves to the case that j' is odd, $j' = j(i)$, and k is such that $r_i = kv_{j'}$. The other cases are similar. Because of the properties of difference vectors the line through q_{i-1} in the direction of r_i, the base segment s, and the vertical line through the right hook form a wedge which is free of grid points so that the orientation of the angular sweep can be reversed. Vector $q_{i-1} + r_i$ lies on or below the base segment, and we know that $r_i + v_{j'}$ exceeds the denominator bound. The vector $r_i - v_{j'-1}$ lies above the base segment and is less than the denominator bound. The only rational number between $r_i - v_{j'-1}$ and r_i that might be less than the denominator bound is $r_i + v_{j'} - v_{j'-1}$. Hence, we need only consider three possibilities: At first we examine if $q_{i-1} + r_i$ falls directly on the base segment. If this is the case, r_i is the searched crossing difference vector. Otherwise, we test $r_i + v_{j'} - v_{j'-1}$ against the denominator bound and take it as the crossing difference vector if possible. If that also fails, vector $r_i - v_{j'-1}$ is guaranteed to be successful and must be perhaps still multiplied several times. Obviously, the turn at q_{i-1} must be zero or to the right, and the turn at q_i must be to the left. Hence, the crossing difference vector describes a part of the shortest path between the two hooks. ❑

We obtain at once:

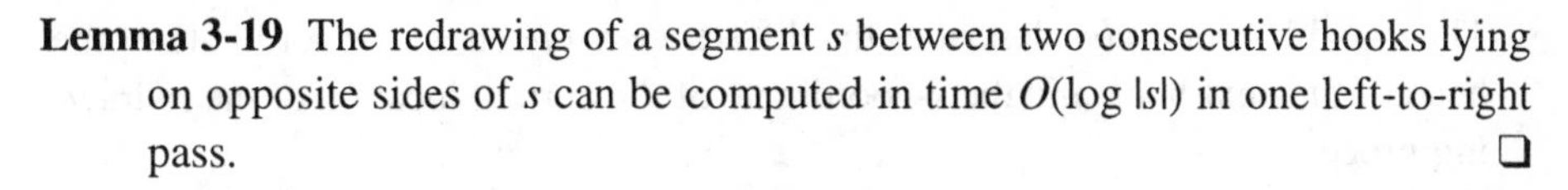

Lemma 3-19 The redrawing of a segment s between two consecutive hooks lying on opposite sides of s can be computed in time $O(\log |s|)$ in one left-to-right pass. ❑

Summarizing the results of Theorem 3-16, Lemma 3-17, and Lemma 3-19, a complete redrawing of a hooked segment with k hooks can be obtained in time $O(k \log |s|)$ by composing the different redrawing parts between each pair of consecutive hooks one after the other.

Theorem 3-20 The redrawing of a segment s with k hooks can be computed in time $O(k \log |s|)$ in one left-to-right pass. ❑

Using a high-level notation, we are now able to formulate the restricted redrawing algorithm which uses the following subalgorithms:

- algorithm *CrossProduct* which for the two-dimensional vectors v_1 and v_2, both with origin (0, 0), computes their cross product $v_1 \times v_2$,
- algorithm *Move* which adds a new difference vector to the redrawing computed so far,
- algorithm *GetNextHook* which provides the next hook for processing,
- algorithm *CrossingMove* which determines the correct difference vector when crossing the base segment from one side to the other, and
- algorithm *Redraw* which is the main routine calling the other algorithms and computing the complete sequence of difference vectors and hence the actual redrawing.

All algorithms are based on the integer arithmetic defined in Section 3.2 and need a collection of global variables. The most important ones together with their meanings are enumerated in the following:

- The flag *away_from_seg* is true if the redrawing is currently bending away from the base segment.
- The flag *below_seg* is true if the redrawing is currently below the base segment.
- The flag *crossing* is true if the redrawing must cross the base segment before reaching the next hook.
- The flag *extend_ok* indicates that a prolongation of the last computed difference vector by a new one is possible if both are collinear.
- The variable $s = \langle x, y \rangle$ views the base segment as a vector with the origin (0, 0).
- The variable *current* denotes a two-dimensional vector describing the current location of the redrawing process.
- The variable *conv* is the array of convergents for the base segment. Each array component *conv*[*index*] is a three-dimensional vector $\langle x, y, z \rangle$ where the vector $\langle x, y \rangle$ is the convergent with index *index* and the component z is the cross product between the base segment and the convergent.

- The variable *remainder* describes the difference between the x-coordinate of the head of the next hook and the x-coordinate of the current location of the redrawing process.
- The variable *last_diff_vec* stores the last computed difference vector.
- The variable *redrawing* denotes the polygonal path $< q_0, q_1, ..., q_n >$ representing the redrawing of a hooked segment as a sequence of points (vectors) where each pair (q_i, q_{i+1}) describes a segment (a difference vector).
- The variable *hooks* represents the non-empty sequence of hooks where each hook is described by a tuple (*head*, *down*). The first component of such a tuple describes the coordinates (x, y) of the head of the hook and the second component whether the hook lies below the base segment. We assume that the hooks are ordered (x, y)-lexicographically.

For the manipulation of sequences (lists) several functions are needed. The symbol $\circ$ denotes the concatenation operator for sequences and the symbol $\Diamond$ the empty sequence. For a sequence L the function *first*(L) positions on the first, *next*(L) on the next, and *last*(L) on the last element of L. The function *eos*(L) tests if the end of L was reached. The function *retrieve*(L) yields a copy of the element at the current position of L, and the function *subst*(L, p) substitutes the element at the current position of L by p.

The algorithm *CrossProduct* computes the cross product of the vectors $v_1 = \langle x_1, y_1 \rangle$ and $v_2 = \langle x_2, y_2 \rangle$ which are both assumed to have origin (0, 0). A cross product can be interpreted as the signed area of the parallelogram formed by the points (0, 0), v_1, v_2, and $v_1 + v_2 = \langle x_1 + x_2, y_1 + y_2 \rangle$. An equivalent but more useful definition views the cross product as the determinant of a matrix:

$$v_1 \times v_2 = \det \begin{pmatrix} x_1 & x_2 \\ y_1 & y_2 \end{pmatrix} = x_1 y_2 - x_2 y_1 = -v_2 \times v_1 .$$

If $v_1 \times v_2$ is positive, then v_1 is clockwise from v_2 with respect to the origin (0, 0). Otherwise v_1 is counterclockwise from v_2. If the cross product is zero, the vectors are collinear and point in either the same or opposite directions. The formulation of the algorithm is obvious.

The algorithm *Move* adds a new difference vector to the redrawing computed so far and updates the difference between the x-coordinates of the head of the next hook and of the current location of the redrawing process (variable *Remainder*). If the new and the previous difference vector are collinear and a hook has not just been passed (*extend_ok* = *true*), the previous difference vector is extended by the new one.

```
algorithm Move
input: A difference vector diff_vec = ⟨x, y⟩
output: A modified redrawing extended by diff_vec (as a side effect)
begin
   remainder := remainder – diff_vec.x;
   if redrawing = ◊ then
      redrawing := redrawing ∘ < (0, 0) > ∘ < (diff_vec.x, diff_vec.y) >
      last_diff_vec = diff_vec;
   else
      last(redrawing);
      q := retrieve(redrawing);
      q := q + diff_vec; (* Points are viewed as vectors. *)
      if (CrossProduct(diff_vec, last_diff_vec) = 0) and extend_ok then
         subst(redrawing, q);
         last_diff_vec = last_diff_vec + diff_vec;
      else
         redrawing := redrawing ∘ < (q.x, q.y) >;
         last_diff_vec = diff_vec;
      end-if;
   end-if;
   current := current + diff_vec;
   extend_ok := true;
end Move.
```

The algorithm *GetNextHook* provides the next hook for processing, checks whether a crossing of the base segment is necessary to reach it, and initializes the variable *remainder* as the difference of the x-coordinates of the heads of the current and the next hook. A special case arises if two hooks, one above and one below the base segment, have heads with the same x-coordinate.

```
algorithm GetNextHook
begin
   away_from_seg := false;
   extend_ok := false;
   next(hooks);
   if eos(hooks) then
      remainder := s.x – current.x;
      crossing := false;
   else
      h := retrieve(hooks);
      remainder := h.head.x – current.x;
      if (remainder = 0) and not h.down then
         next(hooks);
```

```
            below_seg := false;
            index := 1;
            if eos(hooks) then
                remainder := s.x − current.x;
                crossing := false;
            else
                remainder := h.head.x − current.x;
                crossing := below_seg xor h.down;
            end-if;
            if CrossProduct(current, s) > 0 then Move(⟨0, 1⟩) end-if;
        else
            crossing := below_seg xor h.down;
        end-if;
    end-if;
end GetNextHook.
```

The algorithm *CrossingMove* determines the correct crossing difference vector when crossing the base segment from one side to the other and adds it to the redrawing. Three cases have to be distinguished to find the correct crossing difference vector.

```
algorithm CrossingMove
input: Two vectors good and major and an index minor_index of a convergent
begin
    away_from_seg := true;
    below_seg := not below_seg;
    if CrossProduct(current + good, s) = 0 then
        (* Vector current + good is a point lying on s. *)
        Move(good);
    else
        good := good + major − conv[minor_index];
        if good.x ≤ remainder then
            Move(good);
            index := index − 1;
        else
            good := good − major;
            k := remainder div good.x;
            Move(k * good);
            index := minor_index;
        end-if;
    end-if;
end CrossingMove.
```

The algorithm *Redraw* is the main routine calling the algorithms described above and computing the complete sequence of difference vectors and hence the actual redrawing. We assume that the slope of the base segment s is greater than 0° and smaller than 45°. The flag *primary* used in the algorithm is *true* if the convergent *conv*[*index*] currently considered has the same slope relationship with respect to the base segment s as the direction of the moves being produced. Thus, the following cases have to be distinguished:

- If we are below (above) the base segment and turning away from it, then our moves will have slope less (greater) than the base segment, and the even (odd) convergents will be "primary".
- If we are below (above) the base segment and moving towards it, then our moves will have slope greater (less) than the base segment, and the odd (even) convergents will be "primary".

```
algorithm Redraw
begin
   (* Initializations. *)
   first(hooks); h := retrieve(hooks); below_seg := h.down;
   remainder := h.head.x; away_from_seg := true; crossing := false;
   extend_ok := true; redrawing := ◊; index := 2; current := ⟨0, 0⟩;

   (* Computation of the convergents for base segment s
   and of the cross product of s with each convergent. *)
   conv[0] := ⟨0, 0, 0⟩; conv[1] := ⟨0, 1, s.x⟩; conv[2] := ⟨1, 0, -s.y⟩;
   while conv[index].z ≠ 0 do
      partial_quotient := -conv[index – 1].z div conv[index].z;
      conv[index + 1] := conv[index – 1] + partial_quotient * conv[index];
      inc(index);
   end-while;
   (* Repeat the last convergent for convenience later. *)
   conv[index + 1] := conv[index];
   inc(index);

   (* Computation of the redrawing of base segment s. *)
   loop
      increment_index := true;

      (* d = a xor b xor c is true if either one or all three conditions are true. *)
      primary := away_from_seg xor below_seg xor (index mod 2 = 0);

      if away_from_seg then
         if index < 2 then
            if eos(hooks) then exit else GetNextHook() end-if
```

```
        else
            (* Find the largest move that does not exceed remainder. *)
            if primary then base := 0 else base := index - 1 end-if;
            r := (remainder - conv[base].x) div conv[index].x;
            if r > 0 then Move(conv[base] + r * conv[index]) end-if;
            dec(index);
        end-if;
    else (* Redrawing moves towards base segment s. *)
        if remainder = 0 then
            if eos(hooks) then exit else GetNextHook() end-if
        end-if;

        if primary then
            if conv[index].z = 0 then
                (* Parallel to s. *)
                away_from_seg := true;
                below_seg := below_seg xor crossing;
            else if conv[index].x > remainder then
                (* Additional towards move would exceed remainder. *)
                away_from_seg := true;
                dec(index);
                below_seg := below_seg xor crossing;
                increment_index := false;
            else
                (* Find the largest move that does not cross s. *)
                r := CrossProduct(current, s) div conv[index].z;
                if r > 0 then
                    delta := r * conv[index];
                    if delta.x > remainder then
                        (* Reduce the move so that it does not exceed
                        remainder. *)
                        r := remainder div conv[index].x;
                        if crossing then
                            CrossingMove(r * conv[index], conv[index],
                                              index - 1);
                        else
                            away_from_seg := true;
                            if r > 0 then Move(r * conv[index]); end-if;
                            dec(index);
                        end-if;
                        increment_index := false;
```

```
                else if crossing and
                        (delta.x + conv[index].x > remainder) then
                    CrossingMove(delta, conv[index], index – 1);
                    increment_index := false;
                else
                    Move(delta);
                end-if;
            end-if;
        end-if;
    else (* Secondary. *)
        if conv[index].z = 0 then
            away_from_seg := true;
            below_seg := below_seg xor crossing;
        else
            r := CrossProduct(s, current + conv[index + 1])
                    div conv[index].z;
            if r > 0 then
                delta := conv[index + 1] – r * conv[index];
                if delta.x > remainder then
                    away_from_seg := true;
                    below_seg := below_seg xor crossing;
                    increment_index := false;
                else if crossing and (delta.x * 2 > remainder) then
                    CrossingMove(delta, delta, index);
                    increment_index := false;
                else
                    Move(delta);
                end-if;
            end-if;
        end-if;
    end-if;
    if increment_index then inc(index); end-if;
  end-if;
 end-loop;
end Redraw.
```

It is obvious that the algorithms *CrossProduct*, *Move*, *GetNextHook*, and *Crossing-Move* need $O(1)$ time. If s is the segment to be redrawn, k the number of hooks, and n the number of redrawing segments, the main algorithm *Redraw* has worst time complexity $O(k \log|s| + n)$. As shown before, for each of the k hooks the loop is executed at most $\log|s|$ times. The preceding computation of the convergents to s needs $O(\log|s|)$ time.

3.3.5 The Generalized Redrawing Algorithm

The generalized redrawing algorithm treats the general case that the hooked segment to be redrawn has arbitrary position and slope within the realm. Let $t = (v, w)$ be a realm segment, and let α be the counterclockwise angle from the horizontal line through the end point of t with lower y-coordinate to segment t. Then eight cases of different ranges of angles and hence slopes have to be distinguished: (1) $\alpha = 0°$, (2) $0° < \alpha < 45°$ (the case treated in algorithm *Redraw*), (3) $\alpha = 45°$, (4) $45° < \alpha < 90°$, (5) $\alpha = 90°$, (6) $90° < \alpha < 135°$, (7) $\alpha = 135°$, and (8) $135° < \alpha < 180°$. The idea is now not to design seven new specialized redrawing algorithms for the lacking cases but to reduce the cases 4, 6, and 8 to case 2, to reduce case 7 to case 3, and to apply a special simple algorithm to the cases 1 and 5. We use the observation that a redrawing is symmetric (if we disregard auxiliary points) and that it is invariant against topological operations like translation and rotation which do not change the topological relationships between base segment, hooks, and redrawing. Hence, we are allowed to employ translations and mirror operations at the y-axis and at the angle bisector of the first quadrant.

The generalized redrawing algorithm uses the following subalgorithms:

- algorithm *RedrawSegOnAngleBisector* which computes the redrawing of a segment lying on the angle bisector of the first quadrant,
- algorithm *RedrawHorizontalOrVerticalSeg* which computes the redrawing of a horizontal or vertical segment,
- algorithm *SortHooks* which sorts a hook sequence either according to increasing or decreasing x-order for non-vertical segments, or increasing y-order for vertical segments,
- algorithm *Swap* which exchanges the values of two variables, and
- algorithm *Redrawing* which is the main routine calling the other algorithms and computing the redrawing of a segment with arbitrary slope.

The algorithm *RedrawSegOnAngleBisector* computes the redrawing of a segment with slope one. For k hooks and n resultant redrawing segments, the algorithm needs $O(k + n)$ time.

```
algorithm RedrawSegOnAngleBisector
begin
  redrawing := ⟨(0, 0)⟩; x_pos := 0; first(hooks);
  while not eos(hooks) do
    h := retrieve(hooks);
    if (h.head.x = h.head.y) and (x_pos < h.head.x) then
      (* Hook lies on segment and is not identical to point (x_pos, x_pos)
      which would produce a degenerate redrawing segment of length 0. *)
```

```
                redrawing := redrawing ∘ < h.head >;
            else if h.head.x > h.head.y then
                (* Special case: The head of the hook is an auxiliary point. *)
                if x_pos < h.head.x − 1 then
                    (* The point left of the head of the hook on the segment is not yet
                    part of the redrawing and must be added. *)
                    redrawing := redrawing ∘ < (h.head.x − 1, h.head.y) >;
                end-if;

                (* Draw horizontal unit segment. *)
                redrawing := redrawing ∘ < h.head >;

                (* Draw vertical unit segment. *)
                redrawing := redrawing ∘ < (h.head.x, h.head.y + 1) >;
            end-if;
            x_pos := h.head.x; next(hooks);
        end-while;
        if x_pos < s.x then
            redrawing := redrawing ∘ < (s.x, s.y) >;
        end-if;
    end RedrawSegOnAngleBisector.
```

The algorithm *RedrawHorizontalOrVerticalSeg* with the parameter *horizontal* calculates the redrawing of a horizontal (*horizontal* = *true*) or vertical (*horizontal* = *false*) segment. The algorithm needs $O(k + n)$ time.

```
algorithm RedrawHorizontalOrVerticalSeg
input: A parameter horizontal indicating if the segment s to be redrawn is
       horizontal.
begin
    redrawing := ⟨(0, 0)⟩; q := (0, 0); first(hooks);
    while not eos(hooks) do
        h := retrieve(hooks);
        if horizontal then
            q.x := h.head.x;
        else
            q.y := h.head.y;
        end-if;
        redrawing := redrawing ∘ < q >;
        next(hooks);
    end-while;
    redrawing := redrawing ∘ < (s.x, s.y) >;
end RedrawHorizontalOrVerticalSeg.
```

The algorithm *SortHooks* with the three parameters *xorder_sort*, *increasing1*, and *increasing2* sorts a list of hooks in increasing (*increasing1* = *true*) or decreasing (*increasing1* = *false*) x-order (*xorder_sort* = *true*) or y-order (*xorder_sort* = *false*). If the hooks are sorted in x-order / y-order and if two hooks have equal x-coordinates / y-coordinates, these hooks are sorted in increasing (*increasing2* = *true*) or decreasing (*increasing2* = *false*) y-order / x-order. The sorting takes $O(k \log k)$ time. The algorithm *Swap* simply exchanges the values of two variables and needs constant time. A formulation of the two algorithms is omitted.

The algorithm *Redrawing* is the main routine calling the other algorithms and computing the redrawing of a segment with arbitrary slope. If necessary, before the actual redrawing process translation and mirror operations are performed which are afterwards undone.

```
algorithm Redrawing
input: A hooked segment consisting of a segment t = (v, w) and a sequence
        hooks of hooks
output: The redrawing redrawing of t.
begin
  if hooks = ◊ then
    redrawing := ◊;
    return;
  end-if;

  (* Calculate the slope of t and check if t is horizontal, vertical, or inclined.
  If the value of slope_denominator is negative, shift the negative sign to
  slope_numerator. *)
  slope_numerator := t.w.y – t.v.y;
  slope_denominator := t.w.x – t.v.x;
  horizontal_seg := (slope_numerator = 0);
  vertical_seg := (slope_denominator = 0);
  inclined_seg := not horizontal_seg and not vertical_seg;
  if inclined_seg and (slope_denominator < 0) then
    slope_numerator := - slope_numerator;
    slope_denominator := - slope_denominator;
  end-if;

  (* Normalize t and calculate the vector s = ⟨x, y⟩ with origin (0, 0) for which
  the redrawing will be performed, calculate the translation vector, and
  change the hook coordinates correspondingly. *)
  if ((inclined_seg or vertical_seg) and (t.v.y < t.w.y)) or
    (horizontal_seg and (t.v.x < t.w.x)) then
    s := ⟨t.w.x – t.v.x, t.w.y – t.v.y⟩;
    translation_vect := ⟨t.v.x, t.v.y⟩;
```

```
else
   s := ⟨t.v.x − t.w.x, t.v.y − t.w.y⟩;
   translation_vect := ⟨t.w.x, t.w.y⟩;
end-if;
first(hooks);
while not eos(hooks) do
   h := retrieve(hooks); h.head := h.head − translation_vect;
   subst(hooks, h); next(hooks);
end-while;
(* For the redrawing process, the hooks of s have to be sorted. *)
if inclined_seg then
   if slope_numerator > 0 then
      if slope_numerator ≤ slope_denominator then
         (* Sort hooks in increasing x-order and hooks with equal
         x-coordinates in increasing y-order. *)
         SortHooks(true, true, true);
      else
         (* Sort hooks in increasing y-order and hooks with equal
         y-coordinates in increasing x-order. *)
         SortHooks(false, true, true);
      end-if;
   else
      slope_numerator := - slope_numerator;
      if slope_numerator ≤ slope_denominator then
         (* Sort hooks in decreasing x-order and hooks with equal
         x-coordinates in increasing y-order. *)
         SortHooks(true, false, true);
      else
         (* Sort hooks in increasing y-order and hooks with equal
         y-coordinates in decreasing x-order. *)
         SortHooks(false, true, false);
      end-if;
      slope_numerator := - slope_numerator;
   end-if;
else if horizontal_seg then
   (* Sort hooks in increasing x-order. *)
   SortHooks(true, true, true);
else (* Vertical segment. *)
   (* Sort hooks in increasing y-order. *)
   SortHooks(false, true, true);
end-if;
```

```
(* If necessary, perform mirror operations. *)
if inclined_seg and (slope_numerator < 0) then
   (* If the slope of s is less than 0, s and the hooks have to be mirrored
   against the y-axis. *)
   s.x := - s.x;
   slope_numerator := - slope_numerator;
   first(hooks);
   while not eos(hooks) do
      h := retrieve(hooks);
      h.head.x := - h.head.x;
      subst(hooks, h);
      next(hooks);
   end-while;
   mirrored_at_y_axis := true;
else
   mirrored_at_y_axis := false;
end-if;
if inclined_seg and (slope_numerator > slope_denominator) then
   (* If the slope of s is greater than 45°, s and the hooks have to be
   mirrored against the angle bisector of the first quadrant. *)
   Swap(s.x, s.y);
   Swap(slope_numerator, slope_denominator);
   first(hooks);
   while not eos(hooks) do
      h := retrieve(hooks);
      Swap(h.head.x, h.head.y);
      h.down := not h.down;
      subst(hooks, h);
      next(hooks);
   end-while;
   mirrored_at_angle_bisector := true;
else
   mirrored_at_angle_bisector := false;
end-if;

(* Redrawing process. *)
if inclined_seg then
   if slope_numerator ≠ slope_denominator then
      (* 0 < slope_numerator / slope_denominator < 1 *)
      Redraw;
   else
```

```
            (* Slope is equal to 1. *)
            RedrawSegOnAngleBisector;
        end-if;
    else if horizontal_seg then
        RedrawHorizontalOrVerticalSeg(true);
    else (* Vertical segment. *)
        RedrawHorizontalOrVerticalSeg(false);
    end-if;

    (* Mirror back the redrawing, if necessary, and add the translation
    vector. *)
    first(redrawing);
    while not eos(redrawing) do
        q := retrieve(redrawing);
        if mirrored_at_angle_bisector then
            Swap(q.x, q.y);
        end-if;
        if mirrored_at_y_axis then
            q.x := - q.x;
        end-if;
        q := q + translation_vect;
        subst(redrawing, q);
        next(redrawing);
    end-while;

    (* The redrawing path has to be directed in the same way as the original
    segment t. If necessary, the sequence of points of the redrawing has to be
    reversed. *)
    if t.v.y > t.w.y then
        [ Reverse the sequence of points of the redrawing. ]
    end-if;

    (* Restore the original state of the sequence of hooks. *)
    [...]
end Redrawing.
```

The most expensive operations are the algorithm *SortHooks* which needs $O(k \log k)$ time and the algorithm *Redraw* which needs $O(k \log|s| + n)$ time. All other operations need constant or linear time. Hence, the entire algorithm *Redrawing* needs $O(k \log k|s| + n)$ time in the worst case.

3.4 Realms

Realms serve as a discrete geometric basis for the formal definition of spatial data types. They represent a finite, user-definable set of points and *non-intersecting* line segments over a discrete domain (a grid). Given N, a *realm over N*, or *N-realm* for short, is a set $R = P \cup S$ such that

(i) $P \subseteq P_N, S \subseteq S_N$

(ii) $\forall s \in S : s = (p, q) \Rightarrow p \in P \wedge q \in P$

(iii) $\forall p \in P \; \forall s \in S : \neg (p \; \underline{\text{in}} \; s)$

(iv) $\forall s, t \in S, s \neq t : \neg (s \text{ and } t \; \underline{\text{intersect}}) \wedge \neg (s \text{ and } t \; \underline{\text{overlap}})$

Condition (i) means that each point and each end point of a line segment of a realm is a grid point. Condition (ii) emphasizes that each end point of a realm segment is also a point of the realm and condition (iii) that no realm point lies within a realm segment (which means on it without being an end point). Condition (iv) requires that no two realm segments intersect except at their end points. The elements of P and S are called *R-points* and *R-segments*. There is an obvious interpretation of a realm as a *spatially embedded planar graph* with a set P of nodes and a set S of edges.

3.5 Operations on Realms: The Realm Interface

3.5.1 Modelling the Realm Interface

For manipulating realms a *realm interface* is provided whose operations are subdivided into three groups. The fundamental operations on realms are to insert an N-point or N-segment, to split an R-segment by an N-point, and to delete an R-point or R-segment. However, the interface is more complex since we study realms not just as abstract entities but in connection with spatial databases. That means that there are spatial objects in the database depending on realm objects. This dependency needs to be modelled and treated by the operations.

The approach to implement the dependency is the following: We assume that a spatial object (which we in this context call *spatial attribute value*) is stored together with the spatially-referenced object it is associated with in the database and that there is a logical pointer from each segment or point describing the spatial attribute value to the underlying realm object. Furthermore, associated with each realm object (point or segment) is a set of logical pointers; one pointer to each corresponding component of a spatial attribute value in the database. In other words, points and segments in the database are doubly linked with the corresponding points and segments in the realm. Pointers from the database into the realm are realized by *realm object identifiers* from a set *ROID*, pointers from the realm into the database by *spatial component identifiers* from a set *SCID*. It is assumed that each *roid* or *scid* uniquely identifies the corresponding entity and that the implementation guarantees fast access to these components. Additionally, a *scid* also identifies the spatial attribute value as a whole.

This approach stores geometries redundantly with a database object and in the realm. One might save some space by representing attribute values just by structures composed of pointers to realm objects. However, it is crucial for efficient query processing and well worth the extra space to keep the geometries with the objects. In this way one can directly apply spatial data type operations to spatial attribute values whereas otherwise it would always be necessary to access (load pages of) the underlying realm. Note that the two-way linking is necessary in any case, since changes in the realm (e.g., a new intersection point on a segment) need to be propagated to the dependent spatial attribute values.

The management of pointers is modelled formally as follows: Let R be a realm. A *representation of R* is a set of triples $\{(r, roid(r), scids(r)) \mid r \in R\}$ where *roid* is a function giving for a realm object its unique identifier in *ROID* and *scids* is a function returning the set of *SCIDs* of dependent components of spatial attribute values. We also allow the notation $roid(v)$ to assign a new *ROID* to a newly created realm object v.

The realm interface is described by the following signature. Slightly extending standard notations we allow operators to return tuples of values and sets of values – the type of a set of X values is denoted by X*. The first group of operations are those mentioned above:

sorts	*Realm, Point, Segment, RealmObject, ROID, SCID, Rectangle, Bool, Integer*			
ops	*InsertNPoint*:	*Realm* × *Point*	→	*Realm* × *ROID* × (*SCID* × (*Segment* × *ROID*)*)*
	InsertNSegment:	*Realm* × *Segment*	→	*Realm* × (*Segment* × *ROID*)* × (*SCID* × (*Segment* × *ROID*)*)* × *Bool*
	SplitRSegment:	*Realm* × *ROID* × *Point*	→	*Realm* × *ROID* × (*SCID* × (*Segment* × *ROID*)*)* × *Bool*
	Delete:	*Realm* × *ROID*	→	*Realm* × *Bool*

The sort (type) *Realm* refers to a realm representation, *Point* to the set P_N, *Segment* to the set S_N. *RealmObject* is a union type of *Point* and *Segment*; we assume one can recognize whether a given instance is a point or a segment. *ROID* and *SCID* have been discussed above. *Rectangle* denotes the set of axis-parallel rectangles definable over space $N \times N$, that is, *N-rectangles*.

The insertion of line segments makes use of the redrawing algorithm described in Section 3.3 in order to preserve the topology of a set of intersecting line segments defined over a grid. Additionally, collections of points are allowed in realms which are not treated by the redrawing algorithm. The insertion of points and the insertion of line segments are independent of each other. Exceptions are the situations where an *N*-point lies <u>in</u> an *R*-segment, an *N*-segment traverses an *R*-point, and an *N*-point is desired to split an *R*-segment.

The operation *InsertNPoint* takes a realm and an *N*-point. It returns (i) the modified realm, (ii) a *roid* for the inserted point, which could be an old one if the point was in the realm already, and (iii) a (mostly empty) set of segments in the database that need to be split (redrawn) each in two collinear parts because the point to be inserted falls onto their interiors, i.e., the point lies <u>in</u> each of the segments. For each such segment its "address" in the database (SCID) together with a list of two pairs $(s, roid(s))$ (where s is one of the two segments of the splitting) is returned. It is then the task of the DBMS to replace segments by their redrawings.

The operation *InsertNSegment* takes a realm and an *N*-segment. It returns (i) the modified realm, (ii) a list of segments with their *roids* which may contain either the original segment as the only element or a redrawing of this segment, and (iii) a possibly empty set of database segments that must be redrawn. Here the inserted segment may need redrawing because it traverses *R*-points, overlaps *R*-segments, or intersects *R*-segments. Only those *R*-segments need redrawing that are intersected by this segment. For each such *R*-segment its *scid* together with a list of pairs $(s, roid(s))$ (where s is a segment of the redrawing) is returned. The last parameter (iv) indicates whether insertion was performed. It was rejected, if not both end points of the segment were present in the realm.

The operation *SplitRSegment* takes a realm, a *roid* of an *R*-segment, and an *N*-point lying on the proper envelope of the *R*-segment. After the realm construction, this operation allows to additionally insert a point into the realm where the point is intended to lie exactly within the interior of an *R*-segment. Conceptually, the *N*-point splits the *R*-segment into two parts. For this, the *N*-point has to lie on the proper envelope of the *R*-segment, since, in order to place the *N*-point <u>in</u> the *R*-segment, the redrawing of the *R*-segment through the *N*-point is computed. The operation returns (i) the modified realm, (ii) a *roid* for the inserted point, and (iii) a set of segments in the database that need to be redrawn. The last parameter (iv) indicates whether insertion was performed. It was rejected, if the *N*-point did not lie on the proper envelope of the *R*-segment.

The operation *Delete* takes a realm and the identifier of a realm object (point or segment) and removes the object from the realm if this does not violate some predefined integrity constraints. It returns (i) the modified realm and (ii) an indication whether the object was removed. The following conditions are checked: A point is only removed if there is no segment ending in the point, i.e., if it is an isolated point. Any realm object is only removed if its set of *scids* (dependent objects) is empty.

The second group of operations supports the management of the two-way linking between realm objects and components of spatial attribute values in the database:

Register:	$Realm \times ROID \times SCID$	$\rightarrow$	$Realm$
Unregister:	$Realm \times ROID \times SCID$	$\rightarrow$	$Realm$
GetSCIDs:	$Realm \times ROID$	$\rightarrow$	$SCID^*$
GetRealmObject:	$Realm \times ROID$	$\rightarrow$	$RealmObject$

Here *Register* informs a realm object *roid* about a spatial component *scid* depending on it. *Unregister* removes such an information. *GetSCIDs* returns the *scids* of spatial components depending on a given *roid*, *GetRealmObject* returns the geometry. These operations are to be used, for example, as follows: A spatial attribute value is constructed by selecting a number of realm objects in a certain order (this is supported by the last group of operations, see below). After all components have been selected, the representation of this value is built and stored in the database. Then all components are registered with their underlying realm objects. When a spatial attribute value is deleted, the registration is removed for all realm objects. *GetSCIDs* and *GetRealmObject* are general purpose operations to support query processing.

The last group of operations supports the selection of realm objects for the construction of spatial objects:

Window:	$Realm \times Rectangle$	$\rightarrow$	$(RealmObject \times ROID)^*$
Identify:	$Realm \times Point \times Integer$	$\rightarrow$	$ROID \times Bool$

Window returns all realm objects together with their *roid* that are inside or intersect a given rectangular window. *Identify* tries to identify a realm object close to the *N*-point given as a parameter. The number given as a third parameter controls the "pick distance". A *roid* (possibly undefined) is returned together with an indication whether identification was successful. These two operations can be used to retrieve a portion of a realm in order to define spatial objects over it. For example, this portion may be displayed at a graphical user interface. With a pointing device one can select *N*-points which through *Identify* determine realm objects from which the spatial objects can be built.

3.5.2 Algorithms of the Realm Interface

In this subsection we define the semantics of the three most important and most complex realm operations *InsertNPoint*, *InsertNSegment*, and *SplitRSegment* by giving algorithms for them. All the other operations are rather simple so that their meaning should be clear from the explanations above.

The algorithm *InsertNPoint* has to treat the following cases: (i) the point is already present in the realm, (ii) the point is new and does not fall on the interior of any R-segment, and (iii) the point is new and lies within the interior of an R-segment (only exactly one such R-segment is possible). In the last case the R-segment has to be split into two collinear parts.

algorithm *InsertNPoint* (R, p, R', r, SP)
input: A realm $R = P \cup S$ and an N-point p.
output: The modified realm R', a realm object identifier r for p, and a set SP of spatial component identifiers and redrawings for the spatial objects which have to be updated.
begin
$SP := \varnothing$;
if $\exists\, q \in P : p = q$ **then**
(* p was already inserted into R. At most one such R-point may exist. The realm remains unchanged, and only the roid of p is returned. *)
$r := roid(q)$; $R' := R$
else if $\forall\, s \in S : \neg\,(p\ \underline{in}\ s)$ **then**
(* p has not yet been inserted into R and does not lie within the interior of any R-segment. Hence, p can be inserted into R without performing any redrawings. *)
$r := roid(p)$; $R' := R \cup \{(p, r, \varnothing)\}$;
else (* $\exists\, s = (a, b) \in S : p\ \underline{in}\ s$. *)
(* Split (redraw) s into two collinear segments. *)
$s_1 := (a, p)$; $s_2 := (p, b)$;
(* Update realm. *)
$R' := R \setminus \{(s, roid(s), scids(s))\}$;
$r := roid(p)$;
$R' := R' \cup \{(p, r, \varnothing)\}$;
$R' := R' \cup \{(s_1, roid(s_1), \varnothing)\}$;
$R' := R' \cup \{(s_2, roid(s_2), \varnothing)\}$;
$SP := \bigcup_{sc \in scids(s)} \{(sc, \{(s_1, roid(s_1)), (s_2, roid(s_2))\})\}$
end-if;
end *InsertNPoint*.

The algorithm *InsertNSegment* first checks whether the segment was already inserted into the realm. If the segment is new, a second test ascertains whether the end points of the segment are both already present in the realm. If this is not the case, insertion is rejected. This agrees with the graph-theoretic view of a realm: An edge can only exist if its nodes are there. It implies that the user of the realm interface has to make sure that the points are present (in case of doubt just insert them first; this does not hurt). The following cases have now to be distinguished: (i) the segment traverses no R-points and for all R-segments holds that it either meets or is disjoint to them, (ii) the segment traverses R-points, (iii) the segment overlaps R-segments, and (iv) the segment intersects R-segments. Note that the cases (ii), (iii), and (iv) do not exclude each other. The algorithm uses the two predicates *ExistsRPoint* and *ExistsRSegment* with the obvious meaning to check whether a realm object to be created is present already. The notation "[...]" denotes steps of the algorithm described colloquially without using a defined formalism.

algorithm *InsertNSegment* (R, s, R', RD, SP, ok)
input: A realm $R = P \cup S$ and an N-segment $s = (q_1, q_2)$.
output: The modified realm R', a set RD of pairs of R-segments and realm object identifiers either for s or a redrawing of s, a set SP of spatial component identifiers and redrawings for the spatial objects which have to be updated, and a parameter ok which indicates whether the insertion was performed. Insertion was rejected if the end points of s were not present in the realm.
begin
  $SP := \varnothing$;
  $ok := true$;
  if $\exists\, t \in S : s = t$ **then**
    (* s was already inserted into R. At most one such R-segment may exist. The realm remains unchanged, and only the roid of s is returned. *)
    $R' := R$; $RD := \{(s, roid(t))\}$
  else if not $ExistsRPoint(q_1)$ **or not** $ExistsRPoint(q_2)$ **then**
    $ok := false$;
  else if $\forall\, t \in S : (s$ and t are $\underline{\text{disjoint}} \vee s$ and t $\underline{\text{meet}}) \wedge$
      $\forall\, p \in P : \neg\, (p\ \underline{\text{in}}\ s)$ **then**
    $R' := R \cup \{(s, roid(s), \varnothing)\}$; $RD := \{(s, roid(s))\}$
  else
    (* s traverses R-points and/or overlaps R-segments and/or intersects R-segments. *)
    $S_{rd} := \varnothing$; (* The set of R-segments which have to be redrawn. *)
    $P_{traverse}(s) := \{p \in P \mid p\ \underline{\text{in}}\ s\}$;
    $S_{overlap}(s) := \{t \in S \mid s$ and t $\underline{\text{overlap}}\}$;
    $S_{intersect}(s) := \{t \in S \mid s$ and t $\underline{\text{intersect}}\}$;

```
(* Insert hooks. *)
for each p in P_traverse(s) do
   [ Insert a hook h = ⟨p, p⟩ on s. ] (* h is a vector of length 0. *)
end-for;
for each t = (a, b) in S_overlap(s) do
   if a in s then [ Insert a hook h = ⟨a, a⟩ on s. ] end-if;
   if b in s then [ Insert a hook h = ⟨b, b⟩ on s. ] end-if;
end-for;
for each t in S_intersect(s) do
   [ Insert a hook h = ⟨q, p⟩ both on s and on t from the intersection point
   q of s and t to the closest grid point p.
   (* Note that p = q is possible if q is a grid point. *) ];
   S_rd := S_rd ∪ {t};
   for each v in S do
      if [ h and v intersect at p' ] then
         [ Insert a hook h' = ⟨p', p⟩ on v ];
         S_rd := S_rd ∪ {v}
      end-if;
   end-for;
end-for;

(* Redraw hooked segments. *)
[ Redraw s. Let {s_1, ..., s_m} be the R-segments of the redrawing of s. Let
s_i = (p_i, q_i), i ∈ {1, ..., m}.
Redraw all R-segments of S_rd := {t_1, ..., t_n}. Let {t_i,1, ..., t_i,k_i} be the set
of k_i R-segments of the redrawing of t_i. Let t_i,j = (p_i,j, q_i,j), i ∈ {1, ..., n},
j ∈ {1, ..., k_i}. ];

(* Update realm. Insert the end points of the R-segments of the
redrawings and the R-segments themselves if they do not already exist
in the realm. *)
R' := R \ {(t_i, roid(t_i), scids(t_i)) | i ∈ {1, ..., n}};
for each i in 1 .. m do
   if not ExistsRPoint(p_i) then R' := R' ∪ {(p_i, roid(p_i), ∅)};
   if not ExistsRPoint(q_i) then R' := R' ∪ {(q_i, roid(q_i), ∅)};
   if not ExistsRSegment(s_i) then R' := R' ∪ {(s_i, roid(s_i), ∅)}
end-for;
for each i in 1 .. n do
   for each j in 1 .. k_i do
      if not ExistsRPoint(p_i,j) then R' := R' ∪ {(p_i,j, roid(p_i,j), ∅)};
      if not ExistsRPoint(q_i,j) then R' := R' ∪ {(q_i,j, roid(q_i,j), ∅)};
      if not ExistsRSegment(t_i,j) then R' := R' ∪ {(t_i,j, roid(t_i,j), ∅)}
end-for; end-for;
```

$RD := \{(s_i, roid(s_i)) \mid i \in \{1, ..., m\}\};$

$SP := \bigcup_{i=1}^{n} \{(sc, \{(t_{i,j}, roid(t_{i,j})) \mid j \in \{1, ..., k_i\}\}) \mid sc \in scids(t_i)\}$

end-if;
end *InsertNSegment*.

The algorithm *SplitRSegment* first checks whether the point to be inserted lies on the proper envelope of the desired *R*-segment and afterwards whether it was already inserted into the realm. To insert the point, the redrawing of the *R*-segment through the point is computed. A special notation is required for the case that a point *p* lies on the proper envelope of a segment *s* and the redrawing of *s* is intended to go through *p*. In this case a hook is induced, and we take the point (say *q*) on the segment closest to the target point as the start point of the hook and call it *base*(<target>, <segment>), in this case $q = base(p, s)$.

algorithm *SplitRSegment* (*R, rs, p, R', rp, SP, ok*)
input: A realm $R = P \cup S$, a realm object identifier *rs* for an *R*-segment, and an *N*-point *p*.
output: The modified realm *R'*, a realm object identifier *rp* for *p*, a set *SP* of spatial component identifiers and redrawings for the spatial objects which have to be updated, and a parameter *ok* which indicates whether the insertion was performed. Insertion was rejected if the *N*-point did not lie on the proper envelope of the *R*-segment.
begin
  ok := *true*;
  s := *GetRealmObject*(*R, rs*); (* Simplified use. *)
  if $p \notin \bar{E}(s)$ **then**
    ok := *false*;
  else if $\exists\, q \in P : p = q$ **then**
    (* *p* was already inserted into *R*. At most one such *R*-point may exist. The realm remains unchanged, and only the roid of *p* is returned. *)
    rp := *roid*(*q*); *R'* := *R*
  else (* *p* is a new point lying on the proper envelope of *s*. *)
    (* Insert hooks. *)
    [Insert a hook $h = \langle base(p, s), p\rangle$ on *s*.];
    $S_{rd} := \{s\}$; (* The set of *R*-segments which have to be redrawn. *)
    for each *v* **in** *S* **do**
      if [*h* and *v* intersect at *p'*] **then**
        [Insert a hook $h' = \langle p', p\rangle$ on *v*];
        $S_{rd} := S_{rd} \cup \{v\}$
      end-if;
    end-for;

(* Redraw hooked segments. *)
[Redraw all R-segments of $S_{rd} := \{t_1, ..., t_n\}$. Let $\{t_{i,1}, ..., t_{i,k_i}\}$ be the set of k_i R-segments of the redrawing of t_i. Let $t_{i,j} = (p_{i,j}, q_{i,j})$, $i \in \{1, ..., n\}$, $j \in \{1, ..., k_i\}$.];

(* Update realm. Insert p, the end points of the R-segments of the redrawings, and the R-segments themselves if they do not already exist in the realm. *)
$R' := R \setminus \{(t_i, roid(t_i), scids(t_i)) \mid i \in \{1, ..., n\}\}$;
$rp := roid(p)$;
$R' := R' \cup \{(p, rp, \varnothing)\}$;

for each i **in** 1 .. n **do**
  for each j **in** 1 .. k_i **do**
    if not $ExistsRPoint(p_{i,j})$ **then** $R' := R' \cup \{(p_{i,j}, roid(p_{i,j}), \varnothing)\}$;
    if not $ExistsRPoint(q_{i,j})$ **then** $R' := R' \cup \{(q_{i,j}, roid(q_{i,j}), \varnothing)\}$;
    if not $ExistsRSegment(t_{i,j})$ **then** $R' := R' \cup \{(t_{i,j}, roid(t_{i,j}), \varnothing)\}$
  end-for;
end-for;
$SP := \bigcup_{i=1}^{n} \{(sc, \{(t_{i,j}, roid(t_{i,j})) \mid j \in \{1, ..., k_i\}\}) \mid sc \in scids(t_i)\}$
end-if;
end *SplitRSegment*.

An appropriate data structure for representing a realm is a spatial index structure. Hence, the run time complexity of the realm interface operations is in particular dependent on the time requirements of the storage and retrieval operations of the used spatial index structure for realm objects and can thus not be presented here. Additionally, the needed external operations have to support exact-match queries, i.e., direct access to realm objects by their identifiers or by their geometries, and range queries, i.e., the retrieval of all realm objects lying inside or intersecting a given query rectangle. Such a spatial index structure containing a realm representation can also serve as an index into the database since each realm object has an associated list of logical pointers to the spatial attribute values defined over it in the database.

3.6 Realm-Based Structures and Primitives

At this point we can assume that the problems of numerical robustness and topological correctness are solved by the two lower layers. The third layer (see Appendix B) defines structures and relationships between these structures that can be discovered within a realm and that are useful for the definition of spatial data types. For this purpose, a realm is viewed as a planar graph over the grid $N \times N$, and some structures called *R-cycle*, *R-face*, *R-unit*, and *R-block* are defined over this graph.

Informally, an *R-cycle* is a cycle of this graph. An *R-face* is an *R*-cycle possibly enclosing some other disjoint *R*-cycles corresponding to a region or polygon with holes. An *R-unit* is a minimal *R*-face. These three notions support the definition of a *regions* data type. An *R-block* is a connected component of the realm graph; it supports the definition of a *lines* data type. Figure 3-11 gives a complete overview of these *realm-based structures* and of the provided *primitives* and functions defined on them in the remainder of this subsection. The primitives are predominantly binary, overloaded predicates which describe topological relationships between the structures. The functions are numerical and conversion functions.

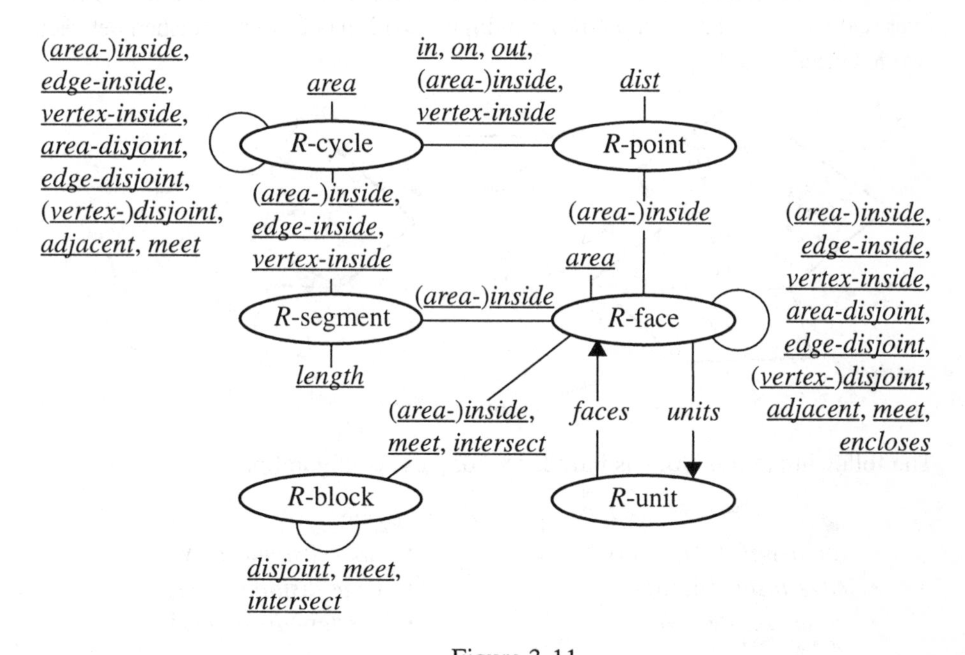

Figure 3-11

An *R-cycle* c is just a cycle in the graph interpretation of a realm, defined by a set of *R*-segments $S(c) = \{s_0, ..., s_{m-1}\}$, such that

(i) $\forall i \in \{0, ..., m-1\} : s_i$ meets $s_{(i+1) \bmod m}$

(ii) No more than two segments from $S(c)$ meet in any point p.

Obviously the following relationships may exist between an *N*-point p and an *R*-cycle c:

(i) p on c $:\Leftrightarrow$ $\exists s \in S(c) : p$ on s

For $p = (x, y)$ let $s_p = ((x, y), (x, n - 1))$ (that is, a vertical segment extending from p upwards to the edge of the grid). Let $S_r(c)$ be the set of segments in $S(c)$ whose right end point, but not the left one, is <u>on</u> s_p (the left end point is the smaller one of the two end points in the (x, y)-lexicographical order). Let $S_i(c)$ be the segments in $S(c)$ that <u>intersect</u> s_p. Then

(ii) p <u>*in*</u> c $:\Leftrightarrow$ $\neg p$ <u>*on*</u> $c \wedge |S_r(c)| + |S_i(c)|$ is odd

(iii) p <u>*out*</u> c $:\Leftrightarrow$ $\neg(p$ <u>*on*</u> $c \vee p$ <u>*in*</u> $c)$

Hence c partitions the set P_N into the three subsets $P_{in}(c)$, $P_{on}(c)$, and $P_{out}(c)$. Let $P(c) := P_{on}(c) \cup P_{in}(c)$.

Cycles are interesting because they are the basic entities for the definition of regions over realms. The relationships shown in Figure 3-12 may be distinguished between two R-cycles c_1 and c_2.

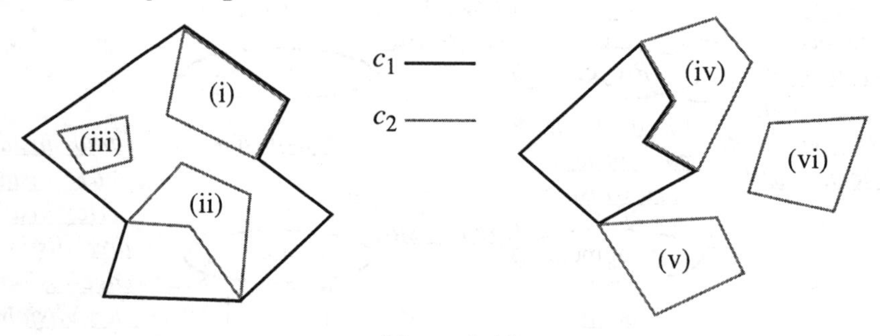

Figure 3-12

The following terminology is introduced for these configurations:

c_2 is
- *(area-)inside* (i, ii, iii)
- *edge-inside* (ii, iii)
- *vertex-inside* (iii)

c_1.

c_1 and c_2 are
- *area-disjoint* (iv, v, vi)
- *edge-disjoint* (v, vi)
- *(vertex-)disjoint* (vi)

The meaning is that (i) c_2 is (with respect to *area*) *inside* c_1, (ii) additionally has no common edges with c_1, (iii) has not even common vertices with c_1. Similarly (iv) c_2 is *disjoint* (with respect to *area*) with c_1, (v) additionally has no common edges with c_1, (vi) additionally has not even common vertices with c_1. *area-inside* is the standard interpretation of the term *inside*, *vertex-disjoint* the standard interpretation of the term *disjoint*. Furthermore, there are two positive notions: c_1 and c_2 are *adjacent* if they are area-disjoint and have common edges; they *meet* if they are edge-disjoint and have common vertices. The predicates are formally defined as follows:

c_1 *(area-)inside* c_2	:⇔	$P(c_1) \subseteq P(c_2)$
c_1 *edge-inside* c_2	:⇔	c_1 *area-inside* $c_2 \wedge S(c_1) \cap S(c_2) = \varnothing$
c_1 *vertex-inside* c_2	:⇔	c_1 *edge-inside* $c_2 \wedge P_{on}(c_1) \cap P_{on}(c_2) = \varnothing$
c_1 and c_2 are *area-disjoint*	:⇔	$P_{in}(c_1) \cap P(c_2) = \varnothing \wedge P_{in}(c_2) \cap P(c_1) = \varnothing$
c_1 and c_2 are *edge-disjoint*	:⇔	c_1 and c_2 are *area-disjoint* $\wedge$ $S(c_1) \cap S(c_2) = \varnothing$
c_1 and c_2 are *(vertex-)disjoint*	:⇔	c_1 and c_2 are *edge-disjoint* $\wedge$ $P_{on}(c_1) \cap P_{on}(c_2) = \varnothing$
c_1 and c_2 are *adjacent*	:⇔	c_1 and c_2 are *area-disjoint* $\wedge$ $S(c_1) \cap S(c_2) \neq \varnothing$
c_1 and c_2 *meet*	:⇔	c_1 and c_2 are *edge-disjoint* $\wedge$ $P_{on}(c_1) \cap P_{on}(c_2) \neq \varnothing$

One can observe similar ways how an *R*-segment s can lie within an *R*-cycle c (Figure 3-13):

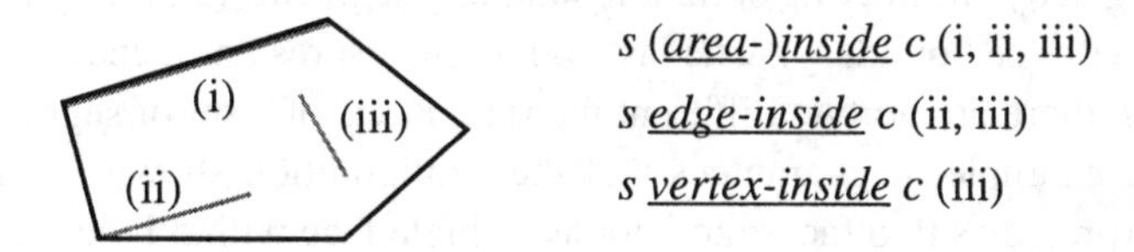

Figure 3-13

For an *R*-point p and an *R*-cycle c we have two possibilities (Figure 3-14):

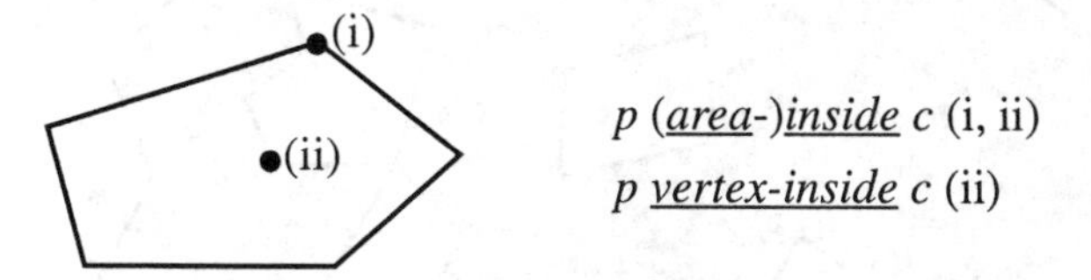

Figure 3-14

Formal definitions are left to the reader.

Based on the concept of *R*-cycles, for the definition of a spatial data type for regions the notions *R-face* and *R-unit* are introduced which describe regions from two different perspectives and which are used equivalently. Both of them make it possible to define polygonal regions with holes. An *R*-unit is a "minimal" *R*-face in the sense that any *R*-face within the *R*-unit is equal to the *R*-unit. Hence *R*-units are the smallest region entities that exist over a realm. We will see that any two *R*-units are area-disjoint and that any *R*-face can be described as a set of *R*-units. In the next chapter a region (data type) will be defined that can either be viewed as a set of *R*-faces or, equivalently, as a set of *R*-units. The first view emphasizes a minimal representation of the boundary of a region whereas the latter view supports the definition of set operations for regions. We will define operations to convert between the two (formal) representations.

An *R-face f* is a pair (c, H) where c is an R-cycle and $H = \{h_1, ..., h_m\}$ is a (possibly empty) set of R-cycles such that the following conditions hold (let $S(f)$ denote the set of segments of all cycles of f):

(i) $\forall i \in \{1, ..., m\} : h_i$ <u>*edge-inside*</u> c

(ii) $\forall i, j \in \{1, ..., m\}, i \neq j : h_i$ and h_j are <u>*edge-disjoint*</u>

(iii) Each cycle in $S(f)$ is either equal to c or to one of the cycles in H (no other cycle can be formed from the segments of f)

The first two conditions allow a hole within a face to touch in a vertex the boundary cycle c or another hole. This is necessary in order to achieve closure under operations (e.g., subtracting face g from face f may lead to a hole in f). On the other hand, to allow two holes to be area-disjoint makes no sense, since then adjacent holes could be merged by eliminating common boundary segments (similarly for adjacency of a hole with the boundary). The last condition ensures uniqueness of representation, that is, there are no two different interpretations of a set of segments as sets of faces. For example, it guarantees that the configuration shown in Figure 3-15 must be interpreted as two faces, and not as a single face with 5 holes (since under the latter interpretation the cycle drawn fat would violate condition (iii)).

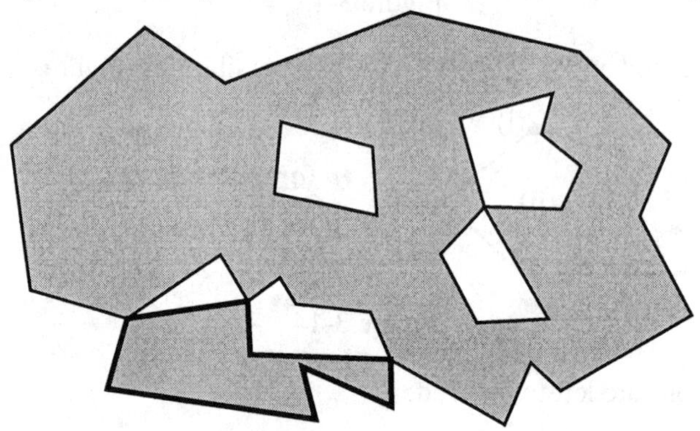

Figure 3-15

With terms defined below, condition (iii) can be rephrased as "an R-face cannot be decomposed into two or more edge-disjoint R-faces". Note that in a given set of faces it is entirely possible for a hole of one face to contain some other faces ("islands").

The grid points belonging to an R-face f are defined as $P(f) := P(c) \setminus \bigcup_{i=1}^{m} P_{in}(h_i)$.

The possible relationships between an R-point p or an R-segment s and an R-face $f = (c, H)$ are:

(i) p (*area-*)*inside* f $:\Leftrightarrow$ p *area-inside* $c \wedge \forall\, h \in H : \neg\, p$ *vertex-inside* h

(ii) s (*area-*)*inside* f $:\Leftrightarrow$ s *area-inside* $c \wedge \forall\, h \in H : \neg\, s$ *edge-inside* h

The various notions of *inside* and *disjoint* can be extended to the comparison of two R-faces $f = (f_0, \bar{F})$ and $g = (g_0, \bar{G})$, for example:

f (*area-*)*inside* g $:\Leftrightarrow$ f_0 *area-inside* $g_0 \wedge$
$\forall\, \bar{g} \in \bar{G} : \bar{g}$ *area-disjoint* $f_0 \vee \exists\, \bar{f} \in \bar{F} : \bar{g}$ *area-inside* $\bar{f}$

This definition is illustrated in Figure 3-16.

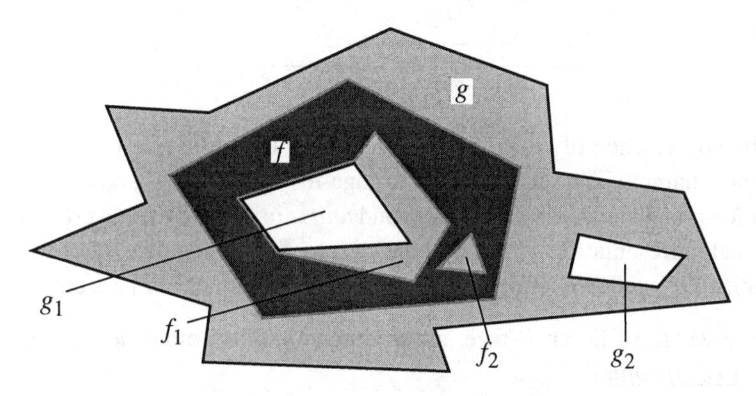

Figure 3-16

f *area-disjoint* g $:\Leftrightarrow$ f_0 *area-disjoint* $g_0 \vee \exists\, \bar{g} \in \bar{G} : f_0$ *area-inside* $\bar{g}\ \vee$
$\exists\, \bar{f} \in \bar{F} : g_0$ *area-inside* $\bar{f}$

f *edge-disjoint* g $:\Leftrightarrow$ f_0 *edge-disjoint* $g_0 \vee \exists\, \bar{g} \in \bar{G} : f_0$ *edge-inside* $\bar{g}\ \vee$
$\exists\, \bar{f} \in \bar{F} : g_0$ *edge-inside* $\bar{f}$

The meaning of the remaining predicates *edge-inside*, *vertex-inside*, (*vertex-*)*disjoint*, *adjacent*, *meet* should be clear; their definitions are left to the reader. We add a primitive *encloses*:

f *encloses* g $:\Leftrightarrow \exists\, \bar{f} \in \bar{F} : g_0$ *area-inside* $\bar{f}$

An R-unit as a minimal R-face is defined as follows. Let $F(R)$ denote the set of all possible R-faces. Let f be an R-face.

f is an *R-unit* $:\Leftrightarrow \forall\, g \in F(R) : g$ *area-inside* $f \Rightarrow g = f$

Figure 3-17 shows an example of a realm with all its R-units u_i and an emphasized R-face which is not an R-unit.

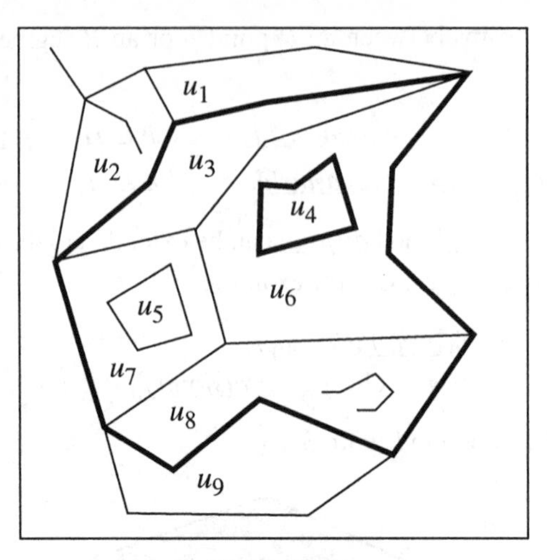

Figure 3-17

Now, the equivalence of two representations of a region over a realm is formally established, namely, as a set of (pairwise) edge-disjoint *R*-faces, and as a set of area-disjoint *R*-units. Operations called *faces* and *units* are defined to convert between the two formal representations. First we consider the conversion of a set of faces into a set of units. We need two lemmas, whose proofs are technical and are only sketched:

Lemma 3-21 Let *f* be an *R*-face and *u* an *R*-unit. Then either *u* area-inside *f* or *u* area-disjoint *f*.

The idea of the proof is that if this is not the case, then one of the cycles of *f*, say *f'* (Figure 3-18), must properly intersect one of the cycles of *u*, say *u'*.

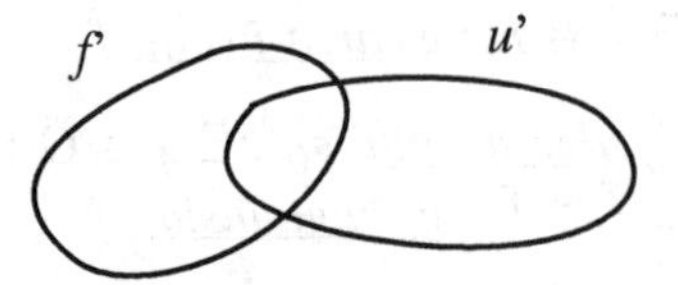

Figure 3-18

But then a part of *f'* lies within *u* and forms a cycle there with a part of *u'*. Hence there would be a face contained in *u* different from *u* which contradicts the definition of an *R*-unit. ❑

Lemma 3-22 Let *f* be an *R*-face and *u* an *R*-unit such that *u* area-inside *f*. Then "subtracting" *u* from *f* results in a set of *R*-faces.

The idea of the proof is the following: If *u* is even edge-inside *f* then removing the area of *u* from *f* either just adds another hole to *f*. If *u*'s outer cycle u_0 has some adjacent parts with *f*'s outer cycle f_0, then a "bay" is formed in f_0 (Figure 3-19). If it is adjacent with a hole f_1 in f, then f_1 will grow (Figure 3-20). If several adjacencies are present, then f may be decomposed into several faces. ❑

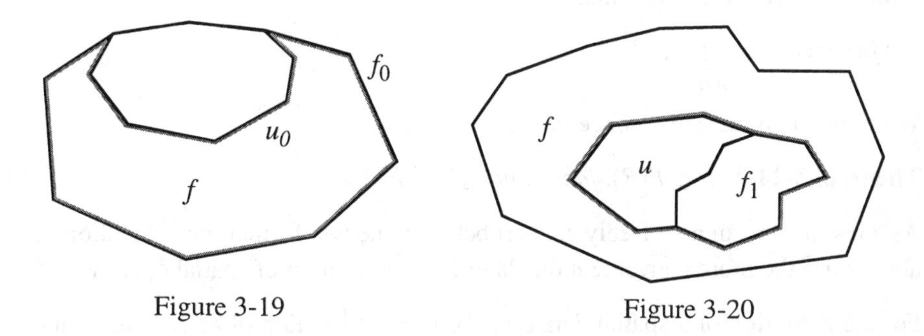

Figure 3-19 Figure 3-20

The second lemma implies that the units inside a face f cover the area of f completely. For, if some area were left, it would form its own face which could again be decomposed into units.

Therefore the following definition correctly decomposes faces into units. Let F be a set of edge-disjoint R-faces, and let $U(R)$ denote the set of all R-units.

$$units(F) := \{u \in U(R) \mid \exists f \in F\text{: } u \ \underline{area\text{-}inside}\ f\}$$

Let T be a set of R-segments, that is, $T \subseteq S$. Then *cycles*(T) denotes the set of all cycles (in the graph interpretation of realm R) that can be formed from segments in T. Furthermore, let $S(F)$ denote the set of all R-segments of a set of R-faces F. We say that a set T of R-segments *describes a set of pairwise edge-disjoint R-faces* :⇔ there exists a set of edge-disjoint R-faces F such that $T = S(F)$. If T describes a set of edge-disjoint R-faces, then a function *regions*(T) is defined to return this set of faces.

We now consider the conversion of a set of units into a set of faces. Let Δ denote the operator for symmetrical set difference, that is, $V \Delta W = (V \setminus W) \cup (W \setminus V)$. Δ forms the union of two sets removing their intersection. The operator is associative and commutative. The basis for the conversion is the following lemma:

Lemma 3-23 Let f and g be two area-disjoint R-faces. Then $S(f) \Delta S(g)$ describes a set of edge-disjoint R-faces.

The basic idea is that the Δ operator just removes segments that are common to both faces. The area-disjointedness condition makes sure that only boundaries between adjacent areas are removed (and not boundaries between a covered region in one face and a hole in the other face). ❑

The lemma can be extended to two sets of faces: Let F, G be two sets of edge-disjoint R-faces such that the faces in $F \cup G$ are pairwise area-disjoint. Then

$S(F) \Delta S(G)$ describes a set of edge-disjoint R-faces. Let the resulting set of R-faces be denoted by $F + G$. Now the conversion from units to faces can be defined as follows. Let U be a set of R-units.

$$faces(U) = \sum_{u \in U} \{u\}$$

We summarize the equivalence in

Theorem 3-24 $\forall F \subseteq F(R)$: $faces(units(F)) = F$ ❑

As a result, we can now freely convert between the two formal representations and always use the more convenient one later in the definition of spatial operations.

For the definition of a spatial data type for lines, the notion of an *R-block* is introduced. A set T of R-segments is called *connected* $:\Leftrightarrow \forall\, r, t \in T\ \exists\, s_1, ..., s_m \in T$: $r = s_1, t = s_m$, and $\forall\, i \in \{1, ..., m-1\}$: s_i and s_{i+1} meet. An *R-block* b is a connected subgraph in the graph interpretation of a realm, defined by its set of R-segments $S(b)$. Two R-blocks b_1 and b_2 are *disjoint* $:\Leftrightarrow \forall\, s_1 \in S(b_1)\ \forall\, s_2 \in S(b_2)$: s_1 and s_2 are disjoint. For an R-point p we consider the angularly sorted cyclic list L_p of R-segments $s \in S(b_1) \cup S(b_2)$ that meet in p. p is called a *meeting point* if L_p is the concatenation of two sublists $L_{p,1}$ and $L_{p,2}$ so that all R-segments of $L_{p,1}$ are elements of $S(b_1)$ and all R-segments of $L_{p,2}$ are elements of $S(b_2)$, or vice versa. In Figure 3-21 p represents and p' does not represent a meeting point.

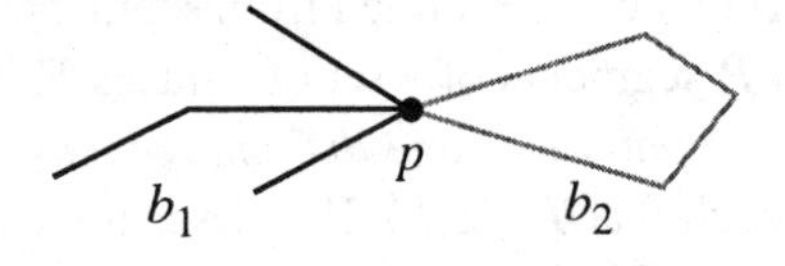

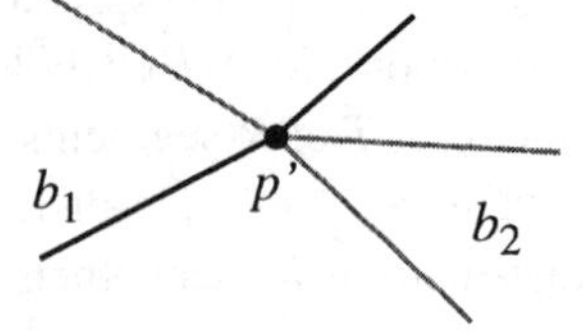

Figure 3-21

Let b_1 and b_2 be two R-blocks.

b_1 and b_2 *meet* $:\Leftrightarrow \exists\, s \in S(b_1)\ \exists\, t \in S(b_2)$: s and t meet in a meeting point $\wedge$
$\forall\, s \in S(b_1)\ \forall\, t \in S(b_2) : s \neq t\ \wedge$
(s and t meet in $p \Rightarrow p$ is a meeting point)

b_1 and b_2 *intersect* $:\Leftrightarrow \forall\, s \in S(b_1)\ \forall\, t \in S(b_2) : s \neq t \wedge \exists\, s \in S(b_1)\ \exists\, t \in S(b_2)$:
s and t meet in $p \wedge p$ is not a meeting point

Again, we have two equivalent representations of a *lines* value, namely, as a set of segments, or as a set of disjoint R-blocks. For a set of segments $T \subseteq S$, $blocks(T)$ denotes its partition into maximal connected components. Then $S(blocks(T)) = T$.

Some primitives relate an R-block b and an R-face f.

b (area-)inside f :⇔ $\forall s \in S(b)$: s area-inside f

b and f meet :⇔ $\forall s \in S(b) : \neg s$ area-inside $f \wedge \exists s \in S(b) \exists t \in S(f) : s$ and t meet

b and f intersect :⇔ $\exists s \in S(b)$: s area-inside f

Embedding N-points and N-segments in the Euclidean plane, we can define some primitives needed for computing numerical values (distances, lengths, etc.) from realm-based structures. The distance $dist(p, q)$ between two N-points p and q and the length $length(s)$ of an N-segment s are computed in the well-known way. The area $area(c)$ inside an R-cycle c with the segment set $S(c) = \{s_0, ..., s_{m-1}\}$ and $s_i = ((x_i, y_i), (x_{(i+1) \bmod m}, y_{(i+1) \bmod m}))$ is calculated by adding up the areas of the trapezia under each segment s_i down to the x-axis of the realm. Segments must be numbered in the order of the cycle so that the areas of the trapezia under the segments on the bottom of the R-cycle are subtracted from the areas of the trapezia under the segments at the top of the R-cycle.

$$\underline{area}(c) := \quad 0.5 \cdot \left| \sum_{i=0}^{m-1} (x_{(i+1)\mathrm{mod}m} - x_i)(y_{(i+1)\mathrm{mod}m} + y_i) \right| .$$

The area inside an R-face $f = (c, H)$ is defined as $\underline{area}(f) := \underline{area}(c) - \sum_{h \in H} \underline{area}(h)$.

3.7 Related Work

The idea of separating geometric primitives from the remainder of spatial modelling has already been proposed by Claire and Guptill [CG82]. They call minimal and basic structures like points and segments *primal spatial elements* and geometric primitives *primal operators*. Primal spatial elements are described as the elementary constructs for the discrete representation of spatial objects. Primal operators perform basic, atomic operations upon these minimal constructs. The main reason for separation is that primal elements and operators can serve as a basis for various higher data structures and are independent of them. The set of primal spatial elements contains besides points and segments half plane segments and raster segments. The set of primal operators contains unary operators like length, area, boundary, and perimeter, and binary operators describing relationships between primal elements. For points and segments binary primal operators are very similar to the geometric primitives presented in Section 3.2.

An alternative approach of introducing a discrete geometric basis for modelling as well as for implementation is the work of Frank and Kuhn [FK86] (later continued in [EFJ89]) which is based on algebraic topology (see Section 2.3.5). Because of the conflict between the infinite precision real numbers of Euclidean geometry and the finite precision number systems of computers (see Section 2.4) Frank and Kuhn only consider the topological structures of point sets underlying spatial objects. Their topological data model is based on simplicial complexes and has a similar purpose as the concept of realms. Essentially, they offer an irregular triangular network partition of the plane as a geometric domain over which spatial objects could be defined. However, the connections are missing to the underlying finite arithmetic as well as to spatial data types based on this model. Also, a triangular partition contains too much information; it is sufficient to keep those points and segments in a geometric domain that are needed for spatial objects. Finally, their model is an abstract one whereas realms are shown within a database context.

Based on the realm concept presented in this chapter, an alternative approach to implement realms has been proposed by Müller *et al.* [MPFDW96]. It is called *virtual* realms (in contrast to *stored* realms), i.e., the realm is not stored explicitly but is generated temporarily, only partially, and as needed by operations that update the database. Whenever a new spatial object O is to be inserted or an existing ROSE algebra object O has to be updated in the database, it or its geometric changes have to become acquainted with all other spatial objects in the database that are virtually defined over the same realm. For this purpose, all spatial objects in the database that intersect the minimum bounding rectangle of O are retrieved and the corresponding parts of the "small" realm relating to these objects are constructed temporarily. After that, interactions between these objects and O are resolved, and the geometric changes made to O and to the retrieved ROSE algebra objects are stored. Finally, the temporarily constructed parts of the realm are discarded.

A main goal of this approach is to reduce the storage space needed for stored realms. But it is not clear whether this goal is achieved, since both approaches require the use of a spatial index. In the case of the stored realm, the realm itself is the index; in the case of the virtual realm an index on spatially-referenced objects of possibly different object classes or types is needed. Since usually an index relates to one object class only, the assumed kind of index poses problems concerning the DBMS architecture. Moreover, both approaches have to access spatial objects in close proximity to the object to be inserted or updated. The stored realm has to propagate updates from the realm to the associated *points*, *lines*, and *regions* objects; the virtual realm loads spatial objects to let them become acquainted with the object to be inserted or updated.

An advantage of virtual realms is that renouncing a stored version of a realm simultaneously eliminates the two-way linking and the redundancy of spatial information occurring both in the realm objects and in the spatial algebra objects. On the other hand a virtual realm is more expensive in terms of CPU time, since additional processing is needed to construct and destroy its main memory representation.

Chapter 4
Realm-Based Spatial Data Types: The ROSE Algebra

Based on realm-based structures and primitives in this chapter a spatial algebra is described and formally defined which is called *ROSE* (*RObust Spatial Extension*) *algebra* [GS95]. It offers general types to represent point, line, and region features together with a comprehensive set of operations. The first section introduces the *realm-based* spatial data types points, lines, and regions of the ROSE algebra and defines the structure of corresponding objects as well as *spatial algebra primitives* on these objects. The second section introduces a flexible type system which allows to precisely describe polymorphic operations that are central to the ROSE algebra. In this type system it is also possible to cleanly model partitions of the plane so that operations can be constrained to be applicable to partitions or regions of partitions. The third section identifies a number of concepts that need to be present in the DBMS data or object model to allow it to cooperate with the ROSE algebra. These concepts are summarized in an abstract *object model interface* (*OMI*). The fourth section describes the operations of the ROSE algebra and formally defines their semantics. The fifth section shows how the ROSE algebra can be integrated with a DBMS data model and query language, using the object-oriented data model O_2 and its query language O_2SQL as an example. The sixth section describes related work.

4.1 Realm-Based Spatial Data Types

The realm-based structures defined in Section 3.6 form the basis for a definition of spatial data types. The fourth definition layer (see Appendix B) introduces basic types called points, lines, and regions which will be part of the ROSE algebra [GS95] defined in Section 4.4. There is a "flat" and a "structured" view of objects of these types.

The "flat" view is the following:

> *For a given realm R, an object of type points is a set of R-points, an object of type lines is a set of R-segments, and an object of type regions is a set of R-units.*

The "structured" view, that we shall assume as the formal definition, is as follows:

> *For a given realm R, an object of type points is a set of R-points, an object of type lines is a set of pairwise disjoint R-blocks, and an object of type regions is a set of pairwise edge-disjoint R-faces.*

We have shown in Section 3.6 that the two views are equivalent. The first view is conceptually very simple and supports a direct understanding of set operations. The second view is "semantically richer" and shows *lines* and *regions* objects as consisting of a number of *components* (blocks or faces). Moreover, it allows one to express relationships between these components and also emphasizes the representation of the boundary in case of regions. Note that a *regions* object may have holes. Holes are important because (i) they allow for an adequate modelling of area features, and (ii) they make it possible to obtain closure under point set operations. Figure 4-1 illustrates the data types.

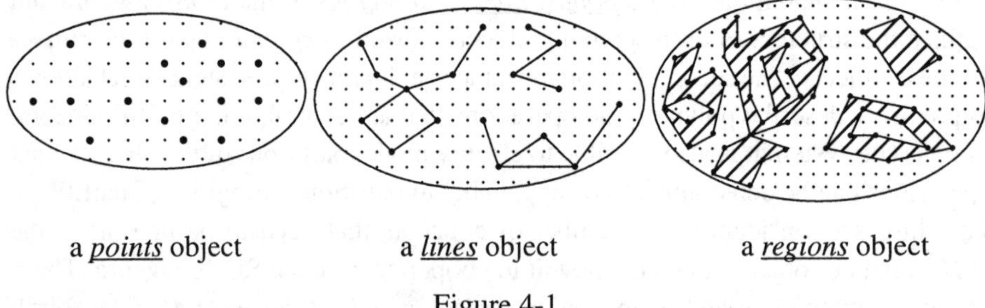

a *points* object a *lines* object a *regions* object

Figure 4-1

It should be obvious that these data types have very nice closure properties. They are closed under the geometric operations *union, intersection*, and *difference* with regard to the same realm. That is, the result of such an operation is a realm-based object as well and corresponds to the definitions of the spatial data types given above. The geometric operations can be reduced to the corresponding set-theoretic ones and are defined as follows. Let P_1, P_2 be two *points* objects, L_1, L_2 two *lines* objects, and R_1, R_2 two *regions* objects. Then

$$\textbf{\textit{union}}\ (P_1, P_2) := P_1 \cup P_2$$
$$\textbf{\textit{union}}\ (L_1, L_2) := blocks(S(L_1) \cup S(L_2))$$
$$\textbf{\textit{union}}\ (R_1, R_2) := faces(units(R_1) \cup units(R_2))$$

For ***intersection*** and ***difference*** the definitions are analogous. Due to the underlying realms, these operations both in theory *and* in practice obey the usual algebraic laws (commutativity, associativity, distributivity, ...).

The realm-based primitives offer a formal basis for the definition of *spatial algebra primitives* of which ***union***, ***intersection***, and ***difference*** have just been introduced. The following further primitives are needed. Let F and G be two regions objects.

F and G are ***area-disjoint***	:⇔	$\forall f \in F \; \forall g \in G$: f and g are *area-disjoint*
F and G are ***adjacent***	:⇔	F and G are ***area-disjoint*** $\wedge$ $\exists f \in F \; \exists g \in G$: f and g are *adjacent*

The meaning of the remaining predicates (***area-***)***inside***, ***edge-inside***, ***vertex-inside***, ***edge-disjoint***, (***vertex-***)***disjoint***, ***meet*** should be clear; their formal definitions are left to the reader. We define two further predicates ***intersect*** and ***encloses***:

F and G ***intersect***	:⇔	$units(F) \cap units(G) \neq \emptyset$
F ***encloses*** G	:⇔	$\forall g \in G \; \exists f \in F$: f *encloses* g

Let P and Q be two points objects.

P and Q are ***disjoint***	:⇔	$P \cap Q = \emptyset$

Let K and L be two lines objects.

K and L are ***disjoint***	:⇔	$\forall k \in K \; \forall l \in L$: k and l are *disjoint*
K and L ***meet***	:⇔	($\forall k \in K \; \forall l \in L$: k and l are *disjoint* $\vee$ k and l *meet*) $\wedge$ ($\exists k \in K \; \exists l \in L$: k and l *meet*)
K and L ***intersect***	:⇔	($\forall k \in K \; \forall l \in L$: k and l are *disjoint* $\vee$ k and l *intersect*) $\wedge$ ($\exists k \in K \; \exists l \in L$: k and l *intersect*)

Let P be a points object, L a lines object, F a regions object, and v, w lines or regions objects.

P (***area-***)***inside*** F	:⇔	$\forall p \in P \; \exists f \in F$: p *area-inside* f
L (***area-***)***inside*** F	:⇔	$\forall l \in L \; \exists f \in F$: l *area-inside* f
L and F ***meet***	:⇔	$\forall l \in L \; \forall f \in F$: $\neg$ l *area-inside* f $\wedge$ $\exists l \in L \; \exists f \in F$: l and f *meet*
L and F ***intersect***	:⇔	$\exists l \in L \; \exists f \in F$: l and f *intersect*
P ***on_border_of*** v	:⇔	$\forall p \in P \; \exists s = (q_1, q_2) \in S(v) : p = q_1 \vee p = q_2$
v ***border_in_common*** w	:⇔	$\exists s \in S(v) \; \exists t \in S(w) : s = t$

4.2 The Type System of the ROSE Algebra

The ROSE algebra that we are going to define is a system of spatial data types together with operations between these types. Many of the operations are applicable to several types. Hence we need a framework and notations to describe polymorphic operations. We would also like to express certain constraints for the applicability of some operations. For example, an adjacency test operation for regions should only be allowed if the two operands are known to come from a set of disjoint regions (that is, a *partition* of the plane). Similarly, an overlay operation should be constrained to two partition operands and not be applicable to arbitrary collections of objects with region attributes. In this section we briefly review a type system powerful enough to express polymorphic operations and the mentioned constraints in a precise manner.

4.2.1 Second-Order Signature

A system of several sets and functions between these sets is called a *many-sorted algebra*. A *many-sorted signature* describes the syntactic aspect of a many-sorted algebra. It consists of two sets of symbols called *sorts* and *operators*; operators are annotated with strings of sorts. Each sort is the *name of a set* of the algebra and each operator the *name of a function*. For example, the symbols lines, regions, and bool may be sorts and $\textbf{intersects}_{\underline{lines}\ \underline{regions}\ \underline{bool}}$ an operator. The annotation with sorts defines the functionality of the operator. A signature defines a set of terms.

Second-order signature, introduced in [Gü93], is a system of two coupled many-sorted signatures where the top-level signature offers *kinds* (sets of types) as sorts and *type constructors* as operators. The terms of this signature define a collection of types, that is, a type system. A simple example is shown below. Each line describes a group of operators (type constructors in this case) with the same functionality.

kinds DATA, GEO, SET

type constructors

	→ DATA	int, real, bool
	→ GEO	points, lines, regions
GEO	→ SET	set

Here int, set, etc. are type constructors which generally have one or more argument kinds and one result kind. A type constructor with zero argument kinds is called a *constant type*. In the example all constructors except for set are constant types. The terms of this signature, and therefore the available types of this type system, can be classified by result kinds. For example, there are exactly three types of kind GEO. The types of kind SET are set(points), set(lines), and set(regions). In this example the set of types is finite, but this is generally not the case.

A second, bottom-level signature uses the types defined by the top-level signature as sorts. Usually one does not write the bottom-level signature directly but rather a *signature specification* which allows one to quantify over kinds and so to define polymorphic operations. For example, we can define:

$\forall$ *data* in DATA.	*data* $\times$ *data*	$\rightarrow$ <u>*bool*</u>	=, <, ≤, ≥, >
$\forall$ *geo* in GEO.	*geo* $\times$ <u>*regions*</u>	$\rightarrow$ <u>*bool*</u>	**inside**

Here *data* and *geo* are *type variables* ranging over the kinds DATA and GEO, respectively. The semantics of such a signature specification is a many-sorted signature which is obtained by substituting for each type variable all types in the respective kind. Hence the first specification says that the comparison operators are defined for two integers, two reals, or two boolean values. The second specification defines an **inside** operator with functionalities <u>*points*</u> $\times$ <u>*regions*</u> $\rightarrow$ <u>*bool*</u>, <u>*lines*</u> $\times$ <u>*regions*</u> $\rightarrow$ <u>*bool*</u>, and <u>*regions*</u> $\times$ <u>*regions*</u> $\rightarrow$ <u>*bool*</u>.

This already completes the description of the basic scheme of second-order signature. Of course, there are also other ways of specifying polymorphic operations. For a discussion and for references see [Gü93].

The basic scheme has been extended in [Gü93] to support the definition of flexible database query languages. Some of these techniques are needed for the definition of the ROSE algebra:

Extensions of the concept of signature. The purpose is to include for a given collection of types (sorts, to be precise) "automatically" product types, union types, list types, and function types. If $s, s_1, ..., s_n$ and t are sorts then

- $(s_1 \times ... \times s_n)$ is a sort (the product sort, denoting tuples of instances of the s_i)
- $(s_1 \cup ... \cup s_n)$ is a sort (the union sort, denoting instances in any of the s_i)
- s^+ is a sort (the sort denoting non-empty lists of instances of s)
- $(s_1 \times ... \times s_n \rightarrow t)$ is a sort (denoting functions from $s_1 \times ... \times s_n$ into t).

With these extensions one can, for example, define the following operations:

$\forall$ *geo* in GEO.

(<u>*set*</u>(*geo*))$^+$	$\rightarrow$ <u>*set*</u>(*geo*)	**union**
<u>*set*</u>(*geo*) $\times$ (*geo* $\rightarrow$ <u>*bool*</u>)	$\rightarrow$ <u>*set*</u>(*geo*)	**select**

Here the **union** operator takes one or more operands that are all sets of geometric objects of the same type and returns a set (the union) of this type. The **select** operator takes an operand of type <u>*set*</u>(*geo*) and a predicate on type *geo* and returns a subset of the operand set fulfilling the predicate.

Specification techniques. Two additional specification techniques are illustrated by the following example:

$\forall$ geo_i in GEO. $(\underline{set}(geo_i))^+$ $\rightarrow$ *data*: DATA **weight**

The notation geo_i is related to operators with a variable number of operands and means that for each substitution of the variable geo_i an instance of the kind GEO is selected independently. Hence *set*(*points*) × *set*(*lines*) × *set*(*lines*) would be one possible operand combination for **weight**. With the quantification "∀ *geo* in GEO" all operands would have to be of the same type (e.g., *set*(*points*)).

The notation "*data*: DATA" is to be read as "some type *data* in DATA" and means that there is a type mapping associated with the **weight** operator. Intuitively the idea is that the operator determines itself the result type within the kind DATA, depending on the given operand types. This is sometimes useful when it is not possible or desirable to describe the result type precisely in the signature. To define the semantics of such an operator one needs to supply a type mapping function (as a part of a second-order algebra, see [Gü93] for details). In this example, the **weight** operator might return a value of type *int* if all operands are sets of points (and return the total number of points), and a value of type *real* otherwise (say, the total area or length). Some examples of meaningful operators with type mappings occur in the ROSE algebra defined below.

Dynamic kinds. (This extension has not yet been covered in [Gü93]). Sometimes it is necessary to modify dynamically the set of instances of a kind, that is, to create new types. For a kind K, the notation *new*(K) creates a new (anonymous) type in K; the value of *new*(K) is a type that can be used in type expressions.

4.2.2 The Type of a Partition

The term *partition* is used to refer to a disjoint subdivision of the plane into regions with associated (non-spatial) attributes. For partitions, one would like to define special operations like testing for adjacency (of two regions of a partition) or overlay (of two partitions, resulting in a new partition). The question is how partitions can be described in a type system so that the operations can be constrained to partition operands.

A partition should be modelled as a set of spatially-referenced objects with associated *regions* attribute values and an additional constraint that for any pair of spatially-referenced objects in one particular partition, their *regions* objects are disjoint. To say this in a more general way, we would like to model and manipulate sets of objects such that for any two distinct objects in such a set a certain condition holds. To consider an example different from partitions, let us assume we would like to model sets of integers with the property that there are no two consecutive integers in the set.

The idea to make this possible in the type system is to introduce *restriction types* and to collect them within a *special kind*. Let d be a data type and p be a binary predicate on d. Then d^p denotes a kind; each type d' in d^p describes a set of values of type d such that for any two distinct elements of d' the predicate p holds. Furthermore, any such type d' is defined to be a *subtype* of d which means that all operations defined for type d are also applicable to instances of type d'.

For the "non-consecutive integer" example, we could introduce a predicate "two-apart" on integers, being true if the difference of the two operands is at least two. Then $\underline{int}^{\text{two-apart}}$ denotes a kind whose element types have carrier sets[15] with the desired property. Hence the set {3, 5, 10} would have a type within this kind whereas for the set {1, 2, 3} there would not exist a type within kind $\underline{int}^{\text{two-apart}}$. The types themselves are anonymous (i.e., no explicit names for them need to be introduced).

We use this as follows: The kind $\underline{regions}^{\textbf{\textit{area-disjoint}}}$ contains all types whose carriers are sets of $\underline{regions}$ objects such that any two distinct objects of the type are ***area-disjoint***. A quantification "$\forall$ *area* in $\underline{regions}^{\textbf{\textit{area-disjoint}}}$" binds the *area* type variable to any such type. Hence an adjacency test can be defined as:

$$\forall\ area \text{ in } \underline{regions}^{\textbf{\textit{area-disjoint}}}. \qquad area \times area \rightarrow \underline{bool} \qquad \textbf{adjacent}$$

Here the quantification selects first one particular partition of the plane as a type *area*. Hence it is guaranteed that any two arguments for the operator **adjacent** are from the same partition and are either area-disjoint or equal. When a new partition is created in query processing, we can use the notation $new(\underline{regions}^{\textbf{\textit{area-disjoint}}})$ to obtain a corresponding new anonymous type for it.

On the side of the database system this should be supported by making it possible to define restriction types and to use them as attribute domains. For example, assume that an operation `area_disjoint` applicable to values of type `regions` has been made known to the DBMS. One might write:

```
type mycountries = restrict (regions, area_disjoint);
class states (name: string; region: mycountries; pop: integer)
```

An insertion of a new object into class `states` should then at least conceptually be viewed as preceded by an insertion of a new `regions` object into the extension of type `mycountries`. It should be checked that the new object is `area_disjoint` with all objects already present.

---

[15] For a type, its set of instances is called the *carrier*.

4.3 The Object Model Interface

Spatial data types as such are rather useless; they need to be integrated into a DBMS data model and query language. On the other hand, their definition should be valid regardless of any particular data model and therefore not depend on it. Hence, spatial data types should not be firmly embedded into a particular DBMS data model. Instead, their definition should be based on an abstract interface to the DBMS data model which we call the *object model interface* (*OMI*). Different DBMS data models can then use the spatial algebra as a provided resource for dealing with geometry. In this section we define an object model interface for the ROSE algebra. In fact, there are two aspects of the interface: (1) There are basic concepts and operations in the object model that are needed to *define* the ROSE algebra, and (2) there are constructs and notations needed to embed the ROSE algebra into the query language, that is, to *use* the ROSE algebra.

In this section the term "object" is used in a database context and describes a spatially-referenced object viewed as a database object.

4.3.1 OMI Concepts for Defining the ROSE Algebra

The concepts that are needed to define the ROSE algebra are the following:

- object types/classes
- collections of objects
- functions for accessing (attribute) values from objects
- data types <u>int</u>, <u>real</u>, <u>bool</u>
- a pool of names (for new objects/functions)
- an object aggregation function
- an object extension function

Object types/classes. We assume that each DBMS data model has some notion of one or more object types or classes. For example, in a relational system, this would be relations; in an object-oriented system we may have object class hierarchies, and objects may have a complex structure. In terms of our type system we model this by a kind OBJ; each DBMS object class is represented as a type <u>obj</u> in OBJ.

Collections of objects. The structures manipulated in (and obtained as a result of) queries may be sets of tuples, nested relations, sequences of object identifiers, graphs, etc. The most simple, universally valid, and data model-independent abstraction is that of a *set of objects*. If a set of objects is not directly available, the DBMS data model must provide functions to transform its structures containing objects into a set of objects, and vice versa. In the type system we have a type constructor <u>set</u> applicable to object types.

Functions for accessing attribute values. The OMI views an object as an abstract entity whose internal structure is hidden. It is assumed that objects may have associated values of standard or spatial data types and that these values can be accessed by means of *attribute functions* of type $obj \rightarrow data$, for any type *obj* in OBJ and data type *data*.

Data types int, real, bool. We assume that standard data types for integers, real numbers, and boolean values exist. Some ROSE operations yield results of these types.

A pool of names. Some operations require (new) names as parameters, in particular for introducing derived attributes (attribute functions). We introduce this pool of names as a type *ident* in a kind IDENT.

Object aggregation function. Some spatial operations construct new objects as "aggregation objects". For that purpose the DBMS data model has to provide a "⊗" (product) function which for two objects o_1 of type obj_1 and o_2 of type obj_2 forms an aggregation object $o_1 \otimes o_2$. The same symbol is used to denote a corresponding type mapping operation; hence there is also a product type $obj_1 \otimes obj_2$, and object $o_1 \otimes o_2$ is of type $obj_1 \otimes obj_2$. On the product type all attribute functions defined on either obj_1 or obj_2 are valid; this should be expressed by the type mapping (defined within the object model). In a relational setting, this corresponds to concatenating two tuples when forming a join; the result tuple has the attributes of both operand tuples.

Object extension function. Sometimes it is necessary to add an attribute to objects of a given object type. For that purpose the DBMS data model must offer an extension function denoted by "⊕". At the instance level, this operation adds a data type value to an object, hence $o \oplus v$ is an object o extended by a value v. At the type level, the given object type *obj* is extended by an attribute function *attr* mapping objects into values of some data type *data*. Hence $obj \oplus (attr, data)$ denotes such an extension type of which $o \oplus v$ is an instance if o has type *obj* and v has type *data*.

4.3.2 OMI Concepts for Embedding the ROSE Algebra into a DBMS Query Language

This part of the object model interface contains requirements about notations and constructs needed in the DBMS query language to allow an embedding and a full use of the ROSE algebra. Facilities are needed to

- denote a (spatial) data type value
- denote a collection of objects together with an attribute (attribute function)
- extend objects by derived (attribute) values
- allow naming of a spatial data type value or a new attribute
- offer a grouping operation.

To motivate why these facilities are needed we give a brief preview of some operations of the ROSE algebra:

$\forall$ *obj* in OBJ. $\forall$ geo, geo_1, geo_2 in GEO.

geo × <u>*regions*</u>	→ <u>*bool*</u>	**inside**
<u>*lines*</u> × <u>*lines*</u>	→ <u>*points*</u>	**intersection**
<u>*set*</u>(*obj*) × (*obj* → geo_1) × geo_2	→ <u>*set*</u>(*obj*)	**closest**
<u>*set*</u>(*obj*) × (*obj* → *geo*) × <u>*ident*</u>	→ <u>*set*</u>(*o*: OBJ)	**decompose**

The meaning of the first two operations should be obvious. The **closest** operator takes a collection of database objects together with a spatial attribute function and a further spatial data type value *v* and returns those objects whose attribute value is closest to *v* (usually one object). The **decompose** operator also takes a collection of objects with a spatial attribute. It produces a new collection of objects as follows: For each object in the operand set its spatial attribute value is decomposed into its components (a component is a point, a block, or a face). If there are *n* components, then *n* copies of the original object are produced each of which has one component as the value of a new attribute. The name of the new attribute is supplied as the third parameter of type <u>*ident*</u>.

We now discuss each of the mentioned facilities in turn and illustrate them in the context of the relational model by (a) showing corresponding notations from geo-relational algebra [Gü88] and (b) by extensions that might be used for SQL. In examples, relations

cities (cname: <u>*string*</u>; center: <u>*points*</u>; pop: <u>*int*</u>)
states (sname: <u>*string*</u>; territory: <u>*regions*</u>; language: <u>*string*</u>)

are used.

Denote a data type value. This is needed to supply operands to operations like **inside** or **intersection**. There are two cases: (i) within the scope of an "object set iteration", and (ii) without object set iteration. In the first case, each object in a set is considered in turn and it suffices to write down the name of an attribute to denote a single data type value.

Q1: Calculate the population (in thousands) of all cities in Germany.

(a) cities **select**[center **inside** Germany] **extend**[pop/1000 {thousands}]

(b) **select** cname, thousands: pop/1000
from cities
where center **inside** Germany

Here, within the scope of a **select** or **extend** operator of the geo-relational algebra or within the where-clause or select-clause of SQL we have an "object set iteration" and an attribute name denotes a data type value.

In the second case (without object set iteration), one would like to refer to a single data type value, in particular to the attribute value of some specific object. A notation is needed to identify a single object and to access one of its attributes. In the geo-relational algebra this is done by an **extract** operator. An error message should appear if none or more than one object is identified by the condition.

Q2: Provide the geometry of the city Hagen (assuming there is only one "Hagen" in the cities relation).

(a) cities **extract**[cname = "Hagen"; center] {Hagen}

(b) **let** Hagen
extract center
from cities
where cname = "Hagen"

Here we have extracted a single <u>*points*</u> value from the cities relation. We have also assigned a name (Hagen) to this value so that it can be used in later queries.

Denote a collection of objects together with an attribute. This is needed for operations like **closest** or **decompose**. Recall the signature for **closest**:

$\underline{set}(obj) \times (obj \rightarrow geo_1) \times geo_2 \quad \rightarrow \underline{set}(obj) \quad$ **closest**

We need a notation to supply the two related operands "$\underline{set}(obj)$" and "$(obj \rightarrow geo_1)$".

Q3: Determine the city or cities closest to Hagen.

(a) cities **select**[cname ≠ "Hagen"] Hagen **closest**[center]

(b) **closest** Hagen
column center
from cities
where cname ≠ "Hagen"

In this example, "cities" corresponds to the "$\underline{set}(obj)$" and "center" to the "$(obj \rightarrow geo_1)$" operand. In geo-relational algebra first the set of objects is written and then the <u>*points*</u> value (the "geo_2" operand); the attribute is given separately in brackets. For an extended SQL the introduction of a "**column** α **from** β" construct is suggested in order to denote a set of objects β with an attribute α. This construct should be viewed as returning the two operands separately as they are needed by the ROSE algebra. In contrast, writing "**select** α **from** β" would yield a set (or multiset) of attribute values, that is, an operand of type $\underline{set}(geo_1)$. This is not what the operator needs; in fact, a set of values is not even available in the ROSE type system given below.

Extend objects by derived (attribute) values. This is needed to make the results of spatial operations available. In geo-relational algebra this is provided by the **extend** operator, in SQL by expressions in the select-clause, as in query Q1.

Allow naming of a spatial data type value or a new attribute. We have already seen two instances of this. In query Q2 a name (Hagen) was assigned to a spatial data type value. An attribute name must also be provided for derived attributes, as in query Q1. Finally, new attribute names are needed by operations that construct new objects such as **decompose**.

Q4: Decompose all states into their basic areas.

(a) states **decompose**[territory {basic_area}]

(b) **decompose into column** basic_area
column territory
from states

Here (a) shows the style for naming the new attribute that would be used in geo-relational algebra (although there was no **decompose** operator). For the extended SQL an "**into column** α" construct has been invented for the same purpose.

Offer a grouping operation. This is needed to support a "fusion" operation (which essentially groups a collection of objects and forms the union of the areas in each group).

Q5: Determine all regions of the states speaking the same language.

(a) states **fusion**[language; territory]

(b) **fusion** territory
from states
group by language

These applications of the fusion operator are really abbreviations of the use of grouping:

(a) states **group_by**[language; group **sum**[territory]]

(b) **select sum**(territory)
from states
group by language

In geo-relational algebra and in SQL such a grouping operation is available; it is used together with a **sum** aggregate function of the ROSE algebra. There may be several attributes for grouping and several aggregate expressions.

4.4 The ROSE Algebra

The fifth and final definition layer (see Appendix B) defines the ROSE algebra itself (ROSE stands for RObust Spatial Extension) [GS95]. It is a *realm-based spatial algebra*, since data types are defined on realms and since operations operate on and produce realm-based spatial objects. All objects occurring as operands are assumed to be defined over the *same* realm.

Defining the ROSE algebra means to give a second-order signature with the types *points*, *lines*, and *regions* as well as types of the object model interface. The algebra then consists of carrier sets for the types and functions for the operations. The carrier sets for the three spatial types have already been defined in Section 4.1. In this section we formally define the functions for all operations.

The type system of the ROSE algebra, as discussed in Section 4.2 and Section 4.3, is summarized in the following specification:

kinds IDENT, DATA, EXT, GEO, OBJ, SET
type constructors

	→ IDENT	*ident*
	→ DATA	*int*, *real*, *bool*, ...
	→ EXT	*lines*, *regions*
	→ GEO	*points*, *lines*, *regions*
OBJ	→ SET	*set*

Kind DATA describes the (standard) data types of the object model interface; there will be other types in addition to the three that are required. There is a kind EXT just

Spatial Predicates Expressing Topological Relationships	**=, ≠, disjoint, meets, inside, area_disjoint, edge_disjoint, edge_inside, vertex_inside, intersects, adjacent, encloses, on_border_of, border_in_common**
Spatial Operators Returning Spatial Data Type Values	**intersection, plus, minus, vertices, common_border, contour, interior**
Spatial Operators Returning Numbers	**no_of_components, dist, diameter, length, area, perimeter**
Spatial Operations on Sets of Spatially-Referenced Objects	**sum, closest, decompose, overlay, fusion**

Figure 4-2

containing types lines and regions which supports the definition of operations not suitable for type points.

The operations of the ROSE algebra are divided into four groups (Figure 4-2). For each group we give an informal introduction, show the signature, and then define the semantics of the operations.

4.4.1 Spatial Predicates Expressing Topological Relationships

These operations compare two spatial objects with respect to their topological relationships and return a boolean value. The predicates' names are self-explanatory.

$\forall$ *geo* in GEO. $\forall$ *ext*, ext_1, ext_2 in EXT. $\forall$ *area* in regions$^{\textit{area-disjoint}}$.

geo × *geo*	→	bool	**=, ≠, disjoint**
geo × regions	→	bool	**inside**
regions × regions	→	bool	**area_disjoint, edge_disjoint, edge_inside, vertex_inside**
$ext_1 \times ext_2$	→	bool	**intersects, meets**
area × *area*	→	bool	**adjacent, encloses**
points × *ext*	→	bool	**on_border_of**
$ext_1 \times ext_2$	→	bool	**border_in_common**

For each operator **op** of the ROSE algebra we define a function $f_{\mathbf{op}}$ which gives the operator's semantics and which has domains and codomain according to the operator's signature entry. An underlying realm *R* is assumed in all definitions. Of course, we rely on the primitives introduced in Section 3.2, Section 3.6, and Section 4.1.

Let v_1, v_2 be two objects of the same type in GEO. Then

$f_{=}(v_1, v_2) := (v_1 = v_2)$
$f_{\neq}(v_1, v_2) := (v_1 \neq v_2)$
$f_{\mathbf{disjoint}}(v_1, v_2) := (v_1 \text{ and } v_2 \text{ are } \textbf{\textit{disjoint}})$

Let v be an object of a type in GEO and F be an object of type regions.

$f_{\mathbf{inside}}(v, F) := (v \textbf{\textit{ inside }} F)$

Let v_1, v_2 be each either a lines or a regions object.

$f_{\mathbf{intersects}}(v_1, v_2) := (v_1 \text{ and } v_2 \textbf{\textit{ intersect}})$
$f_{\mathbf{meets}}(v_1, v_2) := (v_1 \text{ and } v_2 \textbf{\textit{ meet}})$

Let F and G be two *regions* objects of a subtype *area* in *regions*$^{\textit{area-disjoint}}$.

$f_{\textbf{adjacent}}(F, G)$:= (F and G are ***adjacent***)

$f_{\textbf{encloses}}(F, G)$:= (F ***encloses*** G)

The remaining definitions are left to the reader; they all just lift spatial algebra primitives to the ROSE level.

4.4.2 Spatial Operators Returning Spatial Data Type Values

The second group of operations consists of *operators returning atomic spatial objects* as results. The operators **intersection**, **plus**, and **minus** realize the closure properties of the ROSE algebra with respect to geometric intersection, union, and difference of two atomic spatial objects. The **common_border** operator finds the common boundary line(s) of two *regions* or *lines* objects. The **vertices** operator returns the corner points of a *lines* or *regions* object and produces a *points* object. The **contour** operator calculates a *lines* object from a *regions* object's boundary. The **interior** operator is applied to a *lines* object and yields a *regions* object which is composed of all regions that are enclosed by segments of the *lines* object. If F is a *regions* object, **interior**(**contour**(F)) can be used to remove all holes of F; both operators are not inverse to each other.

$\forall$ *geo* in GEO. $\forall$ *ext*, ext_1, ext_2 in EXT.

points × *points*	→ *points*	**intersection**
lines × *lines*	→ *points*	**intersection**
regions × *regions*	→ *regions*	**intersection**
regions × *lines*	→ *lines*	**intersection**
geo × *geo*	→ *geo*	**plus, minus**
$ext_1 \times ext_2$	→ *lines*	**common_border**
ext	→ *points*	**vertices**
regions	→ *lines*	**contour**
lines	→ *regions*	**interior**

Note that the **intersection** operator applied to two *lines* objects does not yield a *lines* object as the set-theoretic intersection of the underlying segment sets (see operator **common_border**) but a *points* object.

Let P and Q be two *points* objects, K and L be two *lines* objects, and F and G be two *regions* objects.

$f_{\textbf{intersection}}(P, Q)$:= ***intersection***(P, Q)

$f_{\textbf{intersection}}(F, G)$:= ***intersection***(F, G)

$f_{\textbf{intersection}}(K, L) := \{p \in R \mid \exists\, s \in S(K)\ \exists\, t \in S(L) : s \text{ and } t \text{ \underline{meet} in } p \wedge p \text{ is not a meeting point}\}$

$f_{\textbf{intersection}}(F, L) := blocks(\{s \in S(L) \mid \exists f \in F : s \text{ \underline{inside} } f\})$

Let v_1 and v_2 be both either two *points* objects, two *lines* objects, or two *regions* objects.

$f_{\textbf{plus}}(v_1, v_2) := \textbf{\textit{union}}(v_1, v_2)$

$f_{\textbf{minus}}(v_1, v_2) := \textbf{\textit{difference}}(v_1, v_2)$

Let K and L be two *lines* objects and F and G be two *regions* objects.

$f_{\textbf{common\_border}}(K, L) := \textbf{\textit{intersection}}(K, L)$

$f_{\textbf{common\_border}}(F, L) := f_{\textbf{common\_border}}(L, F) := blocks(S(F) \cap S(L))$

$f_{\textbf{common\_border}}(F, G) := blocks(S(F) \cap S(G))$

Let v be a *lines* or *regions* object.

$f_{\textbf{vertices}}(v) := \{p \in R \mid \exists\, s \in S(v) : s = (p, q)\}$

Let $F = \{f_1, ..., f_n\} = \{(c_1, H_1), ..., (c_n, H_n)\}$ be a *regions* object.

$$f_{\textbf{contour}}(F) := blocks(\bigcup_{i=1}^{n} S(c_i))$$

Let L be a *lines* object.

$$f_{\textbf{interior}}(L) := regions(\bigcup_{c \in cycles(S(L))} S(c) - \{s \in S(L) \mid \exists\, c \in cycles(S(L)) : s \text{ \underline{edge-inside} } c\})$$

Forming the interior of a *lines* object L is a somewhat more complex operation. First, the union of all segments is computed that occur in any cycles that can be formed from the segments of L. From this set of segments all segments are removed that lie properly within (*edge-inside*) some cycle. Hence only segments of "outer cycles" remain. Since these segments describe a set of edge-disjoint R-faces, the *regions* function can be applied to return a corresponding *regions* object.

4.4.3 Spatial Operators Returning Numbers

The third group of operations contains *spatial operators returning numbers*. The **no_of_components** operator yields the number of components (R-points, R-blocks, or R-faces) of a spatial object. The **dist** operator calculates the minimal distance between any two spatial objects. The **diameter** of a spatial object is defined as the largest distance between any of its components. The **length** operator calculates the length of all segments of a *lines* object. The **area** operator computes the sum of the areas of all faces of a *regions* object. The **perimeter** operator calculates the sum of the length of all cycles of a *regions* object. If we intend to compute only the sum of the length of the outer cycles and not of the holes of a *regions* object, we can use the **contour** operator to eliminate holes first.

$\forall$ *geo*, geo_1, geo_2 in GEO.

geo	$\rightarrow$	*int*	**no_of_components**
$geo_1 \times geo_2$	$\rightarrow$	*real*	**dist**
geo	$\rightarrow$	*real*	**diameter**
lines	$\rightarrow$	*real*	**length**
regions	$\rightarrow$	*real*	**area, perimeter**

Let v and w be objects of types in GEO. Let L be a *lines* object and F be a *regions* object.

$$f_{\mathbf{no\_of\_components}}(v) := card(v)$$

$$f_{\mathbf{diameter}}(v) := max\{\underline{dist}(p, q) \mid p, q \in f_{\mathbf{vertices}}(v)\}$$

$$f_{\mathbf{length}}(L) := \sum_{s \in S(L)} \underline{length}(s)$$

$$f_{\mathbf{area}}(F) := \sum_{f \in F} \underline{area}(f)$$

$$f_{\mathbf{perimeter}}(F) := \sum_{s \in S(F)} \underline{length}(s)$$

Note that the four operators **diameter**, **length**, **area**, and **perimeter** are not invariant against redrawing, i.e., each of these four operations applied before and after a necessary redrawing of one or more segments of a *lines* or *regions* object will yield slightly different results. We want to define the **dist** operator in a way that is invariant against redrawing, since it has besides a numerical aspect also a topological one. Consider a set of spatially-referenced objects with a spatial attribute and a spatial reference object for which the nearest spatially-referenced object has to be computed. If the distance calculations between spatial reference object and spatial attribute value vary depending on possible redrawings, the answer regarding the nearest spatially-referenced object may vary, too, and lead to topological inconsistency. Note the relationship to the **closest** operator discussed below. Therefore we define the distance function as follows. GP will denote the set of grid points associated with a spatial object.

For a *points* object v let $GP(v) := v$, for a *lines* object v let $GP(v) := E(S(v))$ (the union of the envelope points of all segments of v), and for a *regions* object v let $GP(v) := E(S(v)) \cup P_{in}(v)$. Then

$$f_{\mathbf{dist}}(v, w) := \begin{cases} 0, & \text{if } GP(v) \cap GP(w) \neq \varnothing \\ min\{\underline{dist}(p, q) \mid p \in GP(v), q \in GP(w)\}, & \text{otherwise} \end{cases}$$

Although the sets of grid points used in the definition may be very large, this operation can be efficiently implemented, since it can be reduced to distance computations between a point p and a segment s. There it is only necessary to consider those envelope points that are neighbours of the intersection point of s with a perpendicular line going through p.

4.4.4 Spatial Operators on Sets of Spatially-Referenced Objects

Operators of the last group take *sets of spatially-referenced objects* as operands; some of them create new sets of such objects as a result. The **sum** operator aggregates over the values of some spatial attribute of a set of spatially-referenced objects and computes the geometric union of all these values. The **closest** operator yields that element of a set of spatially-referenced objects whose spatial attribute value is nearest to a spatial reference object. The **decompose** operator was already explained in Section 4.3.2; it multiplies each element of a set of spatially-referenced objects according to the number of components of its spatial attribute value and adds one of these components as a new attribute value. The **overlay** operator allows to superimpose one partition of the plane on another one and to combine them into area-disjoint regions. As described in Section 4.2.2, partitions are given as sets of spatially-referenced objects with an attribute of a type in $\underline{regions}^{\textit{area-disjoint}}$. The resulting set of spatially-referenced objects contains one object for each new region obtained as the intersection of a region of the first partition with a region of the second partition. Note that it is not required that the plane is covered completely by the regions of a partition. Thus it is possible that a region of the first partition does not intersect any region of the second partition. In this case it will not be part of any new spatially-referenced object[16] (Figure 4-3).

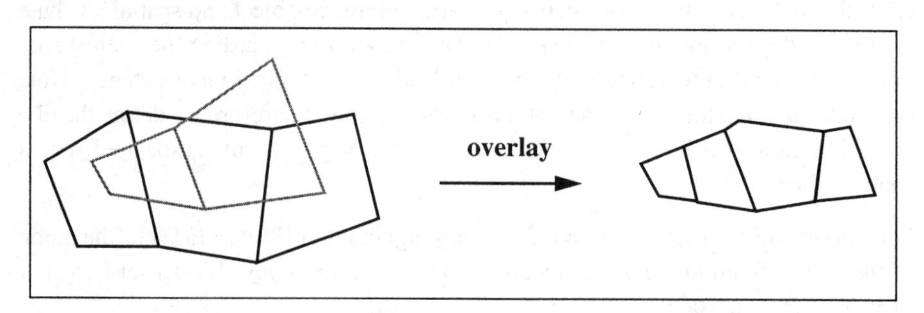

Figure 4-3

The **fusion** operator merges the values of a specified (set of) spatial attribute(s) on the basis of the equality of the values of another (set of) non-spatial attribute(s). For each group of equal non-spatial attribute values a (set of) new spatial object(s) is created as the geometric union of a set of spatial attribute values of the group[17].

In Figure 4-4, we take up the example of Section 2.2.2 (Figure 2-14) where a partition of districts with their land use is given. The task is to compute the regions with the same land use. Neighbour districts with the same land use are replaced by a single region, that is, their common boundary line is erased. Each of the hatched areas on the left is part of a spatially-referenced object describing a district d_i. On the right after the application of the **fusion** operator all areas belonging to the same group g_i form *a single regions* object and are hatched in the same way.

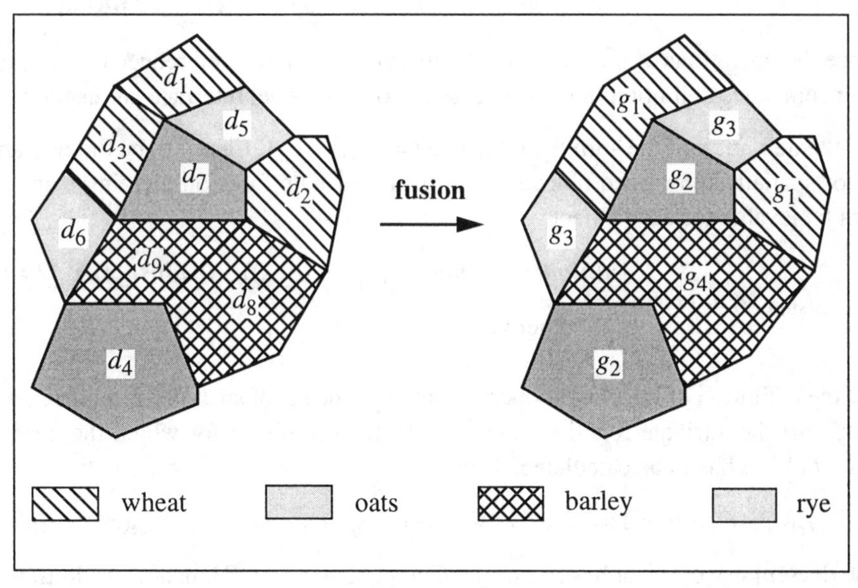

Figure 4-4

[16] This corresponds to the standard join operation. If regions of one partition not intersecting a region of the other partition were in the result, it would be similar to an outer join (see Section 2.2.2).

[17] The **fusion** operator could be extended to allow grouping also by spatial attributes. For efficient implementation this requires a capability of sorting by spatial data type values, which means the ROSE algebra would have to provide a "less-than" operator for each of the three spatial data types imposing a linear order.

The signature for these operations is as follows:

$\forall$ *obj*, obj_1, obj_2 in OBJ. $\forall$ *geo*, geo_1, geo_2 in GEO.
$\forall$ $area_1$, $area_2$ in $\underline{regions}^{\textit{area-disjoint}}$. $\forall$ $data_i$ in DATA. $\forall$ geo_j in GEO.

$\underline{set}(obj) \times (obj \to geo)$	$\to geo$	**sum**
$\underline{set}(obj) \times (obj \to geo_1) \times geo_2$	$\to \underline{set}(obj)$	**closest**
$\underline{set}(obj) \times (obj \to geo) \times \underline{ident}$	$\to \underline{set}(o\text{: OBJ})$	**decompose**
$\underline{set}(obj_1) \times (obj_1 \to area_1) \times \underline{set}(obj_2) \times (obj_2 \to area_2) \times \underline{ident}$		
	$\to \underline{set}(o\text{: OBJ})$	**overlay**
$\underline{set}(obj) \times (obj \to data_i)^+ \times (obj \to geo_j)^+$		
	$\to \underline{set}(o\text{: OBJ})$	**fusion**

Since the operations of this group deal with sets of spatially-referenced objects, for their semantics definition the concepts of the object model interface are needed.

For the definition of the **sum** operator let $O = \{o_1, ..., o_n\}$, for $n \geq 0$, be the operand set of spatially-referenced objects and *attr* the attribute function yielding a spatial data type value for each object.

$$f_{\textbf{sum}}(O, attr) := \begin{cases} \textbf{\textit{union}}(...(\textbf{\textit{union}}(attr(o_1), attr(o_2)), ...), attr(o_n)), \text{ if } O \neq \varnothing \\ \varnothing, \text{ otherwise} \end{cases}$$

For the definition of the **closest** operator let O be the set of spatially-referenced objects, *attr* the attribute function, and *rv* the reference object for which the nearest spatial object has to be calculated. Then

$$f_{\textbf{closest}}(O, attr, rv) := \{o \in O \mid \forall\, o' \in O : f_{\textbf{dist}}(rv, attr(o)) \leq f_{\textbf{dist}}(rv, attr(o'))\}$$

The **decompose** operator has an unspecified result type in OBJ; hence in addition to its semantics function $f_{\textbf{decompose}}$ it needs a type mapping function $\tau_{\textbf{decompose}}$, as described in Section 4.2.1. When an operator **alpha** with a type mapping is used in a query and applied to some operands (say **alpha**(*a*, *b*, *c*)), then this will lead to a call of its semantics function $f_{\textbf{alpha}}(a, b, c)$ during query execution. Additionally, it will lead to a call of the type mapping function $\tau_{\textbf{alpha}}$ during query parsing; the type mapping function is called not with the actual operands (i.e., *a*, *b*, *c*) but instead with *the actual types of these operands*. These types can vary because of the polymorphic specification of operators which is the reason why type mappings are needed at all. The only exception to this rule are operands of type <u>ident</u>; for them not the type <u>ident</u> but the actual identifier is passed to the type mapping function. This is because the main purpose of such operands is the use in type mappings.

$$f_{\mathbf{decompose}}(O, attr, name) := \{o \oplus v \mid o \in O, v \in attr(o)\}$$

$$\tau_{\mathbf{decompose}}(\underline{set}(obj), (obj \to geo), name) := obj \oplus (name, geo)$$

Hence each spatially-referenced object is extended by one of the components of its spatial attribute value; the new object type is an extension of the operand object type by a new attribute *name* of type *geo*. For example, the call in query Q4 (Section 4.3.2) "**decompose**(states, territory, basic_area)" would lead to the following calls of semantics function and type mapping function:

$$f_{\mathbf{decompose}}(states, territory, basic\_area)$$

$$\tau_{\mathbf{decompose}}(\underline{set}(state), (state \to \underline{regions}), basic\_area)$$

The **overlay** operator also needs a type mapping:

$$\begin{aligned} &f_{\mathbf{overlay}}(O_1, attr_1, O_2, attr_2, name) \\ &\quad := \{(o_1 \otimes o_2) \oplus v \mid \exists\, o_1 \in O_1\ \exists\, o_2 \in O_2 : \\ &\qquad f_{\mathbf{intersects}}(attr_1(o_1), attr_2(o_2)) = true \wedge \\ &\qquad v = f_{\mathbf{intersection}}(attr_1(o_1), attr_2(o_2))\} \end{aligned}$$

$$\begin{aligned} &\tau_{\mathbf{overlay}}(\underline{set}(obj_1), (obj_1 \to area_1), \underline{set}(obj_2), (obj_2 \to area_2), name) \\ &\quad := (obj_1 \otimes obj_2) \oplus (name, new(\underline{regions}^{\ \textbf{\textit{area-disjoint}}})) \end{aligned}$$

Here the resulting object type is the product of the two operand types extended by a new attribute *name* of a new type in the kind $\underline{regions}^{\ \textbf{\textit{area-disjoint}}}$.

The **fusion** operator is not formally defined since it is only an abbreviation of a corresponding grouping operation, as described in Section 4.3.2. The semantics definition would rely on a formalization of the semantics of the grouping operation.

4.5 Integration with a DBMS Query Language: O_2SQL/ROSE

The purpose of this section is two-fold: (i) We show the integration of the ROSE algebra with one particular data model and query language, which further illustrates the concepts and requirements of the object model interface. (ii) We demonstrate the "expressive power" of the ROSE algebra (within the context of a query language) by showing some example queries.

For the integration example we select O_2 as one of the state-of-the-art object-oriented database systems with O_2SQL as its current and future standard query language [Ba89, BCD89, BDK92, $O_2$93]. O_2SQL is a functional language that deals with and allows to construct atomic values, tuples, sets and lists, provides operations on these structures, and allows one to define methods on classes. Flat as well as nested structures can be constructed, and all levels of a structure can be accessed. Elements of sets and lists and components of tuples may be of any type or class. The syntax of O_2SQL has an SQL-like style through a *select-from-where* construct corresponding to the three algebraic operations projection, cartesian product, and selection, extended by object-oriented features.

In the sequel we demonstrate the integration of the ROSE algebra with O_2SQL by presenting example queries. The notations regarding class definitions and queries comply with the notations in [BDK92, $O_2$93]. A few notational extensions are necessary. Examples are based on the following simple database which models spatial aspects of Germany. The keyword **public** means that components of a tuple structure are "visible" and can be accessed.

class State
    public type tuple (name : *string*, territory : *regions*)
end;

class City
    public type tuple (name : *string*,
        zipcode : *integer*,
        statistical_data : **tuple** (foundation_date : *integer*,
            population : *integer*,
            unemployment_rate : *real*),
        municipal_area : *regions*)
end;

class Highway
    public type tuple (number : *string*, way : *lines*)
end;

class River
    public type tuple (name : *string*, route : *lines*)
end;

class District
    public type tuple (name : *string*, region : *regions*, land_use : *string*)
end;

A class is a description of a group of objects but not a persistent repository for them in a database. In O_2 only objects associated with names are persistent. We therefore introduce for each class a named collection of objects:

name Cities : **set**(City); **name** States : **set**(State);
name Highways : **set**(Highway); **name** Rivers : **set**(River);
name Districts : **set**(District);

Spatial attributes are defined in the same way as attributes of standard data types, using the spatial data types of the ROSE algebra. Note however, that we have compromised on the typing of *regions* attributes. In the example database, each of these attributes should really have its own type $\underline{area_i}$ within the kind $\underline{regions}^{\textbf{\textit{area-disjoint}}}$ in order to be able to model partitions of the plane. Such a sophisticated typing is not available in O_2. We will therefore assume that for the O_2 integration the definition of the ROSE algebra is slightly changed so that all operators defined on $\underline{area_i}$ types are defined on *regions* instead. This does not change the definition of syntax or semantics of these operators because any value of some type $\underline{area_i}$ in the kind $\underline{regions}^{\textbf{\textit{area-disjoint}}}$ is in fact a *regions* value; it just means that type checking cannot ensure any more that they are applied to partitions.

The syntax of the spatial operations of the ROSE algebra in a query language is not prescribed by the signature of the operations but is part of the process of embedding the operations into the desired query language, i.e., dependent on the extended query language. Here, we select infix syntax for spatial predicates and the two operations **plus** and **minus** and a functional syntax for all other operations.

Q1: List the names and the land use of districts which are neighbours with the same land use.

select **tuple** (dname1: d1.name, dname2: d2.name, land_use: d1.land_use)
from d1 **in** Districts,
d2 **in** Districts
where d1.region **adjacent** d2.region **and** d1.land_use = d2.land_use

All spatial predicates of the ROSE algebra (first group of spatial operations) can be used as selection criteria in the where clause, just like conventional predicates. The result of this query is a set of tuples each formed by the tuple constructor *tuple*. Components of tuples are accessed by the field extraction operator denoted by a dot. Hence here we have the facility of the OMI: *Denote a data type value (within an object set iteration).*

Q2: Which states are enclosed by which other states?

select **tuple** (state1: s1, state2: s2)
from s1 **in** States,
s2 **in** States
where s1.territory **encloses** s2.territory

The result of the query is a set of tuples, each tuple being a pair of *State* objects.

Q3: Determine which highways cross which rivers and list their names, their geometries and their crossings.

```
select  tuple (name: r.name, route: r.route, number: h.number, way: h.way,
              crossing: intersection (r.route, h.way))
from    r in Rivers,
        h in Highways
where   r.route intersects h.way
```

Each tuple of the query result contains an attribute 'crossing' whose value is the intersection of a river and a highway object. (OMI: *Extend objects by derived attribute values, allow naming of a new attribute*.)

Q4: Associate with each state those cities lying inside that state.

```
select  tuple (state: s,
              cities_in_state: select  c
                               from    c in Cities
                               where   c.municipal_area inside s.territory)
from    s in States
```

The result is a set of tuples, each tuple being a pair of a *State* object and a set of *City* objects whose geometries lie inside the geometry of the *State* object.

Q5: Which rivers form partially the boundary line of which states? In which parts do they agree?

```
select  tuple (rname: r.name, sname: s.name,
              border: common_border (s.territory, r.route))
from    s in States,
        r in Rivers
where   s.territory border_in_common r.route
```

Q6: Compute the length of the river and highway network.

```
length (sum (select attribute way from h in Highways)
    plus sum (select attribute route from r in Rivers))
```

Here we have introduced a first extension to O_2SQL to fulfill the requirement of the OMI: *Denote a collection of objects together with an attribute*. The notation is "**select attribute** *attr* **from** *s* **in** *S*" where *attr* is the name of the attribute and *S* the set of spatially-referenced objects.[18]

[18] This is analogous to the "**column** α **from** β" construct for the SQL embedding in Section 4.3.2. The keyword "column" seems to fit with SQL which also speaks of "tables" rather than relations. For O_2 which uses terms like "tuple", a keyword "attribute" appears adequate. Of course, this is just a matter of taste.

It is interesting to observe that in this query first a single *lines* object is formed to which then the length function is applied. Using the built-in **sum** aggregate function of O_2 applicable to sets of real numbers, one might formulate the query as follows:

sum (**select** **length**(h.way) **from** h **in** Highways) +
sum (**select** **length**(r.route) **from** r **in** Rivers)

Actually the result will only be the same if no two highways use the same piece of the highway network. But a more important issue to be discussed here is the view of aggregate functions. The **sum** aggregate function of O_2 used in this last example is applied to a *set of values*. In contrast, the only aggregate function of the ROSE algebra (**sum**) is applied to a *set of database objects with a spatial attribute*. The rationale behind this is to keep the type system of the object model interface as simple as possible. For example, in the relational model sets of values are not available. The ROSE algebra only assumes that collections of database objects and atomic values exist.

Q7: Calculate the perimeter of Bavaria (class *State* is assumed to describe states within Germany).

perimeter (**element** (**select** s.territory **from** s **in** States **where** s.name = "Bavaria"))

The O_2SQL *element* operator extracts the unique element of a singleton set. This is exactly the facility "*denote a data type value (without object set iteration)*" of the OMI. The expression "**element** (...)" denotes the territory of Bavaria.

Q8: Calculate the region outside Bavaria where wheat is cultivated.

```
sum (select   attribute region
       from   d in Districts
       where  d.land_use = "wheat")
minus
element   (select  s.territory
           from    s in States
           where   s.name = "Bavaria")
```

This query yields an atomic spatial object.

Q9: Determine all cities that are located in areas which are completely enclosed by highways.

```
select  c
from    c in Cities
where   c.municipal_area inside
        interior (sum (select attribute way from h in Highways))
```

This query yields a set of *City* objects fulfilling the where condition.

Q10: Check if the highways form a connected network.

no_of_components (**sum** (**select attribute** way **from** h **in** Highways)) = 1

Q11: List the name(s) of the highway(s) being closest to Munich.

define Munich **as**
element (**select** c.municipal_area
from c **in** Cities
where c.name = "Munich");

select h.number
from h **in closest** (**select attribute** way **from** h **in** Highways, Munich)

In the first step a *named query* of O_2 defines Munich as a <u>*regions*</u> object. This is the facility "*allow naming of a spatial data type value*" of the OMI. The **closest** operator takes as operands a class or any other homogeneous set of database objects together with a spatial attribute defined on that object type, and a spatial reference object (in this case Munich). It returns a set of spatially-referenced objects which can be used in a query at all those positions where a set expression is allowed.

Q12: Determine the component regions of the state Niedersachsen (which consists of a main land area as well as several islands in the North Sea).

select s.component
from s **in**
decompose into component
select attribute territory
from s **in** States
where s.name = "Niedersachsen"

The **decompose** operator has three arguments: a class or any other homogeneous set of database objects, a spatial attribute which is defined on that object type and which is to be decomposed, and a name for the new attribute resulting from decomposition. The query yields a set of <u>*regions*</u> objects. Here we have introduced a second extension to O_2SQL to offer the facility "*allow naming of a new attribute*" of the OMI, using a phrase "**into** *attr*", as in Section 4.3.2.

Q13: Partition the state Bavaria with respect to the districts of land use.

overlay into districts_within_Bavaria
(**select attribute** territory **from** s **in** States **where** s.name = "Bavaria",
select attribute region **from** d **in** Districts)

The result of this query is a set of database objects with a new attribute "districts_within_Bavaria". Each partition for the overlay is given as a set of database objects with a <u>*regions*</u> attribute.

Q14: Compute the regions of the same land use.

fusion (Districts; land_use; region)

The **fusion** operator requires three arguments which are syntactically separated by semicolons: a set of spatially-referenced objects, a list of non-spatial attributes used for grouping, and a list of spatial attributes used for geometric union. In the query above the *District* objects are grouped according to equal land use and for each group the geometric union of the *regions* objects of the "region" attribute is formed.

O_2SQL offers a grouping operator **group** so that the query can be formulated without an explicit **fusion** operator:

group d **in** Districts **by** (land_use: d.land_use)
with (region: **sum** (**select attribute** region **from** p **in partition**))

Here the **group** operator is applied to a set of *District* objects. It groups *District* objects by values of their "land_use" attribute and produces for each group one result tuple with two attributes. The first attribute "land_use" receives the value of the "land_use" attribute of the group; the second attribute "region" is determined in the **with**-clause by an expression which computes for each group the geometric union of the "region" attribute values. One can refer to the current group by a keyword **partition**.

4.6 Related Work

Most closely related to this work are the formal definitions of spatial data types (or algebras) given by Güting [Gü88a, Gü88b], Scholl and Voisard [SV89, Vo92], Gargano *et al.* [GNT91], and Worboys and Bofakos [WB93]. In Section 2.2.3 we have seen that these four proposals do not fulfill most of the design criteria presented in Section 1.2. They in particular suffer from restricted data type definitions [GNT91, Gü88a, Gü88b], unnecessarily complicated data type definitions [SV89, Vo92], and complicated definitions of spatial operations [WB93]. Only Worboys and Bofakos define areal objects which may have holes and whose structures have a similar complexity as our *regions* objects.

Details about the formal methods applied by these approaches have been given in Section 2.3. These methods are unfortunately not based on finite resolution and can thus not solve the problem of numerical correctness. Furthermore, they do not treat the geometric consistency problem. All four proposals have connected their spatial types to a fixed data model – Güting and Gargano *et al.* to the relational model, Scholl and Voisard to a complex object algebra, and Worboys and Bofakos to an object-oriented model. Only Scholl and Voisard emphasize the necessity of a clean interface between the spatial algebra and the general object model. We extend their work by offering an abstract object model interface which is independent of any particular data model.

Fernandes *et al.* [FDPW97] describe the integration of the ROSE algebra as a spatial data handling component into the ROCK & ROLL deductive object-oriented database system [BFPWDA95]. They demonstrate that the resulting spatial database system looses none of its general-purpose database functionality and that the requirements posed by the object model interface to incorporate the ROSE algebra conform to current database technology and are not at all a severe burden. The spatial data types of the ROSE algebra are included as primitive types, i.e., their values are treated in the same way as boolean values, number values, and string values.

Chapter 5
Efficient Algorithms for Realm-Based Spatial Data Types

This chapter deals with the implementation of the ROSE algebra [GRS95, Ri95] by providing data structures for its types and new *realm-based geometric algorithms* for its operations. The main techniques used are (parallel) traversal of spatial objects, plane-sweep, and graph algorithms. All algorithms are analysed with respect to their worst case time and space requirements. Due to the realm properties, these algorithms are relatively simple, efficient, and numerically completely robust. The first section emphasizes the difference between a descriptive and an executable algebra, formulates the main design aspects of the new geometric algorithms, and introduces an executable algebra for the given descriptive ROSE algebra. The second section gives a high-level specification of data structures for representing ROSE objects which provides a basis for the subsequent description of algorithms. The third section introduces realm-based geometric algorithms for realizing ROSE operations. The fourth section describes related work.

5.1 Descriptive and Executable ROSE Algebra

Implementing the ROSE algebra [GRS95, Ri95] means to realize its realm-based spatial data types and operations by providing efficient data structures and algorithms defined over a discrete grid. To be precise, we must distinguish between a *descriptive* and an *executable* algebra [BG92, Gü89].

A *descriptive algebra* offers types and operations at a conceptual level which can be used to formulate queries. Its semantics is given by defining a "carrier" set of objects or values for each sort of the algebra and a function for each operator. Such an algebra abstracts completely from representation aspects. Design goals are conceptual clarity, simplicity, and expressiveness. Efficiency plays no role.

An *executable algebra* describes the actual representations and query processing algorithms present in a system. Hence, in such an algebra there is a data structure associated with each sort (or type) and an algorithm, or a procedure realizing it, associated with each operator. Design goals are here efficiency and simplicity.

In a database system, it is the task of the optimizer to translate an expression of the descriptive algebra into an equivalent, efficiently evaluable expression of the executable algebra. A single operator of the descriptive algebra may, in general, be translated to several different operators, or even sequences of operators, of the executable algebra. Reasons may be that many descriptive operators are polymorphic (or more precisely: overloaded), i.e., applicable to several types, and that different algorithms may exist for realizing an operator.

The ROSE algebra, as defined in Chapter 4, is a descriptive algebra. Hence in this chapter we first describe a corresponding executable algebra - essentially polymorphic descriptive operators are decomposed into several executable operators - and then data structures and algorithms to implement it. The main design aspects of the executable ROSE algebra are the following:

- We describe at a very high level, yet precisely, robust, efficient, and realm-based algorithms dealing with the complex geometric entities available in the ROSE algebra. They can be grouped into *parallel traversal*, *plane-sweep*, and *graph algorithms*. For each paradigm, we show a few "prototype" operators and their algorithms and discuss which other operators can be realized similarly and which modifications are necessary. Many algorithms require only linear time, the remaining ones $O(n \log n)$ time where n is a bound on the size of the operand objects.
- All spatial objects processed by the operations are realm-based, i.e., they are defined over a discrete basis and in particular no two segments intersect within their interiors and no point lies within a segment. These properties can be exploited for designing efficient geometric algorithms. For example, many operations can now be realized through a simple parallel traversal for which otherwise more complex and expensive plane sweep algorithms would be needed. Operations that are to be realized by plane-sweep algorithms are now much simpler and more efficient, since they need not discover new intersections and treat special cases. Sweep stations can only be isolated points or end points of segments which are all known in advance. Hence, a static sweep event structure for managing the sweep stations is sufficient.
- Different algorithms processing the same kind of spatial objects usually prefer different internal object representations. For example, plane-sweep algorithms require segments ordered by (x, y)-coordinates whereas a function for computing the area of a <u>regions</u> object might prefer cyclic structures describing the components (e.g., faces, cycles) of the object. In contrast to traditional work on algorithms, the focus is here not on finding the most efficient algorithm for one single problem (operation) together with a corresponding sophisticated data structure, but rather on considering a spatial algebra as a whole and on reconciling the various requirements

posed by different algorithms within a single data structure for each type. The author is not aware that implementations of complete spatial algebras have been described before in a similar manner.

- The implementation is designed for use in a spatial database system. In particular, representations for spatial data types do not use pointer data structures in main memory, but are all embedded into compact storage areas which can be efficiently transferred between a main memory buffer and disk. Data structures are also designed to allow for realm updates.

Realm-based algorithms and data structures of the executable ROSE algebra have been actually implemented within the framework of the so-called *ROSE system* (see Section 6.3).

Both the descriptive and the executable ROSE algebra use *second-order signature* [Gü93] (see also Section 4.2.1) as the underlying formalism. To develop an executable algebra for the given descriptive ROSE algebra means to decompose each polymorphic descriptive operator into corresponding executable, non-polymorphic operators for all possible combinations of data types for the operands. Each executable operator is then realized by an algorithm. For example, it is obvious that an algorithm which examines the disjointedness of two points objects will be different from an algorithm which determines whether two regions objects overlap. Hence, the descriptive operator **disjoint** with the signature specification

$\forall$ *geo* in GEO. *geo* × *geo* → *bool* **disjoint**

is mapped to the three executable operators

points × *points* → *bool* **pp_disjoint**
lines × *lines* → *bool* **ll_disjoint**
regions × *regions* → *bool* **rr_disjoint**

Appendix C lists the signature of the descriptive ROSE algebra and shows its translation into executable operators. For example, the ROSE signature of some spatial predicates is represented as follows:

Descriptive Operator		Executable Operator	PT	PS	G	TC
geo × *geo* → *bool*	=	**pp_equal, ll_equal, rr_equal**	x			$O(n)$
	≠	**pp_unequal, ll_unequal, rr_unequal**	x			$O(n)$
	disjoint	**pp_disjoint, ll_disjoint**	x			$O(n)$
		rr_disjoint		x		$O(n \log n)$

The name of each executable operator is supplied with a prefix indicating the types of the operator's operands. For example, the operator **ll_unequal** is applied to two *lines* objects. The last four columns of this table describe the algorithmic technique used to implement this (group of) executable operators (PT = parallel traversal, PS = plane sweep, G = graph algorithm) and the worst case time complexity (TC). The algorithms are discussed below. There is a gap in the table of Appendix C because efficient algorithms for distance problems (operator **dist** of the third group) have not yet been studied. Moreover, the descriptive operators of the fourth group (Section 4.4.4) have not yet been considered, since the object model interface (Section 4.3) needed for the communication with a DBMS data model has so far not been implemented.

5.2 Data Structures for the Realm-Based Spatial Data Types

Algorithms for the executable ROSE operators need to access, and sometimes to build, the data structures representing objects of the three types *points*, *lines*, and *regions*. Rather than describing these data structures directly in terms of arrays, records, etc., we first introduce a higher level description which offers suitable access and construction operations to be used in the algorithms. Basically, we define an abstract data type for each of the three data structures. In a second step, one can then design and implement the data structure itself.

The specification of an abstract data type consists of a many-sorted signature together with a set of laws, or equations, defining the behaviour of operations. To be precise, we use a slightly different specification method which we call "set-theoretic specification". It simply means that we assign semantics to the sorts and operations of the many-sorted signature directly by defining carrier sets for the sorts and functions for the operations on these carrier sets, i.e., we define an algebra for each of the three data structures representing *points*, *lines*, or *regions* objects, respectively. In other words, we give a concrete mathematical model for the data type instead of a set of laws. In fact, the whole ROSE algebra itself has been defined by the same method.

For most executable operators it turns out to be sufficient to regard a spatial object as an *ordered sequence of elements* where it is possible to access these elements consecutively and to insert a new element into the sequence. Hence this is our basic strategy for modelling the three data structures.

Before we can introduce the algebra *points*, a few notations are repeated. Realms and realm-based spatial objects are defined over a finite discrete space $N \times N$ with $N = \{0, ..., m-1\} \subseteq \mathbf{N}$. $P_N = \{(x, y) \mid x \in N, y \in N\}$ denotes the set of all N-points. Furthermore, an (x, y)-lexicographic order is assumed on P_N which is defined as $p_1 < p_2 \Leftrightarrow x_1 < x_2 \lor (x_1 = x_2 \land y_1 < y_2)$.

algebra *points*

sorts $\underline{points}$, P_N, $\underline{bool}$

ops

new	:		$\rightarrow \underline{points}$
select_first	:	$\underline{points}$	$\rightarrow \underline{points}$
select_next	:	$\underline{points}$	$\rightarrow \underline{points}$
end_of_pt	:	$\underline{points}$	$\rightarrow \underline{bool}$
get_pt	:	$\underline{points}$	$\rightarrow P_N$
insert	:	$\underline{points} \times P_N$	$\rightarrow \underline{points}$

sets $\underline{points} = \{(pos, <p_1, ..., p_n>) \mid pos \geq 0; n \geq 0; \forall\ 1 \leq i \leq n : p_i \in P_N; \forall\ 1 \leq i < n : p_i < p_{i+1}\}$

functions Let $P_p = <p_1, ..., p_n>$, $P = (i, P_p) \in \underline{points}$, and $p \in P_N$.

$$new() = (0, \lozenge)$$

$$select\_first(P) = \begin{cases} (1, P_p) & \text{if } n \geq 1 \\ (0, \lozenge) & \text{otherwise} \end{cases}$$

$$select\_next(P) = \begin{cases} (i+1, P_p) & \text{if } 1 \leq i < n \\ (0, P_p) & \text{otherwise} \end{cases}$$

$$end\_of\_pt(P) = (i = 0)$$

$$get\_pt(P) = \begin{cases} p_i & \text{if } 1 \leq i \leq n \\ \textit{undefined} & \text{otherwise} \end{cases}$$

$$insert(P, p) = \begin{cases} (j, P_p) & \text{if } \exists j \in \{1, ..., n\} : p = p_j \\ (1, <p>) & \text{if } n = 0 \\ (1, <p, p_1, ..., p_n>) & \text{if } p < p_1 \\ (n+1, <p_1, ..., p_n, p>) & \text{if } p > p_n \\ (j+1, <p_1, ..., p_j, p, p_{j+1}, ..., p_n>) & \text{if } \exists j \in \{1, ..., n\text{-}1\} : p_j < p < p_{j+1} \end{cases}$$

end *points*.

The **sorts** and **ops** parts describe the syntax of the algebra, i.e., the signature. The **sets** and **functions** parts give the semantics in terms of carrier set and function definitions. The algebra *points* contains the sorts $\underline{points}$ (to be defined), P_N, and $\underline{bool}$. The carrier set of the sort $\underline{points}$ is defined as the set of all ordered sequences $<p_1, ..., p_n>$ of n N-points together with a pointer indicating a position within the sequence. The symbol $\lozenge$ denotes the empty sequence. Functions manipulate such values, for example, *select_first* positions the pointer *pos* on the smallest element of the point sequence, and *get_pt* yields the point at the current position.

The crucial idea for the representation of the relatively complex $\underline{lines}$ and $\underline{regions}$ objects, which is the basis for most of the algorithms, is to regard them as *ordered sequences of halfsegments*. Let $S_N = \{(p, q) \mid p \in P_N, q \in P_N\}$ again denote the set of N-segments. The equality of two N-segments $s_1 = (p_1, q_1)$ and $s_2 = (p_2, q_2)$ is defined as $s_1 = s_2 \Leftrightarrow (p_1 = p_2 \wedge q_1 = q_2) \vee (p_1 = q_2 \wedge p_2 = q_1)$. W.l.o.g. we normalize S_N by the assumption that $\forall\, s \in S_N : s = (p, q) \Rightarrow p < q$ which enables us to speak of a *left* and a *right end point* of a segment. Let further $H_N = \{(s, d) \mid s \in S_N, d \in \{left, right\}\}$ be the set of *halfsegments*. A halfsegment $h = (s, d)$ consists of an N-segment s and a flag d emphasizing one of the N-segment's end points which is called the *dominating point* of h. If $d = left$ then the left (smaller) end point of s is the dominating point of h, and h is called *left halfsegment*. Otherwise, the right end point of s is the dominating point of h, and h is called *right halfsegment*. Hence, each N-segment s is mapped to two halfsegments $(s, left)$ and $(s, right)$. Let dp be the function which yields the dominating point of a halfsegment.

For two distinct halfsegments h_1 and h_2 with a common end point p, let α be the enclosed angle such that $0° < \alpha \leq 180°$ (an overlapping of h_1 and h_2 is excluded by the realm properties). Let a predicate *rot* be defined as follows: $rot(h_1, h_2)$ is true iff h_1 can be rotated around p through α to overlap h_2 in counterclockwise direction. We can now define a complete order on halfsegments which is basically the (x, y)-lexicographic order by dominating points. For two halfsegments $h_1 = (s_1, d_1)$ and $h_2 = (s_2, d_2)$ it is:

$$\begin{aligned} h_1 < h_2 \;\Leftrightarrow\;& dp(h_1) < dp(h_2) \vee \\ & (dp(h_1) = dp(h_2) \wedge ((d_1 = right \wedge d_2 = left) \vee (d_1 = d_2 \wedge rot(h_1, h_2)))) \end{aligned}$$

We now define the algebra *regions* (the algebra *lines* is almost the same, see below). The carrier set of the sort <u>regions</u> is defined as the set of ordered sequences $< h_1, ..., h_n >$ of n halfsegments where each halfsegment h_i has an attached *set of attributes* a_i whose elements are values of some new sort *attr*. Attribute sets are used in algorithms to attach auxiliary information to segments.

algebra *regions*

sorts <u>regions</u>, H_N, *attr*, <u>bool</u>

ops				
	new	:		→ <u>regions</u>
	select_first	:	<u>regions</u>	→ <u>regions</u>
	select_next	:	<u>regions</u>	→ <u>regions</u>
	end_of_hs	:	<u>regions</u>	→ <u>bool</u>
	get_hs	:	<u>regions</u>	→ H_N
	get_attr	:	<u>regions</u>	→ *attr*
	update_attr	:	<u>regions</u> × *attr*	→ <u>regions</u>
	insert	:	<u>regions</u> × H_N	→ <u>regions</u>

sets $\underline{regions} = \{(pos, <h_1, ..., h_n>, <a_1, ..., a_n>) \mid$
(1) $pos \geq 0, n \geq 0$
(2) $\forall i \in \{1, ..., n\} : h_i \in H_N, a_i \subseteq attr$
(3) $\forall i \in \{1, ..., n\text{-}1\} : h_i < h_{i+1}$ $\}$

functions Let $R_h = <h_1, ..., h_n>$, $R_a = <a_1, ..., a_n>$, $R = (i, R_h, R_a) \in \underline{regions}$, and $h \in H_N$.

$$new() = (0, \Diamond, \Diamond)$$

$$select\_first(R) = \begin{cases} (1, R_h, R_a) & \text{if } n \geq 1 \\ (0, \Diamond, \Diamond) & \text{otherwise} \end{cases}$$

$$select\_next(R) = \begin{cases} (i+1, R_h, R_a) & \text{if } 1 \leq i < n \\ (0, R_h, R_a) & \text{otherwise} \end{cases}$$

$$end\_of\_hs(R) = (i = 0)$$

$$get\_hs(R) = \begin{cases} h_i & \text{if } 1 \leq i \leq n \\ undefined & \text{otherwise} \end{cases}$$

$$get\_attr(R) = \begin{cases} a_i & \text{if } 1 \leq i \leq n \\ undefined & \text{otherwise} \end{cases}$$

$$update\_attr(R, a) = \begin{cases} (i, R_h, <a_1, ..., a_{i\text{-}1}, a, a_{i+1}, ..., a_n>) & \text{if } 1 \leq i \leq n \\ undefined & \text{otherwise} \end{cases}$$

$$insert(R, h) = \begin{cases} (j, R_h, R_a) & \text{if } \exists j \in \{1, ..., n\} : h = h_j \\ (1, <h>, <\varnothing>) & \text{if } R_h = \Diamond \\ (1, <h, h_1, ..., h_n>, <\varnothing, a_1, ..., a_n>) & \text{if } h < h_1 \\ (n+1, <h_1, ..., h_n, h>, \\ \quad <a_1, ..., a_n, \varnothing>) & \text{if } h > h_n \\ (j+1, <h_1, ..., h_j, h, h_{j+1}, ..., h_n>, \\ \quad <a_1, ..., a_j, \varnothing, a_{j+1}, ..., a_n>) \\ \quad \text{if } \exists j \in \{1, ..., n\text{-}1\} : h_j < h < h_{j+1} \end{cases}$$

end *regions*.

Note that the algebra *regions* just offers manipulation of halfsegment sequences. It does not ensure that a sequence indeed represents a correct <u>regions</u> object as defined in the ROSE algebra. The algorithms using this structure are responsible for constructing only sequences that indeed represent <u>regions</u> objects. The algebra *lines* (not presented here) is identical to the algebra *regions* except for all the parts related to attributes which are not needed.

Simple implementations for each of the three data types (algebras) would represent a sequence of *n* points or halfsegments in a linked list or sequentially in an array. The latter representation would also be compatible with the "compact storage area" requirement needed for efficient database loading/storing. In this case, all operations except for *insert* need $O(1)$ time; *insert* requires $O(n)$ time for arbitrary positions and $O(1)$ time for appending an element at the end of the sequence. Such a representation would in fact be quite good for all "parallel traversal" algorithms of the ROSE algebra, because result objects are always constructed in the lexicographic point or halfsegment order and can therefore be built in linear time.

The actual implementation in the ROSE system uses for all three structures an AVL-tree embedded into an array (the array serving as a storage pool for nodes); the elements, i.e. points or halfsegments, are additionally linked in sequence order. With this representation, all operations except *insert* need $O(1)$ time and *insert* $O(\log n)$ time. The requirement to support insertion in $O(\log n)$ time actually does not come from the ROSE algebra but from the connection with realms. Realm updates due to insertion of points or segments into the realm must be propagated to ROSE objects residing in a database (see Section 3.5). This means that the data structures should support replacement of a segment in a *lines* or *regions* object by a chain of segments, i.e., the segment must be deleted and the replacement segments be inserted into the structure. Unfortunately, a consequence of this is that the parallel traversal algorithms cannot construct the resulting objects in linear time any more, but need $O(k \log k)$ for this where k is the size of the result object. This is a case of conflicting requirements, as has already been mentioned previously. On the other hand, deriving the internal structure of a *lines* or *regions* object (e.g., faces and holes) which is needed to complete the construction requires $O(k \log k)$ time anyway.

5.3 Algorithms of the Executable ROSE Algebra

This section introduces *realm-based geometric algorithms* whose characteristic features are numerical robustness, topological correctness, closure properties, and efficiency. Realm-based algorithms are more efficient than their Euclidean counterparts. The design of these algorithms is based on traversal techniques, on the plane-sweep paradigm, and on graph algorithms. We can speak of a *realm-based computational geometry* which deals with spatial objects that are defined over the *same* discrete domain and which assumes that no two segments intersect within their interiors and that no point lies within a segment.

Executable operators are grouped by the applied algorithmic technique. For each group we show and explain some example algorithms. All (groups of) algorithms are analysed with respect to their worst case time and space requirements.

5.3.1 Algorithms with Simple or Parallel Object Traversal

A number of operators of the executable ROSE algebra can be realized by a simple or parallel traversal (scan) through the point or halfsegment sequences of one or two objects. To simplify the description of algorithms, for each possible combination of two spatial data types two operations are introduced which enable a parallel traversal through two ordered sequences of elements (halfsegments, points).

As an example, we consider the two operations for two regions objects given by their halfsegment sequences. The operation *rr_select_first*(R_1, R_2, *object*, *status*) selects the first halfsegment of each of the regions objects R_1 and R_2 (compare to the function *select_first* of algebra *regions*) and positions a logical pointer on both of them. The parameter *object* with possible values {*none*, *first*, *second*, *both*} indicates which of the two object representations contains the smaller halfsegment. If the value of *object* is *none*, no halfsegment is selected, since R_1 and R_2 are empty. If the value is *first* (*second*), the smaller halfsegment belongs to R_1 (R_2). If it is *both*, the first halfsegments of R_1 and R_2 are identical. The parameter *status* with possible values {*end_of_none*, *end_of_first*, *end_of_second*, *end_of_both*} describes the state of both halfsegment sequences. If the value of *status* is *end_of_none*, both objects still have halfsegments. If it is *end_of_first* (*end_of_second*), R_1 (R_2) is empty. If it is *end_of_both*, both object representations are empty.

The operation *rr_select_next*(R_1, R_2, *object*, *status*) searches for the next smaller halfsegment of R_1 and R_2; parameters have the same meaning as for *rr_select_first*. Obviously, this is realized by *select_next* operations of the two objects.

Both operations together allow one to scan in linear time two object representations like one ordered sequence. Analogous operations can be defined for two lines objects (*ll_select_first*, *ll_select_next*) and a lines and a regions object (*lr_select_first*, *lr_select_next*). For the comparison of halfsegments with points, the dominating points of the halfsegments are used so that points and lines objects (*pl_select_first*, *pl_select_next*) and correspondingly points and regions objects (*pr_select_first*, *pr_select_next*) can be treated in a similar way.

In the sequel we discuss algorithms for the operations:

points × regions	→ bool	**pr_on_border_of**
points × points	→ points	**pp_plus**
lines × lines	→ bool	**ll_intersects, ll_disjoint**

Operator **pr_on_border_of** determines whether all points of a points object lie on the faces' boundaries of a regions object. Hence the algorithm checks whether for each point p of a points object P (denoted by $p \in P(P)$) a halfsegment h of a regions object R (denoted by $h \in H(R)$) exists whose dominating point is equal to p.

algorithm pr_on_border_of
input: A points object P and a regions object R
output: *true*, if $\forall\, p \in P(P)\ \exists\, h \in H(R) : p = dp(h)$
false, otherwise
begin
  pr_select_first(P, R, object, status);
  while (*object* ≠ *first*) **and** (*status* = *end_of_none*) **do**
    pr_select_next(P, R, object, status);
  end-while;
  return (*object* ≠ *first*) **and** (*status* ≠ *end_of_second*)
end pr_on_border_of.

The while-loop of the algorithm is executed as long as no point is found which is in P but not a dominating point of a halfsegment of R and as long as none of the object sequences is exceeded. For the predicate to be *true*, termination of the while-loop must not have occurred because a point was found which is not on the boundary of R (*object* ≠ *first*). This implies that termination is due to reaching the end of one or both sequences, and the predicate is *true* if this was not the regions sequence alone (*status* ≠ *end_of_second*).

Operator **pp_plus** forms the union of two points objects. The algorithm just scans the point sequences of the two objects and merges them into a new points object.

algorithm pp_plus
input: Two points objects P_1 and P_2
output: A points object P_{new} containing all points of P_1 and P_2
begin
  P_{new} := *new()*;
  pp_select_first(P_1, P_2, *object, status*);
  while *status* ≠ *end_of_both* **do**
    if *object* = *first* **then** p := *get_pt*(P_1)
    else if *object* = *second* **then** p := *get_pt*(P_2)
    else if *object* = *both* **then** p := *get_pt*(P_1)
    end-if;
    P_{new} := *insert*(P_{new}, p);
    pp_select_next(P_1, P_2, *object, status*);
  end-while;
  return P_{new}
end pp_plus.

Operator **ll_intersects** examines whether two lines objects L_1 and L_2 intersect. According to the definition of the ROSE algebra it yields *true* if both objects have no common (half)segments but at least one common point which is not a *meeting point* but an intersection point. Point p is a *meeting point* if the angularly sorted list of halfsegments of L_1 and L_2 with the same dominating point p can be subdivided into two sublists so that one list contains only halfsegments of L_1 and the other list only halfsegments of L_2. The idea is now to walk around p, scanning the segments, and to count the number of "object changes" in this ordered list of halfsegments of L_1 and L_2. Point p is a meeting point if this number is less than or equal to two; otherwise an intersection point has been found.

```
algorithm ll_intersects
input: Two lines objects L1 and L2
output: true, if no common segment exists, but a common point which is not
               a meeting point
        false, otherwise
begin
  ll_select_first(L1, L2, object, status);
  if object = first then act_dp := dp(get_hs(L1))
  else if object = second then act_dp := dp(get_hs(L2))
  end-if;
  old_obj := object; found := false; count := 0;
  while (status = end_of_none) and (object ≠ both) do
    ll_select_next(L1, L2, object, status);
    if (status ≠ end_of_both) and (object ≠ both) and not found then
      if object = first then
        new_dp := dp(get_hs(L1))
      else if object = second then
        new_dp := dp(get_hs(L2))
      end-if;
      if new_dp ≠ act_dp then (* new point *)
        act_dp := new_dp;
        count := 0;
        old_obj := object;
      else if object ≠ old_obj then (* object switch *)
        count := count + 1;
        old_obj := object;
        found := found or (count > 2);
      end-if;
    end-if;
  end-while;
  return found and (object ≠ both);
end ll_intersects.
```

The while-loop of the algorithm terminates if either the end of one of the objects has been reached or a common halfsegment has been found. In the latter case the result value is *false* (*object* $\neq$ *both*), in the first case the decision is based on whether at least one intersection point has been found or not (*found*).

Operator **ll_disjoint** examines whether two lines objects L_1 and L_2 are disjoint. According to the definition of the ROSE algebra it is not sufficient only to test for common halfsegments because the operator yields also *false* if there are segments of both objects which have common points. Due to the realm properties such a common point can only be an end point of two segments of L_1 and L_2. Hence, there must be two halfsegments $h_1 \in H(L_1)$ and $h_2 \in H(L_2)$ with the same dominating point. Since in the halfsegment order all halfsegments with the same dominating point lie one behind the other, a parallel object traversal can check whether two consecutive halfsegments from different objects have the same dominating point. In such a case, L_1 and L_2 are not disjoint.

```
algorithm ll_disjoint
input: Two lines objects L1 and L2
output: true, if ∀ h1 ∈ H(L1) ∀ h2 ∈ H(L2) : dp(h1) ≠ dp(h2)
        false, otherwise
begin
  ll_select_first(L1, L2, object, status);
  if object = first then act_dp := dp(get_hs(L1))
  else if object = second then act_dp := dp(get_hs(L2))
  end-if;
  old_obj := object;
  found := false;
  while (object ≠ both) and (status = end_of_none) and not found do
    ll_select_next(L1, L2, object, status);
    if (status ≠ end_of_both) and (object ≠ both) then
      if object = first then
        new_dp := dp(get_hs(L1))
      else
        new_dp := dp(get_hs(L2))
      end-if;
      if object ≠ old_obj then
        found := (new_dp = act_dp);
      end-if;
      act_dp := new_dp;
      old_obj := object;
    end-if;
  end-while;
  return (object ≠ both) and not found;
end ll_disjoint.
```

The while-loop is executed as long as no common halfsegment has been found (*object* ≠ *both*), the end of both objects has not been reached (*status* = *end_of_none*) and no common point has been discovered (**not** *found*). Hence, the operator returns *true* if the end of at least one object has been reached and neither a common half-segment nor a common point has been found.

points × *points*	→	*bool*	**pp_equal, pp_unequal, pp_disjoint**
lines × *lines*	→	*bool*	**ll_equal, ll_unequal, ll_meets, ll_border_in_common**
regions × *regions*	→	*bool*	**rr_equal, rr_unequal, rr_border_in_common**
lines × *regions*	→	*bool*	**lr_border_in_common**
regions × *lines*	→	*bool*	**rl_border_in_common**
points × *lines*	→	*bool*	**pl_on_border_of**
points × *points*	→	*points*	**pp_intersection, pp_minus**
lines × *lines*	→	*points*	**ll_intersection**
lines × *lines*	→	*lines*	**ll_plus, ll_minus, ll_common_border**
regions × *regions*	→	*lines*	**rr_common_border**
lines × *regions*	→	*lines*	**lr_common_border**
regions × *lines*	→	*lines*	**rl_common_border**
lines	→	*points*	**l_vertices**
regions	→	*points*	**r_vertices**
points	→	*int*	**p_no_of_components**
lines	→	*real*	**l_length**
regions	→	*real*	**r_area, r_perimeter**

The algorithms for the other operators whose signature is given above are similar. We explain them briefly. Operator **pp_equal** checks if two *points* objects contain the same elements. It yields *true* if both objects are completely traversed and if no point has been found which is contained in only one of the objects. The operators **ll_equal** and **rr_equal** for the comparison of *lines* and *regions* objects, respectively, are similar; they compare halfsegments. Operator **pp_disjoint** returns *true* if the point sequences of two *points* objects have no common points or if during the parallel traversal the end of one object is reached. Operator **ll_meets** is similar to **ll_intersects**. No equal segments may exist, and additionally each common point of two *lines* objects must be a meeting point. Operator **pl_on_border_of** is analogous to **pr_on_border_of**. The operators **ll_border_in_common**, **rr_border_in_common**, **lr_border_in_common**, and **rl_border_in_common** yield *true* if both operands have at least one common halfsegment. This is directly supported by the *select* procedures which applied in a loop are called until a common halfsegment is found (*object* = *both*) or the end of one of the objects has been reached.

Operator **pp_intersection** determines all common points of two *points* objects (*object* = *both*) and inserts them into a new *points* object. Operator **ll_intersection** computes all common points of two *lines* objects which are no meeting points. A corresponding algorithm is a combination of the algorithms for **pp_intersection** and **ll_intersects**. Operator **ll_plus** which forms the geometric union of two *lines* objects is similar to **pp_plus**. The operators **pp_minus** applied to two *points* objects and **ll_minus** applied to two *lines* objects must take the order of their operands into account. In a loop all those elements are determined that are contained in the first operand but not in the second operand (*object* = *first*). The loop can be finished if the end of the first operand has been reached (*status* = *end_of_first* or *status* = *end_of_both*). The four operators **ll_common_border**, **rr_common_border**, **lr_common_border**, and **rl_common_border** create a new object containing the common halfsegments of their operands. Their algorithms are similar to **pp_intersection**.

Operator **p_no_of_components** calculates the number of points of a *points* object which can be determined by a simple traversal and by incrementing a counter for each point. Something similar holds for the operators **l_vertices** and **r_vertices**. During a scan all dominating points of the halfsegments of a *lines* object or *regions* object, respectively, are determined and inserted into a new *points* object. The definition and implementation of the *points* type ensures that each point can be inserted only once. The operators **l_length** and **r_perimeter** compute the sum of the length of all segments of a *lines* and *regions* object, respectively. In a loop, their algorithms consider only left halfsegments and add their lengths. The operator **r_area** calculates the area of a *regions* object and uses the formula known from Section 3.6. For each segment the information is needed if the interior of the object lies above or below the segment (see Section 5.3.2).

The complete list of operators that can be treated by simple or parallel traversal is indicated by column PT in Appendix C. For all predicates and for operations returning numbers realized by PT algorithms, the worst case time complexity is $O(n)$, where n is the total number of points or halfsegments in the one or two operands. For operations returning new spatial objects the time bound is $O(n + k \log k)$ where k is the number of points or halfsegments in the result object. $O(n)$ time is needed for scanning the operands and $O(k \log k)$ for constructing the result. Since $k = O(n)$, this is always bounded by $O(n \log n)$.

5.3.2 Algorithms Using the Plane Sweep Paradigm

Plane-sweep [Me84, NP82, PS85] is a popular technique of computational geometry for solving geometric set problems which transforms a two-dimensional problem into a sequence of one-dimensional problems which are easier than the original two-dimensional one. A vertical *sweep line* sweeping the plane from left to right stops at special points called *event points* which are generally stored in a queue called *event point schedule*. The event point schedule must allow one to insert new event points discovered during processing; these are normally the initially unknown intersections of line segments. The state of the intersection of the sweep line with the geometric structure being swept at the current sweep line position is recorded in vertical order in a data structure called *sweep line status*. Whenever the sweep line reaches an event point, the sweep line status is updated. Event points which are passed by the sweep line are removed from the event point schedule. Note that in general an efficient fully dynamic data structure is needed to represent the event point schedule and that in many plane-sweep algorithms an initial sorting step is needed to produce the sequence of event points in x-order (or (x, y)-lexicographic order).

In the special case of realm-based computational geometry where no two segments intersect within their interiors, the event point schedule is static (because new event points cannot exist) and given by the ordered sequence of points or halfsegments of the operand objects. No further explicit event point structure is needed. Also, no initial sorting is necessary since the plane-sweep order of points and segments is the base representation of objects anyway.

If a left (right) halfsegment of a <u>*regions*</u> object is reached during a plane-sweep, its segment component is stored into (removed from) the segment sequence of the sweep line status sorted by the order relation *above*. A segment s lies *above* a segment t if the intersection of their x-intervals is not empty and if for each x of the intersection interval the y-coordinate of s is greater than the one of t (except possibly for a common end point where the y-coordinates are equal). Points and halfsegments of <u>*lines*</u> objects are used to query the sweep line status. A point p lies *above* a non-vertical segment t if the x-coordinate of p lies within the x-interval of t and the y-coordinate of p is greater than the y-coordinate of t at $p.x$. If t is vertical, the x-coordinates of p and t must be identical and the y-coordinate of p must be greater than all y-coordinates of t.

The sweep line status can be described by an algebra as follows:

algebra *sweep line status*

sorts *status*, S_N, P_N, *attr*, <u>*bool*</u>

ops	*new_sweep*	:		$\rightarrow$ *status*
	add_left	:	*status* $\times S_N$	$\rightarrow$ *status*
	del_right	:	*status* $\times S_N$	$\rightarrow$ *status*
	pred_of_s	:	*status* $\times S_N$	$\rightarrow$ *status*
	pred_of_p	:	*status* $\times P_N$	$\rightarrow$ *status*

$$\begin{array}{lll} \mathit{current\_exists} & : \mathit{status} & \rightarrow \underline{\mathit{bool}} \\ \mathit{pred\_exists} & : \mathit{status} & \rightarrow \underline{\mathit{bool}} \\ \mathit{get\_attr} & : \mathit{status} & \rightarrow \mathit{attr} \\ \mathit{set\_attr} & : \mathit{status} \times \mathit{attr} & \rightarrow \mathit{status} \\ \mathit{get\_pred\_attr} & : \mathit{status} & \rightarrow \mathit{attr} \end{array}$$

sets $\mathit{status} = \{ (\mathit{pos}, < s_1, ..., s_n >, < a_1, ..., a_n >) \mid$
(1) $\mathit{pos} \geq 0, n \geq 0,$
(2) $\forall\, i \in \{1, ..., n\} : s_i \in S_N, a_i \subseteq \mathit{attr},$
(3) $\forall\, i \in \{2, ..., n\} : s_i \; \mathit{above} \; s_{i-1} \}$

functions Let $S_s = < s_1, ..., s_n >$, $S_a = < a_1, ..., a_n >$, and $S = (i, S_s, S_a) \in \mathit{status}$.

$$\mathit{new\_sweep}() = (0, \lozenge, \lozenge)$$

$$\mathit{add\_left}(S, s) = \begin{cases} (j, S_s, S_a) & \text{if } \exists\, j \in \{1, ..., n\} : s = s_j \\ (1, < s >, < \varnothing >) & \text{if } S_s = \lozenge \\ (1, < s, s_1, ..., s_n >, < \varnothing, a_1, ..., a_n >) & \text{if } s_1 \; \mathit{above} \; s \\ (n+1, < s_1, ..., s_n, s >, < a_1, ..., a_n, \varnothing >) & \text{if } s \; \mathit{above} \; s_n \\ (j+1, < s_1, ..., s_j, s, s_{j+1}, ..., s_n >, < a_1, ..., a_j, \varnothing, a_{j+1}, ..., a_n >) & \text{if } \exists\, j \in \{1, ..., n\text{-}1\} : s \; \mathit{above} \; s_j \wedge s_{j+1} \; \mathit{above} \; s \end{cases}$$

$$\mathit{del\_right}(S, s) = \begin{cases} (0, < s_1, ..., s_{j-1}, s_{j+1}, ..., s_n >, < a_1, ..., a_{j-1}, a_{j+1}, ..., a_n >) & \text{if } \exists\, j \in \{1, ..., n\} : s = s_j \\ (0, S_s, S_a) & \text{otherwise} \end{cases}$$

$$\mathit{pred\_of\_s}(S, s) = \begin{cases} (j, S_s, S_a) & \text{if } \exists\, j \in \{1, ..., n\text{-}1\} : s \; \mathit{above} \; s_j \wedge (s_{j+1} \; \mathit{above} \; s \vee s_{j+1} = s) \\ (0, S_s, S_a) & \text{otherwise} \end{cases}$$

$$\mathit{pred\_of\_p}(S, p) = \begin{cases} (j, S_s, S_a) & \text{if } \exists\, j \in \{1, ..., n\text{-}1\} : p \; \mathit{above} \; s_j \wedge \neg\, (p \; \mathit{above} \; s_{j+1}) \\ (0, S_s, S_a) & \text{otherwise} \end{cases}$$

$$\mathit{current\_exists}(S) = (1 \leq i \leq n)$$

$$\mathit{pred\_exists}(S) = (1 < i \leq n)$$

$$\mathit{get\_attr}(S) = \begin{cases} a_i & \text{if } 1 \leq i \leq n \\ \mathit{undefined} & \text{otherwise} \end{cases}$$

$$\mathit{set\_attr}(S, a) = \begin{cases} (i, S_s, < a_1, ..., a_{i-1}, a, a_{i+1}, ..., a_n >) & \text{if } 1 \leq i \leq n \\ \mathit{undefined} & \text{otherwise} \end{cases}$$

$$\mathit{get\_pred\_attr}(S) = \begin{cases} a_{i-1} & \text{if } 1 < i \leq n \\ \mathit{undefined} & \text{otherwise} \end{cases}$$

end *sweep line status.*

The sweep line status structure and its operations are described as an algebra with an ordered sequence of segments as a carrier set where each segment s_i has an attached set of attributes a_i of some sort *attr* and where a pointer indicates the position within the sequence. For the sweep line status an efficient internal dynamic structure like the AVL tree can be employed (and is used in the ROSE system) which realizes each of the operations *add_left*, *del_right*, *pred_of_s*, and *pred_of_p* in worst case time $O(\log n)$ and the other operations in constant time.

In the sequel for all algorithms we assume that all those halfsegments of a <u>*regions*</u> object R have an associated attribute *InsideAbove* where the area of R lies above or left of its segments. This *segment classification* can be computed by a plane-sweep algorithm (shown below) which views all segments intersecting the current sweep line from bottom to top. It is obvious that the lowest segment obtains the attribute *InsideAbove*, the following does not, the third again obtains it, etc. Whether the attribute *InsideAbove* is associated with a segment depends on the assignment of the attribute to the immediate preceding segment in the sweep line status.

algorithm *InsideAttribute*
input: A <u>*regions*</u> object $R = (i, < h_1, ..., h_n >, < a_1, ..., a_n >)$
output: $R' = (i', < h_1, ..., h_n >, < a_1', ..., a_n' >)$ with $a_i' = a_i$ if h_i is a right halfsegment, $a_i' = a_i \cup \{InsideAbove\}$ if h_i is a left halfsegment and the area of R lies above or left of the segment component of h_i, and $a_i' = a_i \setminus \{InsideAbove\}$ otherwise (for $1 \le i \le n$)

```
begin
  S := new_sweep();
  R := select_first(R);
  while not end_of_hs(R) do
    h := get_hs(R); (* Let h = (s, d) *)
    attr := get_attr(R);
    if d = left then
      S := add_left(S, s);
      if not pred_exists(S) then
        S := set_attr(S, {InsideAbove});
        R := update_attr(R, attr ∪ {InsideAbove});
      else
        pred_attr := get_pred_attr(S);
        if InsideAbove ∈ pred_attr then
          S := set_attr(S, ∅);
          R := update_attr(R, attr \ {InsideAbove});
        else
          S := set_attr(S, {InsideAbove});
          R := update_attr(R, attr ∪ {InsideAbove});
        end-if;
```

```
            end-if;
        else
            S := del_right(S, s);
        end-if;
    R := select_next(R);
    end-while;
end InsideAttribute.
```

This kind of segment classification which must be calculated only for left halfsegments is called at the end of the construction of a _regions_ object. The while-loop is executed once for each halfsegment of a _regions_ object. The most expensive operations within the loop are *add_left* which inserts the segment component of a halfsegment into the sweep line status S and *del_right* which removes it from S. Both operations take $O(\log m)$ time where m is the number of elements of S. Since $m \leq n$, algorithm *InsideAttribute* requires $O(n \log n)$ time for an object with n halfsegments.

The following example (Figure 5-1) will demonstrate the algorithm (and some concepts introduced so far). We consider the following _regions_ object R:

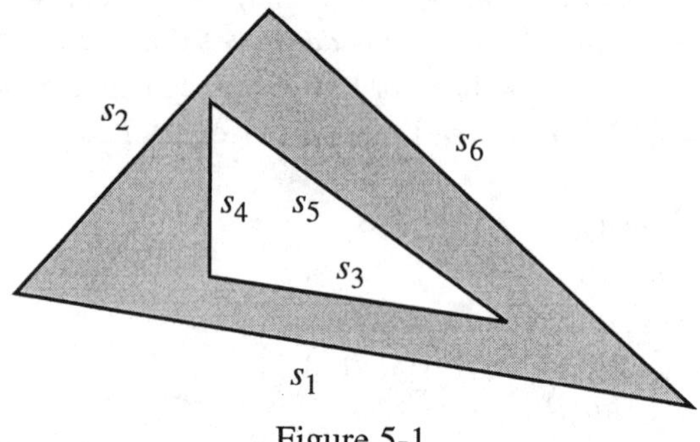

Figure 5-1

Let $h_i^l = (s_i, left)$ and $h_i^r = (s_i, right)$ denote the left and right halfsegments belonging to the segments s_i for $1 \leq i \leq 6$. We assume that no attributes have been assigned to the halfsegments of R. Then R has the following structure after the call of *select_first*:

$$R = (1, < h_1^l, h_2^l, h_3^l, h_4^l, h_4^r, h_5^l, h_2^r, h_6^l, h_5^r, h_3^r, h_6^r, h_1^r >, < \varnothing, ..., \varnothing >)$$

After initialization with *new_sweep* the sweep line status S has the structure $S = (0, \lozenge, \lozenge)$. Then within the while-loop the first halfsegment of R is considered. Since h_1^l is a left halfsegment, the corresponding segment component s_1 is inserted into S by *add_left*: $S = (1, < s_1 >, < \varnothing >)$. Obviously, s_1 has no predecessor in S so that it obtains the attribute *InsideAbove* (abbreviated by I): $S = (1, < s_1 >, < \{I\} >)$. The attribute is also associated with the halfsegment h_1^l in R (*update_attr*), and the logical pointer is positioned on the next halfsegment by *select_next*:

$$R = (2, < h_1^l, h_2^l, h_3^l, h_4^l, h_4^r, h_5^l, h_2^r, h_6^l, h_5^r, h_3^r, h_6^r, h_1^r >, < \{I\}, \varnothing, ..., \varnothing >)$$

The second halfsegment of R, h_2^l, is again a left halfsegment. Segment s_2 is inserted after s_1 into S because s_2 *above* s_1 holds. Since the attribute *InsideAbove* has been attached to the predecessor s_1 of s_2, s_2 does not obtain the attribute. In R this attribute is not set for halfsegment h_2^l. In this way, the algorithm continues and if a right halfsegment is considered, the corresponding segment component is removed from S. Figure 5-2 shows in the first column the halfsegment currently processed within the while-loop and in the second column the effect of this process on the sweep line status.

After the execution of algorithm *InsideAttribute*, the regions object R has the following structure:

$$R = (0, \; < h_1^l, h_2^l, h_3^l, h_4^l, h_4^r, h_5^l, h_2^r, h_6^l, h_5^r, h_3^r, h_6^r, h_1^r >,$$
$$< \{I\}, \varnothing, \varnothing, \{I\}, \varnothing, \{I\}, \varnothing, \varnothing, \varnothing, \varnothing, \varnothing, \varnothing >)$$

Halfsegment	Sweep line status S		
-	(0,	$\lozenge$,	$\lozenge$)
h_1^l	(1,	$< s_1 >$,	$< \{I\} >$)
h_2^l	(2,	$< s_1, s_2 >$,	$< \{I\}, \varnothing >$)
h_3^l	(2,	$< s_1, s_3, s_2 >$,	$< \{I\}, \varnothing, \varnothing >$)
h_4^l	(3,	$< s_1, s_3, s_4, s_2 >$,	$< \{I\}, \varnothing, \{I\}, \varnothing >$)
h_4^r	(0,	$< s_1, s_3, s_2 >$,	$< \{I\}, \varnothing, \varnothing >$)
h_5^l	(3,	$< s_1, s_3, s_5, s_2 >$,	$< \{I\}, \varnothing, \{I\}, \varnothing >$)
h_2^r	(0,	$< s_1, s_3, s_5 >$,	$< \{I\}, \varnothing, \{I\} >$)
h_6^l	(4,	$< s_1, s_3, s_5, s_6 >$,	$< \{I\}, \varnothing, \{I\}, \varnothing >$)
h_5^r	(0,	$< s_1, s_3, s_6 >$,	$< \{I\}, \varnothing, \varnothing >$)
h_3^r	(0,	$< s_1, s_6 >$,	$< \{I\}, \varnothing >$)
h_6^r	(0,	$< s_1 >$,	$< \{I\} >$)
h_1^r	(0,	$\lozenge$,	$\lozenge$)

Figure 5-2

The first class of plane-sweep algorithms considers the relationships between a *points* or *lines* object and a *regions* object. The algorithm scheme is to insert only the segments of the *regions* object into the sweep line status and to use the elements of the *points* and *lines* object, respectively, as query elements. The operations of this class have the following signature:

points × *regions*	→ *bool*	**pr_inside**
lines × *regions*	→ *bool*	**lr_inside, lr_intersects, lr_meets**
regions × *lines*	→ *bool*	**rl_intersects, rl_meets**
regions × *lines*	→ *lines*	**rl_intersection**

As examples, we show the algorithms for **pr_inside** and **rl_intersection** (see algorithms below). The algorithms for the other operations are similar. The algorithm **pr_inside** checks whether all points of a *points* object P lie within the areas of a *regions* object R. A point of P may coincide with an endpoint of a segment of R. Both objects are traversed in parallel during a plane-sweep. The segment components of the left halfsegments of R together with the associated attribute *InsideAbove* are inserted into the sweep line status, the segment components of the right halfsegments are removed. If a point p of P does not coincide with a dominating point of a halfsegment of R, the existence of a segment in the sweep line status immediately below p is checked. If no segment is found, then p definitely lies outside of R. Otherwise, it must be checked if the attribute *InsideAbove* has been assigned to the segment. If this is the case, then p lies inside of R, otherwise outside.

```
algorithm pr_inside
input: A points object P and a regions object R
output: true, if all points of P lie in the area of R
        false, otherwise
begin
  S := new_sweep();
  inside := true;
  pr_select_first(P, R, object, status);
  while (status ≠ end_of_first) and inside do
    if (object = both) or (object = second) then
      h := get_hs(R); (* Let h = (s, d). *)
      attr := get_attr(R);
      if d = left then
        S := add_left(S, s);
        if InsideAbove ∈ attr then
          S := set_attr(S, {InsideAbove});
        end-if
      else
        S := del_right(S, s);
```

```
            end-if
        else
            S := pred_of_p(S, get_pt(P));
            if current_exists(S)
                then inside := (InsideAbove ∈ get_attr(S))
                else inside := false
            end-if
        end-if;
        pr_select_next(P, R, object, status);
    end-while;
    return inside;
end pr_inside.
```

The while-loop of the algorithm is executed at most $l+m$ times (l the number of points of P, m the number of halfsegments of R). The loop terminates when all points of P have been examined or when a point has been found which does not lie in R. The insertion of a left halfsegment into and the removal of a right halfsegment from the sweep line status needs $O(log\ m)$ time. A point which coincides with the dominating point of a halfsegment can be ignored, since it lies definitely within R. For all other points the preceding segment in the sweep line status has to be searched which also needs $O(\log m)$ time. Altogether, the worst case time complexity of **pr_inside** is $O((l + m) \log m)$.

The algorithm for **rl_intersection** in a similar way produces a new *lines* object which contains all segments lying within R. It is crucial for the correctness of this algorithm that we can be sure that a complete (half)segment lies within R, if its dominating point lies within an area of R. This is because the boundary of R cannot intersect the interior of the segment due to the realm properties. This algorithm requires $O((l + m) \log m + k \log k)$ where k is the size of the result object and l and m the size of the *lines* and *regions* operand, respectively.

```
algorithm rl_intersection
input: A lines object L and a regions object R
output: A new lines object L_new containing all halfsegments of L whose
        segment components lie in R
begin
    L_new := new();
    S := new_sweep();
    lr_select_first(L, R, object, status);
    while status = end_of_none do
        if (object = both) or (object = second) then
            h := get_hs(R); (* Let h = (s, d). *)
            attr := get_attr(R);
```

```
            if d = left then
                S := add_left(S, s);
                if InsideAbove ∈ attr then
                    S := set_attr(S, {InsideAbove});
                end-if
            else
                S := del_right(S, s);
            end-if
        end-if;
        if object = both then
            h := get_hs(L);
            L_new := insert(L_new, h);
        else if object = first then
            h := get_hs(L); (* Let h = (s, d). *)
            S := pred_of_s(S, s);
            if current_exists(S) and (InsideAbove ∈ get_attr(S)) then
                L_new := insert(L_new, h);
            end-if;
        end-if;
        lr_select_next(L, R, object, status);
    end-while;
    return L_new;
end rl_intersection.
```

For all other operations of this class, the time complexity is $O((l + m) \log m)$ if m is the size of the <u>regions</u> operand and l the size of the other operand. Of course, for $n = l + m$, $O(n \log n)$ is a simpler upper bound for all operations. Let L be a <u>lines</u> and R be a <u>regions</u> object. The algorithm for **lr_inside** applied to L and R is similar to the one for **pr_inside**. During a plane-sweep (the event points are given by a parallel traversal of the operands) the segment components of the left (right) halfsegments of R are inserted into (removed from) the sweep line status S. For each left halfsegment h of L which does not correspond to a left halfsegment of R, the sweep line status must be checked whether a segment of R with attribute *InsideAbove* lies immediately below the segment component of h. If this is not the case, the predicate yields the value *false*.

An algorithm for **lr_intersects** applied to L and R must search for at least one segment of L lying within the areas of R. The search can be terminated if a segment of L has been found lying within R. Operator **lr_meets** yields the value *true* if no segment of L lies within R and if L and R have at least one common point. The fulfilment of the first condition implies that common points are automatically meeting points. The algorithm is similar to the one for **lr_intersects**. But as long as no common point has been found, additionally the dominating point of the last considered halfsegment together with an identifier of the object in which it is contained must be stored. If it corresponds with the dominating point of the current halfsegment and if the stored and the current object identifier are distinct, the second condition for **lr_meets** is satisfied.

The second class of plane-sweep algorithms considers the relationships between two *regions* objects.

regions × *regions*	→ *bool*	**rr_disjoint, rr_inside, rr_area_disjoint, rr_edge_disjoint, rr_edge_inside, rr_vertex_inside, rr_intersects, rr_meets, rr_adjacent, rr_encloses**
regions × *regions*	→ *regions*	**rr_intersection, rr_plus, rr_minus**

Note that here the immediate application of the technique introduced above is impeded by the fact that *regions* objects may have holes. Hence, for the algorithms of this class we introduce the concepts of *overlap numbers* and *segment classification*. A point of the realm grid obtains the *overlap number k* if it is covered by (or part of) *k* *regions* objects. For example, for two intersecting simple polygons the area outside of both polygons gets overlap number 0, the intersecting areas get overlap number 2, and the other areas get overlap number 1. Since a segment of a *regions* object separates space into two parts, an inner and an exterior one, each segment is associated with a pair (*m*/*n*) of overlap numbers, a *lower* (or right) one *m* and an *upper* (or left) one *n*. The lower (upper) overlap number indicates the number of overlapping *regions* objects below (above) the segment. In this way, we obtain a *segment classification* of a fixed set of *regions* objects and speak of (*m*/*n*)-segments.

For two *regions* objects (we only consider binary operators here) $m, n \leq 2$ holds; of the nine possible combinations only seven describe valid segment classes. This is because a (0/0)-segment contradicts the definition of a *regions* object, since then at least one of both *regions* objects would have two holes or an outer cycle and a hole with a common border. Similarly, (2/2)-segments cannot exist, since then at least one of the two *regions* objects would have a segment which is common to two outer cycles of the object. Hence, possible (*m*/*n*)-segments are (0/1)-, (0/2)-, (1/0)-, (1/1)-, (1/2)-, (2/0)-, and (2/1)-segments. Examples of (*m*/*n*)-segments are given in Figure 5-3.

In order to get a deeper understanding of overlap numbers and segment classification, we generalize these concepts to a fixed number of *k* *regions* objects $R_1, ..., R_k$ and present an algorithm which demonstrates the computation of the segment classification of these *k* objects. The algorithm is not relevant for practice, since we consider only binary operators here. Furthermore, it is frequently not necessary and too

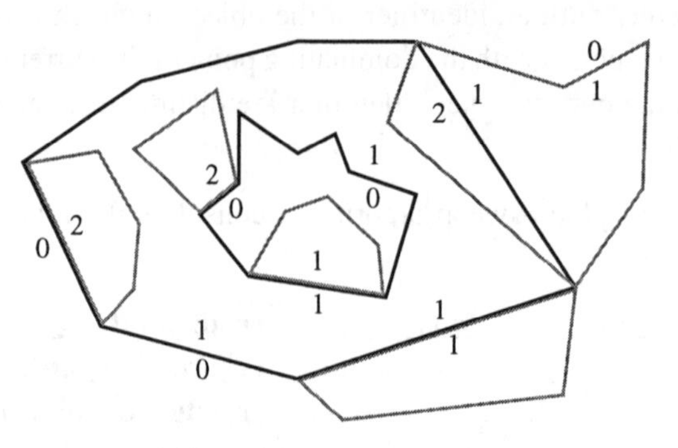

Figure 5-3

expensive to compute the segment classification for all halfsegments because many algorithms can be terminated early. At first, a common ordered halfsegment sequence is constructed from the k objects which contains all halfsegments once. Each halfsegment obtains two *local* overlap numbers indicating in how many regions objects the segment component of the halfsegment is a lower and an upper boundary. This halfsegment sequence can be constructed by the following algorithm scheme:

Step 1. Perform algorithm *InsideAttribute*(R_i) for all regions objects R_i ($1 \leq i \leq k$)

Step 2. Select the first halfsegment from all k objects (*select_first*(R_i)) ($1 \leq i \leq k$) and initialize a new value R of type regions[19] (*new*(R)).

Step 3. Let h be the smallest of the selected halfsegments. Count in how many objects that contain h attribute *InsideAbove* has been set and in how many objects that contain h it has not been set. Let a be number of attached and b be number of non-attached attributes.

Step 4. Insert h into R (*insert*(R, h)) and mark h with the local overlap numbers b and a (*update_attr*(R, (b/a))).

Step 5. Select the next halfsegments of all regions objects containing h, if these halfsegments exist (*select_next*(R_i)).

Step 6. If there is an object whose end has not been reached, continue with step 3.

The final task now is to "lift" the local overlap numbers, i.e., to determine the number of object areas in which each segment is located. This number is added to the local overlap numbers of a segment in order to achieve its segment classification.

[19] As a rule, R will be no regions object as defined in the ROSE algebra but as defined in the algebra *regions*. For each right halfsegment in R there will be a corresponding left halfsegment, and vice versa.

The strategy of the algorithm *SegmentClassification* is similar to that of algorithm *InsideAttribute*. For a segment its immediate predecessor in the sweep line status is retrieved whose upper overlap number is added to the local overlap numbers of the segment. The segment classification is only computed for left halfsegments.

algorithm *SegmentClassification*
input: $R = (i, < h_1, ..., h_n >, < a_1, ..., a_n >) \in$ <u>*regions*</u>. R is an ordered halfsegment sequence with local overlap numbers for each halfsegment as described above.
output: $R' = (i', < h_1, ..., h_n >, < a_1', ..., a_n' >) \in$ <u>*regions*</u> where a_i' contains the segment classification of the segment component of a left halfsegment h_i.

```
begin
  S := new_sweep();
  R := select_first(R);
  while not end_of_hs(R) do
    h := get_hs(R); (* Let h = (s, d). *)
    {(b/a)} := get_attr(R); (* Let (b/a) be the local overlap numbers of h. *)
    if d = left then
      S := add_left(S, s);
      if not pred_exists(S) then
        S := set_attr(S, {(b/a)});
        R := update_attr(R, {(b/a)});
      else
        {(m/n)} := get_pred_attr(S);
        b_new := n;
        a_new := n + a – b;
        S := set_attr(S, {(b_new/a_new)});
        R := update_attr(R, {(b_new/a_new)});
      end-if
    else
      S := del_right(S, s);
    end-if
    R := select_next(R);
  end-while;
  return R;
end SegmentClassification.
```

As an example for the plane-sweep algorithms of the second class we show the algorithm for **rr_inside** (see algorithm below) which tests whether a <u>*regions*</u> object R_1 is completely contained in a <u>*regions*</u> object R_2. This means that all segments of R_1 must lie within the area of R_2 but no segment (and hence no hole) of R_2 may lie within R_1. If we consider the objects R_1 and R_2 as halfsegment sequences together with the segment classes, the predicate **rr_inside** is true if (i) all halfsegments that are *only* element of R_1 have segment class (1/2) or (2/1), since only these segments lie within R_2, (ii) all halfsegments that are *only* element of R_2 have segment class (0/1) or (1/0), since these definitely do not lie within R_1, and (iii) all *common* halfsegments have segment class (0/2) or (2/0), since the areas of both objects lie on the same side of the halfsegment. In the case of a (1/1)-segment the areas would lie side by side so that R_1 could not be contained by R_2. In the algorithm, whenever a segment is inserted into the sweep line status, first the pair (m_p/n_p) of overlap numbers of the predecessor is determined (it is set to (*/0) if no predecessor exists). Then the overlap numbers (m_s/n_s) for this segment are computed. Obviously $m_s = n_p$ must hold; n_s is also initialized to n_p and then corrected.

```
algorithm rr_inside
input: Two regions objects R1 and R2
output: true, if R1 lies within R2
        false, otherwise
begin
  S := new_sweep();
  inside := true;
  rr_select_first(R1, R2, object, status);
  while (status ≠ end_of_first) and inside do
    if (object = first) or (object = both)
      then h := get_hs(R1); (* Let h = (s, d). *)
      else h := get_hs(R2); (* Let h = (s, d). *)
    end-if;
    if d = right then
      S := del_right(S, s);
    else
      S := add_left(S, s);
      if not pred_exists(S)
        then (mp/np) := (*/0)
        else {(mp/np)} := get_pred_attr(S)
      end-if;
      ms := np;
      ns := np;
      if ((object = first) or (object = both)) and
        (InsideAbove ∈ get_attr(R1))
        then ns := ns + 1
        else ns := ns − 1
      end-if;
      if ((object = second) or (object = both)) and
        (InsideAbove ∈ get_attr(R2))
```

```
                then n_s := n_s + 1
                else n_s := n_s − 1
            end-if;
            S := set_attr(S, {(m_s/n_s)});
            if object = first then
                inside := ((m_s/n_s) ∈ {(1/2), (2/1)})
            else if object = second then
                inside := ((m_s/n_s) ∈ {(0/1), (1/0)})
            else
                inside := ((m_s/n_s) ∈ {(0/2), (2/0)})
            end-if;
        end-if;
        rr_select_next(R1, R2, object, status);
    end-while;
    return inside;
end rr_inside.
```

If R_1 has l and R_2 m halfsegments, the while-loop is executed at most $n = l + m$ times, since each time a new halfsegment is visited. The most expensive operations within the loop are the insertion and the removal of a segment into and from the sweep line status. Since at most n elements can be contained in the sweep line status, the worst case time complexity of the algorithm is $O(n \log n)$ which is also valid for all other operations of this class.

The other operations mostly require slight modifications of the algorithm above. The algorithm for **rr_edge_inside** forbids common segments, the algorithm for **rr_vertex_inside** even common points, a problem which to handle is a little bit more complicated. The operation **rr_area_disjoint** yields *true* if both objects have no common areas and only allows (0/1)-, (1/0)-, and (1/1)-segments. The operation **rr_edge_disjoint** additionally forbids common segments (no (1/1)-segments) and **rr_disjoint** even common points which needs a little bit more effort. The operation **rr_adjacent** which checks the neighbourhood of two *regions* objects is equal to **rr_area_disjoint** but additionally requires the existence of at least one (1/1)-segment. The operation **rr_meets** which checks whether two *regions* objects meet in a point is equal to **rr_edge_disjoint** but additionally requires the existence of at least one common point. The operation **rr_intersects** is *true* if two *regions* objects have a common area which means that there exist some segments of segment class (0/2), (1/2), (2/0), or (2/1).

The following three operations produce a new *regions* object. The intersection of two *regions* objects (operation **rr_intersection**) implies the search for all segments with segment classification (0/2), (1/2), (2/0), and (2/1). For the union of two *regions* objects (operation **rr_union**) all (0/1)-, (1/0)-, (0/2)-, and (2/0)-segments are collected. The computation of the difference of two <u>*regions*</u> objects R_1 and R_2 (operation **rr_minus**) requires all (0/1)- and (1/0)-segments of R_1, all (1/2)- and (2/1)-segments of R_2, and all common (1/1)-segments.

The operation **rr_encloses** yields *true* for two <u>*regions*</u> objects R_1 and R_2 if each face and hence each segment of R_2 is contained in a hole of R_1. Note that this condition does not mean that R_1 and R_2 are area-disjoint, since it is possible that another face of R_1 lies within R_2. Here a method is used which gives the overlap numbers a different interpretation: We do not consider the overlapping of object areas but the overlapping of the single cycle areas of an object. In this way, the exterior of R_1 gets the number 0, the area of a face of R_1 the number 1, and a hole the number 2. If a hole of R_1 contains another face of the same object, this face gets the number 3 and a hole of this face the number 4, etc. If we compute such a segment classification for R_1, then R_1 encloses R_2 if all segments of R_2 lie on a level with even overlap number (greater than 0).

An algorithm for this operation must have the information where the inner side of a cycle is in order to be able to assign an attribute *CycleAbove* to a segment. For this purpose we can use the results of algorithm *InsideAttribute*. If for a segment of an outer cycle the attribute *InsideAbove* has been set, the attribute *CycleAbove* can be set too, because the inner sides of the object area and of the outer cycle coincide. If for a segment of a hole cycle the attribute *InsideAbove* has been set, the attribute *CycleAbove* cannot be set, because the inner sides of the object area and of the hole cycle differ. The problem can therefore be reduced to the question which segments belong to an outer cycle and which segments to a hole cycle of a <u>*regions*</u> object. The solution of this problem is part of the next subsection.

5.3.3 Graph Algorithms

A realm can be interpreted as a *spatially embedded planar graph*. Hence, a <u>*lines*</u> or a <u>*regions*</u> object defined over such a realm can be regarded as a planar subgraph $G = (V, E)$ where the vertex set V is the set of all end points of the segments and the edge set E is the set of all segments of the object. Note that such an embedded planar graph represents not only the usual incidence relationships between nodes and

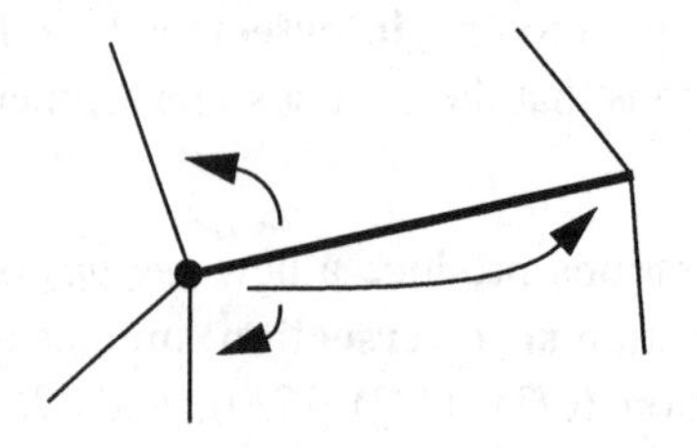

Figure 5-4

edges, but also the *neighbourhood relationship among segments incident to the same node*. This graph-theoretic view offers two primitive operations, illustrated in Figure 5-4, that are crucial for the algorithms discussed in this section: For a given halfsegment, (i) find its two neighbours incident to the same node with respect to the counterclockwise order, and (ii) find the "partner halfsegment" representing the same segment (which is equivalent to following an edge of the graph).

Basically, the data structure needed to support these two primitives in $O(1)$ time is an adjacency list for each node containing the outgoing edges in counterclockwise order. As it happens, the halfsegment sequence representing a lines or regions object

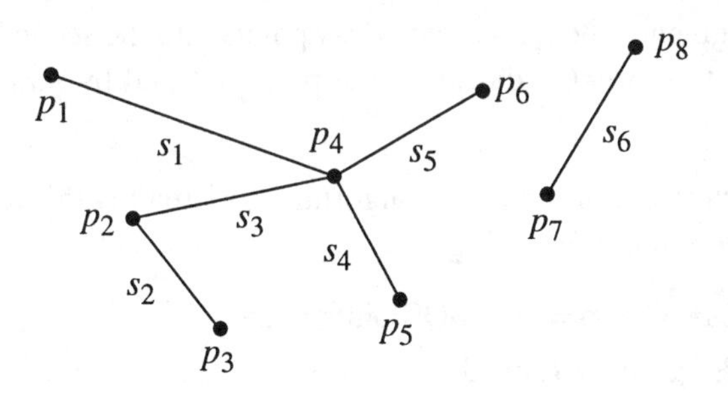

Array *Edge*

index	*h*	*pred*	*succ*	*link*	*node_index*
1	h_1^l	1	1	5	1
2	h_2^l	3	3	4	2
3	h_3^l	2	2	6	2
4	h_2^r	4	4	2	3
5	h_1^r	8	6	1	4
6	h_3^r	5	7	3	4
7	h_4^l	6	8	9	4
8	h_5^l	7	5	10	4
9	h_4^r	9	9	7	5
10	h_5^r	10	10	8	6
11	h_6^l	11	11	12	7
12	h_6^r	12	12	11	8

Array *Node*

index	*on_stack*
1	*false*
2	*false*
3	*false*
4	*false*
5	*false*
6	*false*
7	*false*
8	*false*

Figure 5-5

is already close to the desired structure because it contains all halfsegments with the same dominating point as a compact subsequence in counterclockwise order (this fact has already been used in algorithm **ll_intersects**). What is needed additionally is a pointer from each halfsegment to the partner halfsegment. For convenience, we also doubly link the halfsegments around a node.

Figure 5-5 shows a *lines* object and its graph representation in two arrays *Edge* and *Node*. This is essentially the temporary representation of a *lines* or *regions* object used in the ROSE system as a basis for graph algorithms. In array *Edge*, field *h* contains the halfsegment. The fields *pred* and *succ* contain the indexes of the preceding and succeeding halfsegments in the counterclockwise order; *link* is the index of the partner halfsegment. The field *node_index* points into the second array *Node*. In array *Node*, field *on_stack* indicates if the point indexed by *index* has been pushed onto a stack.

The data structure definition and an algorithm for creating this temporary representation are shown in the following:

```
const  MaxComp = ...; NoOfPoints = ...;
type   EdgeRec = record
                    h                  : H_N;
                    pred, succ         : cardinal;
                    link, node_index   : cardinal;
                 end;
       NodeRec = record
                    on_stack           : boolean;
                 end;
var    Edge = array [1..MaxComp] of EdgeRec;
       Node = array [1..NoOfPoints] of NodeRec;

algorithm InitEdgeAndNodeArray
input: A lines object L (or a regions object R)
output: The two arrays Edge and Node
begin
   top_V := 0;
   top_E := 0;
   old_dp := (m, m); (* outside of the realm *)
   L := select_first(L);
   while not end_of_hs(L) do
      top_E := top_E + 1;
      h := get_hs(L); (* Let h = (s, d). *)
      Edge[top_E].h := h;
      act_dp := dp(h);
```

```
        if (act_dp ≠ old_dp) or (top_V = 0) then
            (* New or first point reached. *)
            top_V := top_V + 1;
            Node[top_V].on_stack := false;
            Edge[top_E].node_index := top_V;
            Edge[top_E].succ := top_E;
            Edge[top_E].pred := top_E;
        else
            (* The same dominating point. *)
            Edge[top_E].node_index := top_V;
            (* Produce doubly-linked ring. *)
            Edge[top_E].pred := top_E − 1;
            Edge[top_E].succ := Edge[top_E − 1].succ;
            Edge[top_E − 1].succ := top_E;
            Edge[Edge[top_E].succ].pred := top_E;
        end-if;
        if d = right then
            [ Compute index i of the corresponding left halfsegment of the array
            Edge in the range 1 to top_E by using binary search. ]
            Edge[top_E].link := i;
            Edge[i].link := top_E;
        end-if;
        old_dp := act_dp;
        L := select_next(L);
    end-while;
    return Edge, Node;
end InitEdgeAndNodeArray.
```

In algorithm *InitEdgeAndNodeArray*, the while-loop is executed once for each half-segment. All operations within the loop need constant time except for linking a right with its corresponding left halfsegment which requires $O(\log n)$ time for binary search where n is the number of halfsegments of the *lines* or *regions* object. Hence the whole algorithm has time complexity $O(n \log n)$. The search range for a binary search can be restricted since a left halfsegment must be located before the corresponding right one in the halfsegment sequence. After initialization of the arrays, for an index of an element we can find its predecessor, successor, opposite halfsegment, and node information in constant time.

This graph-theoretic view is used to realize the executable operators **l_interior**, **r_contour**, **l_no_of_components**, and **r_no_of_components** which have the following signature:

lines	→	*regions*	**l_interior**
regions	→	*lines*	**r_contour**
lines	→	*int*	**l_no_of_components**
regions	→	*int*	**r_no_of_components**

Operation **l_interior** determines a *regions* object formed from the areas enclosed by segments of a *lines* object, and **r_contour** returns a *lines* object formed from the segments of only the outer cycles of the faces of a *regions* object (holes are omitted). The other two operations return the number of components which is the number of blocks (connected components) for a *lines* object and the number of faces for a *regions* object. As examples, the algorithms for **r_contour**, **l_interior**, and **l_no_of_components** are shown.

The main problem of the algorithm for **r_contour** is the assignment of the segments to the correct *outer cycles* and *hole cycles* which according to the face definition is unique (see Section 3.6). According to that definition, the *regions* object in Figure 5-6(a) consists of two faces rather than of a single face with a hole.

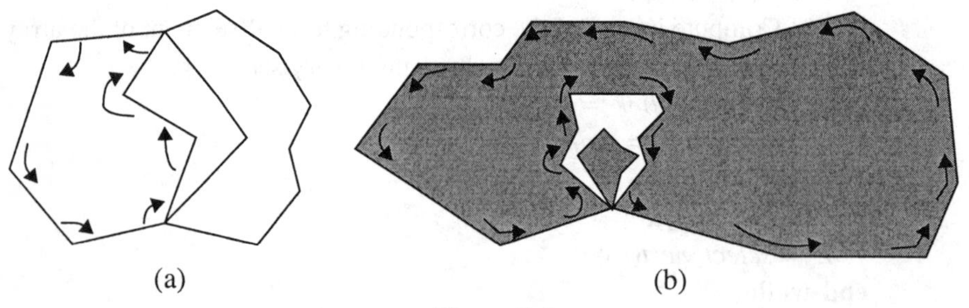

Figure 5-6

An important observation is that for the *first* halfsegment of any cycle (with respect to the order of halfsegments) we can decide whether it belongs to an outer cycle or a hole. It is a left halfsegment and belongs to an outer cycle if and only if the attribute *InsideAbove* has been set, otherwise it belongs to a hole.

We adopt the following strategy: If for a given left halfsegment it is known that it belongs to an outer cycle, then we traverse the graph forming a *minimal* cycle containing that segment. This works as follows: For the given halfsegment, get the partner halfsegment (i.e. follow the edge). From the partner, go around that node to the *predecessor* in the counterclockwise order. Follow that edge, etc. As soon as the node of the initial halfsegment is reached again, a complete cycle has been found and its halfsegments can be marked as outer halfsegments.

This strategy works fine for the *regions* object in Figure 5-6(a) where it correctly determines the left face. However, in Figure 5-6(b) the cycle would also include the hole segments. Therefore the strategy is refined as follows: If the first halfsegment belongs to an outer cycle, then try to form a minimal cycle traversing the graph as described above. Put each encountered halfsegment on a stack and mark its node as being *on_stack*. As soon as a node is encountered which is on the stack already, two cases are possible:

- *Case 1.* This is the node of the initial halfsegment. Then a complete outer cycle has been found. Remove all halfsegments from the stack, marking them as outer segments, and also from the graph. Repeat the procedure for the remaining halfsegments.
- *Case 2.* This is not the initial node. Then a hole cycle has been found. Remove halfsegments from the stack until the current node is found there, marking them as hole segments. Remove these halfsegments also from the graph. Then continue building the outer cycle. - Before removing halfsegments from the stack one must store the next halfsegment of the outer cycle in order to avoid continuing with some other face that may lie in the hole, as shown in Figure 5-6(b).

If the first halfsegment belongs to a hole, then try to form a *maximal* cycle by going always to the *successor* around a node. Apart from that, proceed in the same way as for outer cycles. However, if here a node is encountered which is not the initial one, then a cycle belonging to another hole has been found sharing a vertex with the hole cycle of the initial halfsegment.

The following algorithm *CycleClassification* classifies the segments of a <u>regions</u> object as outer or hole segments, following the strategy just discussed. Here the type *EdgeRec* is extended by the fields *visited* and *inside_above*. The first field is initialized by the value *false*; the latter field is true if a halfsegment of the <u>regions</u> object has the attribute *InsideAbove*. A variable *empty* indicates if the stack is empty; it is implicitly changed by the stack operations *push* and *pop*. "Remove *Edge*[*j*] from the graph" means remove the edge from the cycle of segments around its node. This algorithm requires *O*(*n* log *n*) time for a <u>regions</u> object with *n* halfsegments due to the included preprocessing step for computing the *Edge* and *Node* arrays; apart from that it needs only *O*(*n*) time.

algorithm *CycleClassification*
input: A <u>regions</u> object *R*
output: A modified <u>regions</u> object *R* whose halfsegments obtain the attribute
    HoleSegment if they belong to a hole and *OuterSegment* otherwise.
begin
  InitEdgeAndNodeArray(*R*);
  for *i* := 1 **to** [number of halfsegments in *Edge*] **do**
    if not *Edge*[*i*].*visited* **then**
      if *Edge*[*i*].*inside_above* **then** (* Outer cycle. *)

```
                Node[Edge[i].node_index].on_stack := true;
                push(i); Edge[i].visited := true;
                first_node_index := Edge[i].node_index;
                l := Edge[i].link; push(l); Edge[l].visited := true;
                repeat
                  j := Edge[l].node_index;
                  if not Node[j].on_stack then
                    Node[j].on_stack := true;
                    j := Edge[l].pred;
                    push(j); Edge[j].visited := true;
                    l := Edge[j].link;
                    push(l); Edge[l].visited := true;
                  else if j = first_node_index then
                    while not empty do (* Outer cycle. *)
                      j := pop();
                      [ Remove Edge[j] from the graph. ];
                      [ Set attribute OuterSegment for Edge[j]. ];
                      Node[Edge[j].node_index].on_stack := false
                    end-while
                  else (* Hole cycle. *)
                    rem := Edge[l].pred;
                    count := 0;
                    repeat
                      k := pop();
                      [ Remove Edge[k] from the graph. ];
                      [ Set attribute HoleSegment for Edge[k]. ];
                      Node[Edge[k].node_index].on_stack := false;
                      if Edge[k].node_index = j then
                        count := count + 1
                      end-if
                    until (j = Edge[k].node_index) and (count = 2);
                    push(rem); Edge[rem].visited := true;
                    l := Edge[rem].link;
                    push(l); Edge[l].visited := true;
                  end-if
                until empty;
              else (* Hole cycle. *)
                [ Proceed analogously. ]
              end-if
            end-if
          end-for;
        end CycleClassification.
```

We can now formulate the algorithm **r_contour** which computes the contour of a *regions* object by using the first algorithm. After cycle classification has been done, this is trivial and needs only $O(n)$ additional time. The total time for **r_contour**, as presented, is $O(n \log n)$.

algorithm r_contour
input: A *regions* object R
output: A *lines* object L containing the halfsegments of all outer cycles of R.
begin
  L := *new*();
  CycleClassification(R);
  R := *select_first*(R);
  while not *end_of_hs*(R) **do**
    attr := *get_attr*(R);
    if *OuterSegment* $\in$ *attr* **then**
      h := *get_hs*(R);
      L := *insert*(L, h);
    end-if;
    R := *select_next*(R);
  end-while;
  return L;
end r_contour.

The algorithm for **l_interior** is made up of two steps: First, all *maximal* cycles are extracted from a *lines* object. This implies that all segments that do not belong to any cycle are eliminated. Figure 5-7 shows one of four maximal cycles of an example *lines* object. Second, all cycles that lie within another cycle are removed. The whole algorithm needs $O(n \log n)$ time.

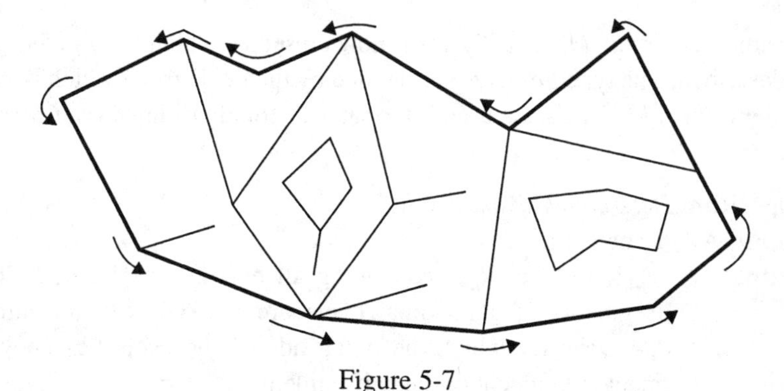

Figure 5-7

For the first algorithm step, we use a strategy similar to that of algorithm *CycleClassification*. It basically works as follows: For a given halfsegment get the partner halfsegment (i.e. follow the edge). In order to find a *maximal* cycle, from the partner go around that node to the *successor* in the counterclockwise order. Follow that edge, etc. As soon as a node of the halfsegments visited so far is reached again, a maximal cycle has been found and its segments can be transferred into a new <u>*regions*</u> object. The algorithm has, of course, also to take segments into account that do not belong to any cycle. Hence, we refine the strategy as follows:

- *Step 1.* Determine the smallest halfsegment of the graph representation of the <u>*lines*</u> object and put it on a stack.
- *Step 2.* Determine the partner halfsegment and put it on the stack, too.
- *Step 3.* Three cases are possible:
  - *Case 1.* The node of the halfsegment on top of the stack has not been put on the stack before, and the halfsegment has a successor. Put the successor on the stack and continue with step 2.
  - *Case 2.* The node of the halfsegment on top of the stack has not been put on the stack before, and the halfsegment has no successor. That means that the halfsegment cannot belong to any cycle. Remove halfsegments from the stack and at the same time from the graph until a halfsegment is found which has a further successor or until the stack is empty. In the first case put this successor on the stack and continue with step 2, otherwise continue with step 1.
  - *Case 3.* The node of the halfsegment on top of the stack has already been put on the stack before. All halfsegments on the stack enclosed by the two halfsegments with the same node on top and within the stack belong to the same maximal cycle. Remove them from the stack and from the graph and transfer them to a new <u>*regions*</u> object. If the stack is empty, continue with step 1, otherwise with step 3.

An algorithm *DetermineMaximalCycles* is now presented which follows the strategy just described. The type *EdgeRec* is extended by the field *visited* which is initialized by the value *false*. The stack used is the same as for algorithm *CycleClassification*.

algorithm *DetermineMaximalCycles*
input: A <u>*lines*</u> object *L*
output: A <u>*regions*</u> object R_{new} containing all maximal cycles of *L*. Those halfsegments of a maximal cycle are marked with the attribute *InsideAbove* for which the inner side of the cycle lies above the segment component of the halfsegment.
begin
R_{new} := *new*();
InitEdgeAndNodeArray(*L*);

```
for i := 1 to [ number of halfsegments in Edge ] do
  if not Edge[i].visited then
    Node[Edge[i].node_index].on_stack := true;
    push(i);
    l := Edge[i].link; push(l);
    repeat
      j := Edge[l].node_index;
      if not Node[j].on_stack then
        Node[j].on_stack := true;
        if l ≠ Edge[l].succ then
          (* Case 1: Successor exists. Stack can be extended. *)
          j := Edge[l].succ; push(j);
          l := Edge[j].link; push(l);
        else
          (* Case 2: No successor exists. Reduce stack. *)
          repeat
            l := pop(); (* Remove halfsegment. *)
            Node[Edge[l].node_index].on_stack := false;
            Edge[l].visited := true;
            [ Remove Edge[l] from the graph. ];

            l := pop(); (* Remove partner halfsegment. *)
            Edge[l].visited := true;
            [ Remove Edge[l] from the graph. ];
            if not empty then
              l := pop(); push(l);
            end-if;
          until empty or (l ≠ Edge[l].succ);
          if not empty then
            Node[Edge[l].node_index].on_stack := false;
          end-if;
        end-if;
      else
        (* Case 3: Node is already on stack. Cycle has been found. *)
        top_node := j;
        repeat
          l := pop(); (* Remove halfsegment. *)
          Node[Edge[l].node_index].on_stack := false;
          Edge[l].visited := true;
          R_new := insert(R_new, Edge[l].h);
          (* Determine inner side of cycle. *)
          if Edge[l].h.d = right then
```

```
                $R_{new}$ := update_attr($R_{new}$, {InsideAbove});
              end-if;
              [ Remove Edge[l] from the graph. ];

              l := pop(); (* Remove partner halfsegment. *)
              Edge[l].visited := true;
              $R_{new}$ := insert($R_{new}$, Edge[l].h);
              (* Determine inner side of cycle. *)
              if Edge[l].h.d = left then
                $R_{new}$ := update_attr($R_{new}$, {InsideAbove});
              end-if;
              [ Remove Edge[l] from the graph. ];
            until empty or (top_node = Edge[l].node_index);
            if not empty then
              Node[Edge[l].node_index].on_stack := false;
            end-if;
          end-if;
        until empty;
      end-if;
    end-for;
    return $R_{new}$;
  end DetermineMaximalCycles.
```

For the second algorithm step, a plane sweep is used to remove any cycles enclosed by other cycles, that is, we search for *outermost cycles*. Only those halfsegments belong to an outermost cycle whose segment components are either (0/1)- or (1/0)-segments. The following algorithm *DetermineOutermostCycles* uses the results of the previous algorithm and computes these outermost cycles.

```
  algorithm DetermineOutermostCycles
  input: A regions object R as the result of algorithm DetermineMaximal-
         Cycles.
  output: A regions object $R_{new}$ containing only the outermost cycles of R.
  begin
    $R_{new}$ := new();
    S := new_sweep();
    R := select_first(R);
    while not end_of_hs(R) do
      h := get_hs(R); (* Let h = (s, d). *)
      if d = left then
        S := add_left(S, s);
        if not pred_exists(S)
          then (m/n) := (*/0)
```

```
            else {(m/n)} := get_pred_attr(S)
          end-if;
          b := n;
          a := n;
          if InsideAbove ∈ get_attr(R)
            then inc(a);
            else dec(a);
          end-if;
          if (b/a) ∈ {(0/1), (1/0)} then
            R_new := insert(R_new, h);

            (* Insert the corresponding right halfsegment h' of h into R_new. *)
            h'.s := h.s;
            h'.d := right;
            R_new := insert(R_new, h');
          end-if;
        else
          S := del_right(S, s);
        end-if;
        R := select_next(R);
      end-while;
      return R_new;
    end DetermineOutermostCycles.
```

Now we can simply formulate the algorithm for **l_interior**:

```
algorithm l_interior
input: A lines object L.
output: A regions object R_new formed from the areas enclosed by segments
        of L.
begin
  R := DetermineMaximalCycles(L);
  R_new := DetermineOutermostCycles(R);
  return R_new;
end l_interior.
```

Computing the (connected) components in a *lines* object (**l_no_of_components**) can be done by a simple depth-first traversal [AHU83]. For this purpose, the type *EdgeRec* is extended by a field *visited* which is initialized by the value *false*. The algorithm for **l_no_of_components** each time searches for the smallest halfsegment which has not yet been visited and increments a counter for a new found block. Then for such a halfsegment, algorithm *MarkBlock* is called which recursively marks all halfsegments as visited that belong to the same block.

```
algorithm l_no_of_components
input: A lines object L.
output: The number of components (blocks) of L.
begin
   count := 0;
   InitEdgeAndNodeArray(L);
   for i := 1 to [ number of halfsegments in Edge ] do
      if not Edge[i].visited then
         (* New block found. *)
         count := count + 1;
         MarkBlock(i);
      end-if;
   end-for;
   return count;
end l_no_of_components.
```

```
algorithm MarkBlock
input: Index i of the halfsegment in Edge which has to be processed.
output: The changed array Edge (as a side effect).
begin
   Edge[i].visited := true;
   (* Process partner halfsegment. *)
   if not Edge[Edge[i].link].visited then
      MarkBlock(Edge[i].link);
   end-if;
   (* Process neighbour halfsegment. *)
   if not Edge[Edge[i].succ].visited then
      MarkBlock(Edge[i].succ);
   end-if;
end MarkBlock.
```

Determining the number of components (faces) in a _regions_ object (algorithm **r_no_of_components**) is also a by-product of cycle classification. A counter is introduced which counts the number of outer cycles found by the algorithm. The algorithms for **l_no_of_components** as well as for **r_no_of_components** require $O(n)$ time once the graph representation has been constructed.

5.3.4 Special Algorithms for Distance Problems

The **diameter** operator of the descriptive ROSE algebra determines the maximal extent of a spatial object, that is, the maximal distance between any two of its vertices.

The implementation of the corresponding three executable operators **p_diameter**, **l_diameter**, and **r_diameter** with the signature

<u>*points*</u>	→	<u>*real*</u>	**p_diameter**
<u>*lines*</u>	→	<u>*real*</u>	**l_diameter**
<u>*regions*</u>	→	<u>*real*</u>	**r_diameter**

uses a special strategy different from the three techniques mentioned before. The computation of all distances between any two points of a spatial object is too time-consuming. To reduce the number of elements, we determine the convex hull of the object, since the diameter of the convex hull is equal to the diameter of the whole object [PS85]. An algorithm which calculates the convex hull of the point set of a simple polygon in linear time can be found in [Me84]. An algorithm which computes the diameter of a convex polygon in linear time is shown in [PS85]. The combination of these two algorithms is used in the ROSE system to realize the three diameter operations in $O(n)$ time for an object with n points or halfsegments.

Efficient algorithms for the nine executable operators of the descriptive operator **dist** (see Appendix C) have not yet been studied.

5.3.5 Using Filter Techniques: The Bounding Box

Exact methods (shown so far) can be rather complex and time-consuming. In order to avoid running the more expensive algorithms on the actual halfsegments, it is frequently worthwhile first to approximate spatial objects by simpler geometric structures and to gain knowledge by performing simple operations on these approximations called *filters*. Such strategies are well-known (e.g., [OM88, Gü94b]). As an example of a filter we consider the *object bounding box* (*OBB*) of a (non-empty) spatial object which is defined as the smallest, axis-parallel rectangle enclosing the object. The bounding box of a <u>*points*</u>, <u>*lines*</u>, or <u>*regions*</u> object with n points or halfsegments, respectively, can be computed in time $O(n)$ by a simple object traversal and is usually stored within the object representation.

The algorithm scheme for an executable operator is then as follows: First, perform an appropriate OBB test. This takes time $O(1)$. We now have to distinguish two cases. The first case is that the OBB test is a necessary but insufficient filter condition. If the condition is unsatisfied, terminate the algorithm; otherwise, apply the exact method. The second case is that the OBB test is a sufficient but unnecessary filter condition. If the condition is satisfied, terminate the algorithm; otherwise, apply the exact method.

For the operator **pp_equal**, for instance, which tests if two <u>*points*</u> objects are equal, there is the necessary but insufficient filter condition that the OBBs of the two objects are identical. If the two OBBs are not identical, at least one point exists which is contained in only one of the objects so that the two *points* objects cannot be equal. A comparison of the single points by applying the exact method is then unnecessary. If the two OBBs are identical, it is of course not sure whether the two objects are really equal. Then additionally the exact method has to be carried out. The OBB test is especially applicable to all spatial predicates. For example, for two *regions* objects R_1 and R_2 the result of **rr_disjoint**(R_1, R_2) is certainly *true* if the objects fulfil the sufficient (but unnecessary) condition that their OBBs are (vertex-)disjoint; the result of **pr_inside**(P, R) for a *points* object P and a *regions* object R is certainly *false* if the OBB of P is not (area-)inside of the OBB of R.

If the use of OBBs does not lead to a clear result, the procedure can be refined by comparing the *component bounding boxes* (*CBB*) of spatial objects which of course makes only sense for *lines* and *regions* objects. As an example we consider the operator **ll_disjoint**. Let L_1 and L_2 be two *lines* objects with the OBBs obb_1 and obb_2. Let $\{cbb_{1,1}, ..., cbb_{1,b_1}\}$ be the CBBs of L_1, and $\{cbb_{2,1}, ..., cbb_{2,b_2}\}$ be the CBBs of L_2. The first approach of an algorithm scheme is then as follows:

- *Step 1.* Check if obb_1 and obb_2 are disjoint. If this is the case, the algorithm yields *true* and terminates, since then L_1 and L_2 are disjoint, too.
- *Step 2.* Otherwise, check the condition "$\forall\, i \in \{1, ..., b_1\}\ \forall\, j \in \{1, ..., b_2\}$: $cbb_{1,i}$ and $cbb_{2,j}$ are disjoint", and mark all k pairs of CBBs which do not fulfil the condition. If the condition is fulfilled, the algorithm yields *true* and terminates, since then L_1 and L_2 are disjoint, too.
- *Step 3.* Otherwise, take all marked, critical pairs of CBBs which do not fulfil step 2 and perform the exact method of **ll_disjoint** for the blocks belonging to these pairs.

While the first step only needs constant time, the other two steps are quite more expensive. In the second step $b_1 \cdot b_2$ comparisons of CBBs are necessary. Assuming that L_1 consists of n_1 and L_2 of n_2 halfsegments, in the worst case there can be $n_1/2$ and $n_2/2$ blocks, respectively, which implies a clear worsening compared to the $O(n_1 + n_2)$-algorithm **ll_disjoint** in Section 5.3.1. Here the decision has to be made whether first to compare CBBs or at once to perform a parallel traversal with all halfsegments of both objects. Such a decision can, for instance, rest on the number of necessary comparisons in the worst case. The comparison of the CBBs should only be carried out if

$$c_1 \cdot (b_1 \cdot b_2) < n_1 + n_2$$

The selection of the constant c_1 must take into account that the comparison of CBBs is a factor more expensive then the comparison of the dominating points of two halfsegments and that the third step has to be carried out, if the CBB test is not fulfilled.

The third step can be even still more expensive, since in the worst case there can be $b_1 \cdot b_2$ critical pairs. Assuming that the halfsegments of the *lines* objects are distributed evenly to the blocks, theoretically we additionally obtain $(b_1 \cdot b_2)$ $(n_1/b_1 + n_2/b_2) = n_1 \cdot b_2 + n_2 \cdot b_1$ comparisons of dominating points. The exact method is applied to the blocks of the k critical pairs if

$$c_2 \cdot k \cdot (n_1/b_1 + n_2/b_2) < n_1 + n_2$$

The constant c_2 takes the necessary access operations to the critical pairs of CBBs into account.

In summary, we obtain the following modified algorithm scheme for **ll_disjoint**:

- *Step 1*. Check if obb_1 and obb_2 are disjoint. If this is the case, the algorithm yields *true* and terminates, since then L_1 and L_2 are disjoint, too.
- *Step 2*. Otherwise, if for a given c_1 the condition $c_1 \cdot (b_1 \cdot b_2) < n_1 + n_2$ is fulfilled, continue with step 3, otherwise with step 6.
- *Step 3*. Check the condition "$\forall\, i \in \{1, ..., b_1\}\ \forall\, j \in \{1, ..., b_2\}$: $cbb_{1,i}$ and $cbb_{2,j}$ are disjoint", and mark all k pairs of CBBs which do not fulfil the condition. If the condition is fulfilled, the algorithm yields *true* and terminates, since then L_1 and L_2 are disjoint, too.
- *Step 4*. Otherwise, if for a given constant c_2 the condition $c_2 \cdot k \cdot (n_1/b_1 + n_2/b_2) < n_1 + n_2$ is fulfilled, continue with step 5, otherwise with step 6.
- *Step 5*. For the blocks belonging to the critical pairs of CBBs carry out the exact algorithm for **ll_disjoint**. If for only one pair the result is *false*, the algorithm as a whole yields *false*, otherwise *true*. Terminate the algorithm in both cases.
- *Step 6*. Solve the task by a parallel traversal over *all* halfsegments of both objects. Terminate the algorithm.

Note that this algorithm scheme does not guarantee that always the most efficient strategy is selected. On the one hand, the decision criteria are too rough and imprecise; this could be eliminated by a refinement of the criteria. On the other hand, it is for example possible that the steps 1 to 4 have been executed and nevertheless step 6 has to be carried out so that all intermediate tests have only caused additional costs without leading to a usable result.

For plane-sweep algorithms the decision criteria have to be modified. For example, the operation **lr_intersects**(L, R) needs time $O((n_L + n_R) \log n_R)$ if n_L and n_R are the numbers of halfsegments of L and R, respectively. Hence, the condition $c_1 \cdot (b_1 \cdot b_2) < n_1 + n_2$ has to be changed to the condition $c_1' \cdot (b_L \cdot b_R) < (n_L + n_R) \log n_R$ where

b_L is the number of blocks of L and b_R the number of faces of R. This modification is also transferable to the other decision criteria and run times.

Filter techniques especially on the basis of CBBs have only found restricted use for implementing operators of the executable ROSE algebra. Several operations operate only on one single object. In this case bounding boxes are not advantageous. Other operations for algorithmical reasons must treat spatial objects as a whole.

5.3.6 Algorithms for Operations on Sets of Database Objects

The descriptive operators of the fourth group (Section 4.4.4) that are defined on sets of database objects have not yet been implemented, since the object model interface (OMI) (Section 4.3) needed for the data model-independent communication between the ROSE algebra and a DBMS data model has so far not been realized. By using the implemented OMI, a DBMS should on the one hand be able to access and execute the operations of the ROSE algebra, hand over the database objects and spatial objects needed as operands for the operations, and accept the results. On the other hand with the aid of the implemented OMI the operations of the ROSE algebra should on their own be able to access and traverse sets of database objects.

In order to demonstrate a possible communication mechanism between ROSE algebra and DBMS data model and in order to show how an operator of this group could be realized, some assumptions about the OMI are made. We assume that at the executable level the OMI provides some types and operations that allow access to a specified set or, more precisely, sequence of database objects (spatially-referenced objects). Because only identifiers of database objects and not the database objects themselves are needed, a type *obj-id* for identifiers of database objects is introduced. A value of type *obj-id* contains the unique and stable identifier of a single database object. A value of type *set*(*obj-id*) contains the unique and stable identifier of a specified set (sequence) of database objects.

For the manipulation of a sequence of database objects the OMI is assumed to provide the following operations: Let *CreateObjSet* initialize and open, *OpenObjSet* open, *CloseObjSet* close, and *DestroyObjSet* destroy such a sequence. Let furthermore, *SelectFirstObj* select the first and *SelectNextObj* select the next database object of a sequence. Predicate *EndOfObj* is to indicate if the end of a sequence has been reached. The identifier of a selected database object of a sequence is retrieved by the operation *GetObj*, and to insert an identifier of a database object into a sequence, operation *AddObjToSet* is used. At last, the OMI provides an attribute function for accessing an attribute value of a database object.

Among the sixteen executable operators of this group as an example we show a possible algorithm for the operator **pp_closest** which has the following signature:

<u>set</u>(<u>obj-id</u>) × (<u>obj-id</u> → <u>points</u>) × <u>points</u> → <u>set</u>(<u>obj-id</u>) **pp_closest**

The **pp_closest** operator yields all those elements of a set of database objects whose *points* attribute value is nearest to a reference *points* object. During query execution the DBMS calls this algorithm supplied with the identifier of a set of database objects, an attribute function, and a reference point as operands. Then the algorithm gets the control and traverses the set of database objects by operations of the OMI.

algorithm pp_closest
input: A set identifier *set_id* of type *set*(*obj-id*), an attribute function *PointAttr*: *obj-id* → *points*, and a reference *points* object *P*.
output: A set of database objects whose *points* attributes have the smallest distance to *P*.

```
begin
  result_set_id := CreateObjSet();
  d_min := ∞;
  OpenObjSet(set_id);
  SelectFirstObj(set_id);
  while not EndOfObj(set_id) do
    obj_id := GetObj(set_id);
    P_obj := PointAttr(obj_id);
    d := pp_dist(P, P_obj);
    if d < d_min then
      (* Initialize new object set. *)
      CloseObjSet(result_set_id);
      DestroyObjSet(result_set_id);
      result_set_id := CreateObjSet();
      AddObjToSet(result_set_id, obj_id);
      d_min := d
    else if d = d_min then
      (* Insert object into the (preliminary ?) solution set. *)
      AddObjToSet(result_set_id, obj_id);
    end-if;
    SelectNextObj(set_id);
  end-while;
  CloseObjSet(set_id);
  CloseObjSet(result_set_id);
  return result_set_id;
end pp_closest.
```

5.4 Related Work

The importance of a *finite precision / finite resolution computational geometry*, as described in this chapter, which is defined on a uniform, discrete grid such that points, end points of line segments, vertices of polygons etc. have integer coordinates instead of floating-point coordinates, has been emphasized by Greene and Yao [GY86] as well as Yao [Ya92]. But there has only been little research on this topic which has been summarized in Section 2.5. To the author's knowledge, geometric algorithms over a discrete domain for more complex structures like those of the ROSE algebra have not been described in the literature before.

Chapter 6

Implementing Concepts: Realm System and ROSE System

The illustration, verification, and visualization of concepts through *systems* turns out to be very important for the underpinning of ideas, theories, and algorithms, and for the demonstration of their applicability in practice. Furthermore, it allows both the designer and the user to gain more insight into the problems. Hence, we have made an effort to realize the ideas presented in this book. The concept of realms and the algorithms and data types of the executable ROSE algebra have been implemented within the framework of two software systems called the *Realm system* [Sc95b] and the *ROSE system* [Ri95]. This chapter gives a brief overview of their architecture and functionality. The first section introduces both systems as modular components whose handling is supported by two so-called *application environment interfaces* and by two editors called the *Realm editor* and the *ROSE editor.* The second section sketches the architecture and modularization of the Realm system in more detail and gives an overview of the functions of the realm editor. The third section does the same for the ROSE system and the ROSE editor. The fourth section deals with the coupling of both systems by putting the ROSE system on top of the Realm system. The fifth section describes related work.

6.1 Providing Realm System and ROSE System as Modular Components

The concept of realms (Chapter 3) and the algorithms and data types of the executable ROSE algebra (Chapter 5) have been implemented as modular components within the framework of two software systems called the *Realm system* [Sc95b] and the *ROSE system* [Ri95]. Both systems are available for study or use as stand-alone Modula-2 libraries and are running on Unix systems. They are in principle suitable for use in database systems since all objects have compact representations. However, for a serious integration it is necessary to solve the communication problem of these systems with a DBMS and the problem of managing very large ROSE objects in a way that is compatible with the DBMS object and storage management. A goal pursued in the future is to provide the ROSE algebra as an external computation server, that is, as a *geo-server*, which is able to offer services for geometric computations and to communicate with a database server or some other kind of server over a network (see Section 7.2).

Realm system and ROSE system are both provided as modular, compact, and closed software components with cleanly defined interfaces. Both systems can be used either separately or connected together (applying the second alternative the ROSE system is put on top of the Realm system). For demonstration purposes and in order to simplify the database-independent handling of both systems for the user, two *application environment interfaces* (*AEI*) have been introduced as utilities. They support the user to become acquainted with the capabilities of both systems and to be able to use them for own simple applications. Facilities are provided to store and load collections of realm objects and spatial objects, respectively, in a persistent and database-independent way, to traverse these collections once from the beginning to the end, to provide interactive or other mechanisms for generating such collections, and to retrieve single objects. The management of alphanumeric information is not supported and belongs to the user's tasks. Two editors called the *Realm editor* and the *ROSE editor* offer window- and graphic-oriented user interfaces for displaying and interactively manipulating realm objects and spatial objects, respectively.

6.2 The Realm System

6.2.1 Modularization and Architecture

The *Realm system* is a collection of software modules and contains

- the module *IntegerArithmetic* which realizes an integer arithmetic that is error-free with respect to overflow according to Section 3.2,
- the module *RobustGeoPrimitives* which provides data types for *N*-points and *N*-segments and robust geometric primitives on these types according to Section 3.2,
- the module *Redrawing* which computes the redrawing of an arbitrary segment in the realm according to Section 3.3,
- the module *RealmInterface* which realizes the operations of the realm interface according to Section 3.5,
- the module *RealmStorageInterface* which offers some special external storage operations for realms, and
- the module *SpatialIndexStructure* which provides an independent interface to an external spatial index structure.

The module *RealmAEI* realizes the *Realm Application Environment Interface* (*Realm-AEI*) which contributes to a better handling of the Realm system. Operations are provided to import a set of realm objects given in a special text file format into the external realm representation, to export an external realm representation into such a specially formatted text file, and to create application data. The implementation of the Realm-AEI is based on the realm interface.

The module *RealmEditor* implements the realm editor and can be viewed as a special user application on top of the Realm system and the Realm-AEI. It allows to view and interactively manipulate realm objects.

Figure 6-1 shows the architecture of the Realm system, the Realm-AEI, and the realm editor.

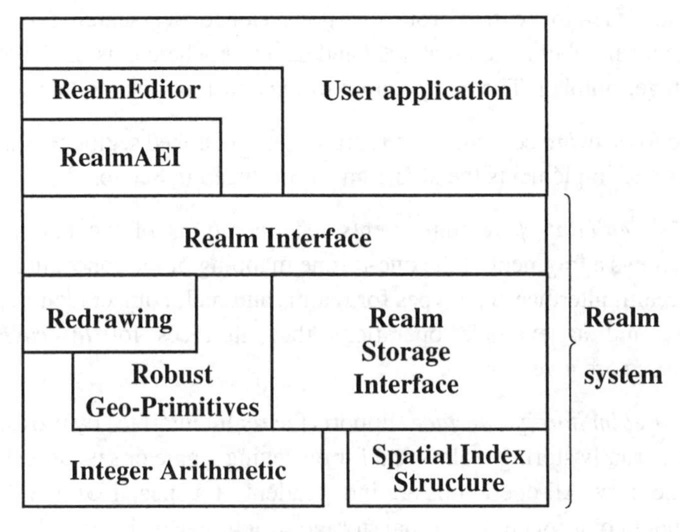

Figure 6-1

6.2.2 Implementation Aspects

The module *IntegerArithmetic* implements a special integer arithmetic which allows to construct and to use "arbitrarily" large integer numbers with standard and a few extended operations and which is error-free with respect to overflow. Each integer number is represented as an $(n+1)$-tuple (array) $(v_0, v_1, ..., v_n)_b$ of cardinals to the base b. Hence, each tuple component (field) v_i can at most have the value $b - 1$. The component v_0 describes the sign of the integer number, and the value for base b is chosen as the square root of the largest, representable computer integer number. This strategy has the advantage that it allows very fast multiplications without overflow by using the computer arithmetic and that it avoids a slower software version of a multiplication algorithm based on additions. But it has the drawback that the possible representation range of a component is not exhausted so that this strategy is more space-consuming. The standard operations like arithmetic and comparison operations are realized by the familiar pencil-and-paper algorithms known from school. The most complex operation is integer division which contains some tricky algorithmic steps and has to treat some special cases. The used algorithm is a variation of an algorithm from Knuth [Kn81].

The module *RobustGeoPrimitives* provides data types for N-points and N-segments and robust geometric primitives on and between these types. Additionally, a type for rectangles over an N-realm, called *N-rectangle*, is introduced with corresponding operations. From Appendix A we know that, if $n = |N|$ is the resolution or size of a realm, then numbers in the range $[-2n^3, 2n^3]$ must be representable by the integer arithmetic in order to be free from overflow. Hence, we search for the largest allowed integer number l such that $n \leq l$ and $|2l^3| \leq m$ where m is the largest, representable integer number. The computation of l is a task of this module.

The module *Redrawing* computes the redrawing of a hooked segment as a sequence of segments and implements the algorithms introduced in Section 3.3.

The module *RealmInterface* implements the operations of the realm interface. Figure 6-2 shows a fragment of the one-to-one mapping of the conceptual to the implemented realm interface. The types for realms and realm object identifiers (*roids*) are opaque, and as example operations the interfaces for *InsertNPoint* and *InsertNSegment* are given.

The module *RealmStorageInterface* supports the realm interface by providing operations for externally storing realms and for retrieving single or sets of realm objects into main memory. All operations are independent of a special DBMS data model and independent of a concrete external storage structure. But in general, the external storage structure will be a spatial index structure.

The module *SpatialIndexStructure* provides a general interface to spatial index structures like an R-tree [Gu84] or an LSD-tree [HSW89] with elementary operations for storing, retrieving, and indexing geometric entities by rectangles. While the definition of the module is given in advance, the implementation or coupling of a concrete spatial index structure is task of the database administrator.

The module *RealmAEI* simplifies the use of the Realm system by offering some file and conversion operations and is furthermore an interface for data exchange with other applications. There are operations to generate a realm from a specially formatted text file representation (import) which contains either points and segments with integer coordinates that obey the realm properties or which contains application data with integer coordinates that consist of arbitrary points and possibly intersecting line segments. Other operations map a realm to the special text file representation (export), generate random application data with integer coordinates and store

```
TYPE
  (* Opaque types for realms and for realm object identifiers ad-
     dressing R-points and R-segments. *)
  Realm;
  ROID;

  (* Type representing the redrawing of an N-segment. The redrawing
     of an N-segment is given as a list of consecutive N-segments
     where each N-segment 'Seg' is associated with its correspon-
     ding realm object identifier 'Roid'. *)
  RedrawingList = POINTER TO RedrawingListType;
  RedrawingListType = RECORD
                        Seg  : NSegment;
                        Roid : ROID;
                        Next : RedrawingList;
                      END;

  (* Type representing the redrawings of spatial components (N-seg-
     ments) in spatial objects. Spatial components are addressed
     by their spatial component identifiers (SCIDs). Identifiers
     of spatial components with the same redrawing are contained
     in the same scid list 'Scids'.*)
  SpatialCompRedrawList = POINTER TO SpatialCompRedrawType;
  SpatialCompRedrawType = RECORD
                            Scids     : SCIDList;
                            Redrawing : RedrawList;
                            Next      : SpatialCompRedrawList
                          END;

  RealmObjects = (RealmPoint, RealmSegment);
  RealmObject = RECORD
                  CASE Kind : RealmObjects OF
                    RealmPoint   : Point   : NPoint;
                  | RealmSegment : Segment : NSegment;
                  END;
                END;

  (* Type representing a list of realm objects 'RObj' together with
     their realm object identifiers 'Roid'. *)
  RealmObjectAndRoidList = POINTER TO RealmObjectAndRoidListType;
  RealmObjectAndRoidListType = RECORD
                                 RObj : RealmObject;
                                 Roid : ROID;
                                 Next : RealmObjectAndRoidList;
                               END;

PROCEDURE InsertNPoint
  (VAR GeoRealm   : Realm                 (* in/out *);
       Point      : NPoint                (* in     *);
   VAR Roid       : ROID                  (* out    *);
   VAR Redrawings : SpatialCompRedrawList (* out    *));

PROCEDURE InsertNSegment
  (VAR GeoRealm           : Realm                 (* in/out *);
       Segment            : NSegment              (* in     *);
   VAR RedrawingOfSeg     : RedrawingList         (* out    *);
   VAR Redrawings         : SpatialCompRedrawList (* out    *);
   VAR EndPointsInserted  : BOOLEAN               (* out    *));
```

Figure 6-2

them in the special text file representation, and convert a text file containing application data with real coordinates to a text file containing the same application data but with integer coordinates.

The module *RealmEditor* will be explained in more detail in Section 6.2.3.

For further details concerning the whole architecture, the study of [Sc95b] is recommended.

6.2.3 The Realm Editor

The *realm editor* offers a window- and graphic-oriented user interface for visualizing realms and for constructing, retrieving, manipulating, and identifying realm objects. It can be viewed as a special user application on top of the Realm system and the Realm-AEI and allows for example to display realm objects graphically, to zoom in and zoom out in realms, to scroll through the realm representation, to select realm objects by using the mouse, and to apply the operations of the realm interface and the Realm-AEI with graphical feedback.

The layout of the editor (shown in Figure 6-3) consists of two parts. The left part is a graphic window of square size and serves as a graphical representation area for all or a selected collection of objects of a given realm. The right part is a command and

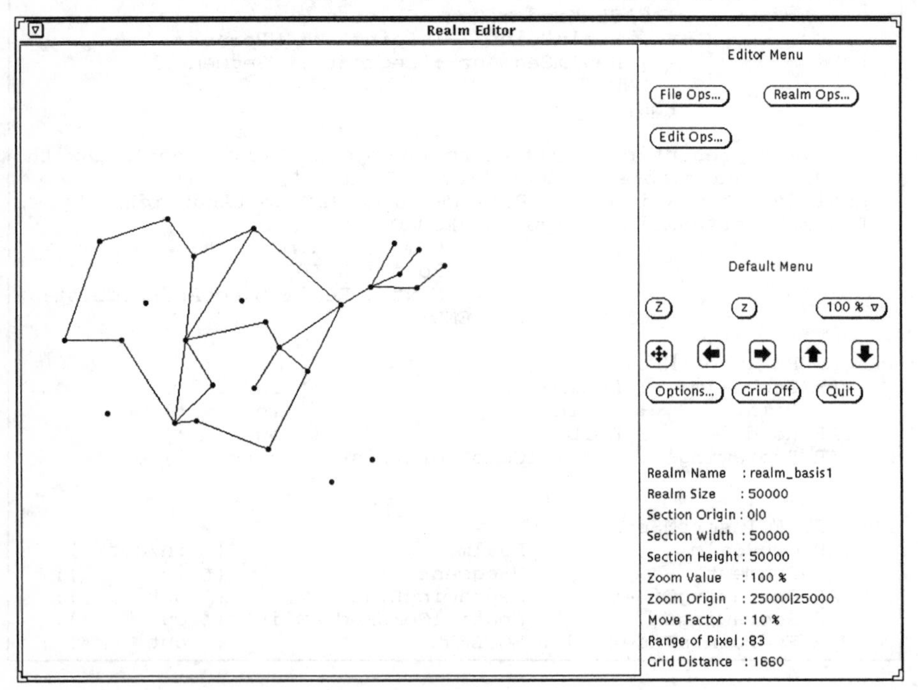

Figure 6-3

control panel and contains an application-specific menu at the top (*Editor Menu*), application-independent zoom and scrolling functions in the middle (*Default Menu*), and application-independent control informations like realm name, realm size, origin, width, and height of the displayed section of the realm, move factor, range of a pixel, and grid distance at the bottom. The layout of the editor window as well as the application-independent functionalities are provided by a so-called *visualization kernel* which forms the basis of the realm editor and the ROSE editor. The application-specific functionalities are dependent on the two editors.

The *Editor Menu* of the realm editor contains file, realm, and editing operations. Menu item *File Ops* calls the file operations of the Realm-AEI (Figure 6-4). Five file operations are provided: Menu item *Import* allows to load a collection of realm objects stored in a specially formatted text file into the external storage structure of a new realm. Menu item *Export* stores the geometry contained in the external storage structure of a realm into a specially formatted text file. Menu item *Import Application Data* is similar to menu item *Import* but here the collection of objects may be an arbitrary set of points and possibly intersecting line segments which has for instance been produced by another application program. Since application data, in general, do not satisfy the realm properties, redrawings of segments may be necessary. Menu item *Create Application Data* generates random application data with integer coordinates, i.e., a collection of arbitrary points and segments which can then be imported with menu item *Import Application Data*. Application data with real coordinates must first be converted to data with integer coordinates. This is performed by menu item *Convert Data*. Each real number is simply multiplied with a multiple of the decimal number 10. The three special text file formats needed for the file operations are public; their description is omitted here. Each operation requires a number of parameters which after selecting the operation are underlined and have to be specified. The parameters relate to the number of points and seg-

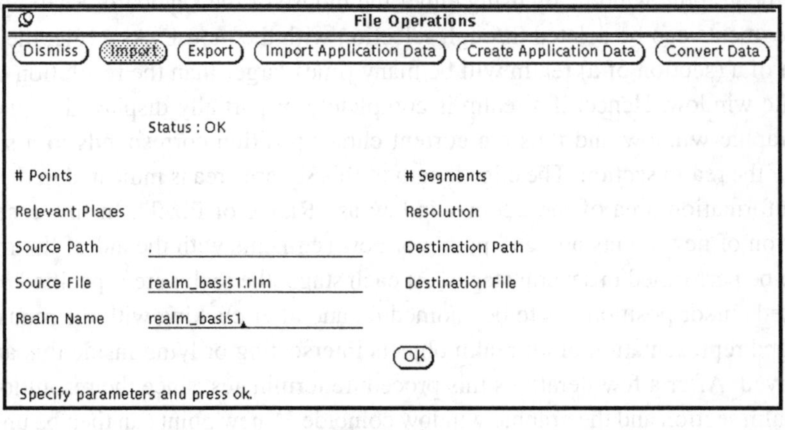

File Operations

Dismiss | Import | Export | Import Application Data | Create Application Data | Convert Data

Status : OK

\# Points | # Segments

Relevant Places | Resolution

Source Path . | Destination Path

Source File realm_basis1.rlm | Destination File

Realm Name realm_basis1

Ok

Specify parameters and press ok.

Figure 6-4

ments to be created, to the number of relevant decimal places when converting data, to the name and the resolution of a realm, and to the source and destination path of a source and destination text file, respectively.

Menu item *Realm Ops* offers realm operations for creating, opening, closing, and destroying the external storage structure of a selected realm (Figure 6-5) and hence implicitly calls the operations of the realm interface. For creating a new realm, its resolution and its name have to be entered. The opening of a realm requires its name and the desired access mode (i.e., only reading access or reading and writing access). Afterwards, the geometry of all contained objects is automatically displayed in the graphic window. When closing a realm, its graphical representation disappears. Only closed realms can be destroyed.

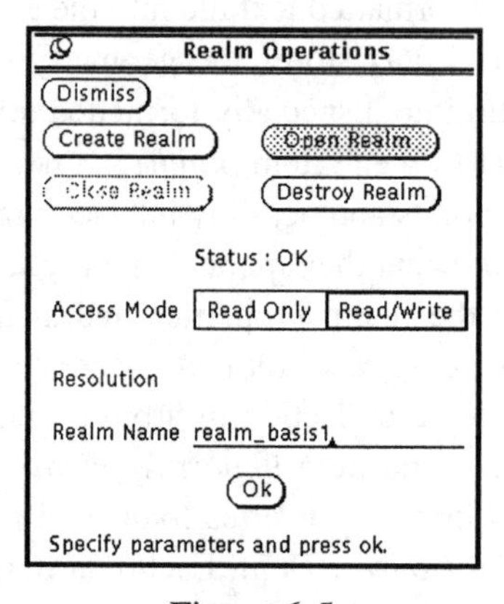

Figure 6-5

Menu item *Edit Ops* allows access to the objects of an opened realm with reading and writing access and offers six editing operations (four insertion and two deletion modes) for them (Figure 6-6). New points and segments can be interactively inserted in the graphic window by using either the mouse or the keyboard. Realm points and segments can be deleted through selection with the mouse. As a rule, the resolution of a (section of a) realm will be many times larger than the resolution of the graphic window. Hence, if a realm is completely or partially displayed, a pixel of the graphic window and thus the current cursor position corresponds to a square area of the realm section. The edge length of this square area is indicated in the control information area of the editor window as "Range of Pixel". In this case, the insertion of new points and end points of new segments with the aid of the mouse has to be performed in several stages. At each stage, the realm area specified by the selected cursor position has to be zoomed in, and after clicking with the mouse, an enlarged representation of all realm objects intersecting or lying inside this area is displayed. After a few iterations this procedure terminates, since the resolutions of the realm section and the graphic window coincide. A new point can then be unique-

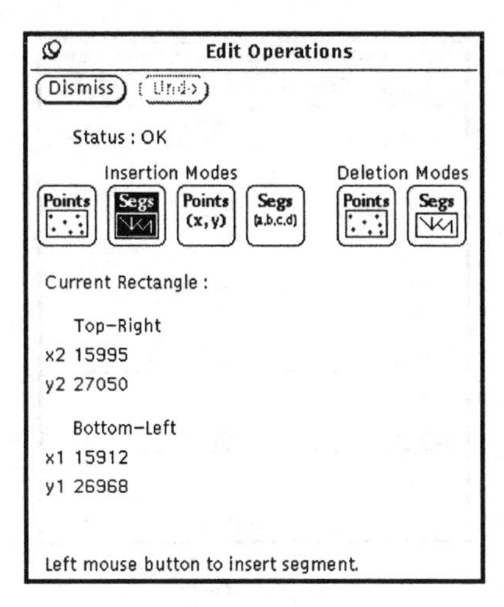

Figure 6-6

ly inserted. The realm area identified by the current cursor position is indicated by the coordinates of its bottom-left and top-right corner. If the coordinates of both corners are the same, cursor positions can be uniquely mapped on grid points, and a point can be inserted. Afterwards, the original realm section is shown again, and the just inserted point or segment is emphasized by colour or by a special line style. The *Undo* button undoes the last performed editing operation.

The *Default Menu* of the editor window allows to define, change, and shift the displayed section of a realm as well as to alter default parameters. As soon as a realm has been opened, it is completely drawn in the graphical representation area. The current zoom value which is always shown in the control information area is then 100% so that all objects of the realm are visible. It is now possible to confine oneself to a realm section of interest and hence to present an enlarged view of a limited set of realm objects (zoom-in function). On the other hand, the set of visible realm objects can be increased by extending the current size of the realm section and hence reducing the scale of the realm representation (zoom-out function). A section can also be moved over the realm space.

To enlarge, to reduce, or to shift a section is feasible in two ways, either by pressing buttons or graphically by manipulating a rectangle representing a realm section with the mouse. The use of buttons works as follows: The visualization kernel manages a sequence of ten different zoom factors (see table in Figure 6-7) which are given in percent, ordered by size, and changeable by the user. If, for instance, the zoom factor is 400%, only a quarter of the realm space is displayed which corresponds to one of four quadrants. The question which quadrant is shown in the graphic window is dependent on the zoom origin (explanation see below). The two zoom buttons of the editor window labelled “Z” and “z” realize the zoom-in and zoom-out function,

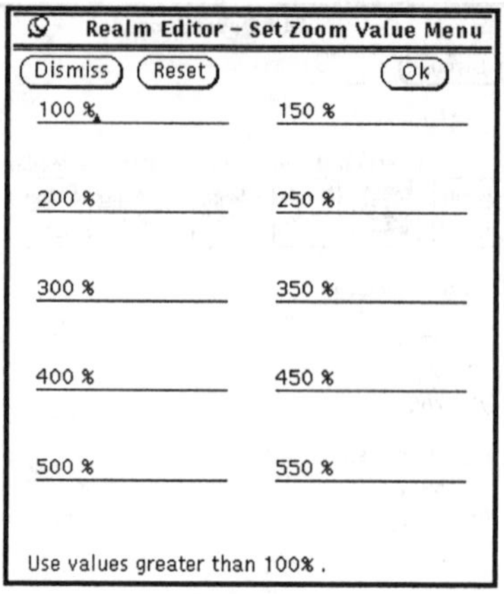

Figure 6-7

respectively, and choose the successor and predecessor, respectively, of the current zoom factor. A button labelled with the current zoom factor contains a pull-down menu and allows the direct selection of one of the ten predefined zoom factors. In each case, the result is an immediate update of the graphical realm representation. A shift of the visible realm section can be carried out with the four arrow buttons each directed to one of the four points of the compass, provided that an enlarged representation has been chosen. The shift is determined by a variable move factor whose value is always shown in the control information area. The move distance is dependent on the move factor and the width and height of the currently selected realm section.

The graphical definition, change, and shift of a section by using the mouse is initiated by clicking the button with the crossed arrows which has the effect that an additional window (Figure 6-8) appears which always displays the whole realm. A rectangle on the realm representation which is always a square indicates the currently selected realm section. Changing the zoom factor by pressing buttons has influence on the extent of the square, and vice versa. The square can now be manipulated by the mouse. Pressing the left mouse key near a side of the square and dragging the mouse in or away from it causes the square and thus the visible realm section to become smaller or larger, respectively, and thus the zoom factor to become larger or smaller. Pressing the middle mouse key within the square and then dragging the mouse changes the position of the square and hence the position of the visible realm section whereas its extent remains unchanged. The button labelled "Reset" undoes all manipulations on the square. After the user has manipulated and positioned the square as desired, a press on the button labelled "Ok" updates the graphical realm representation in the editor window according to the new selected section.

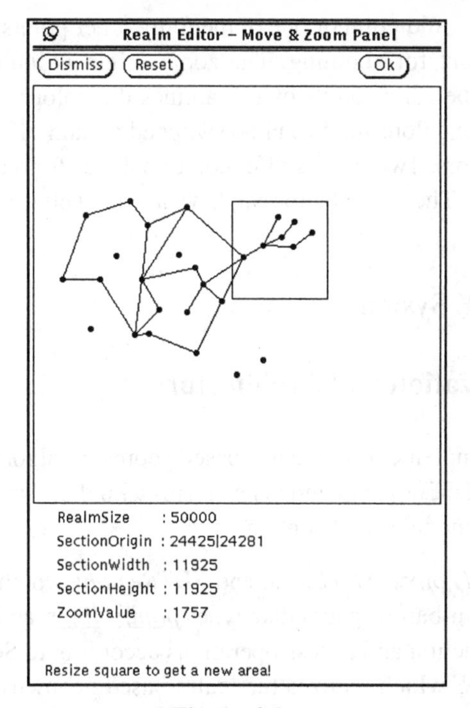

Figure 6-8

The button labelled "Grid on" or "Grid off" in the Default Menu fades in or fades out a grid as background in the graphical representation area which can be helpful for insertion or deletion operations and which should not be confused with the realm grid. Two adjacent grid points have a fixed distance of 10 pixels. The consequence is that the distance between two adjacent grid points measured in realm coordinates can have different size which is dependent on the extent of the realm section. The grid distance in realm coordinates is shown in the control information area. Default parameters can be changed by clicking the menu item *Options*. A window (Figure 6-9) appears showing the default parameters and their current values. The default parameters relate to zoom origin, colour graphic, and move factor. The zoom origin

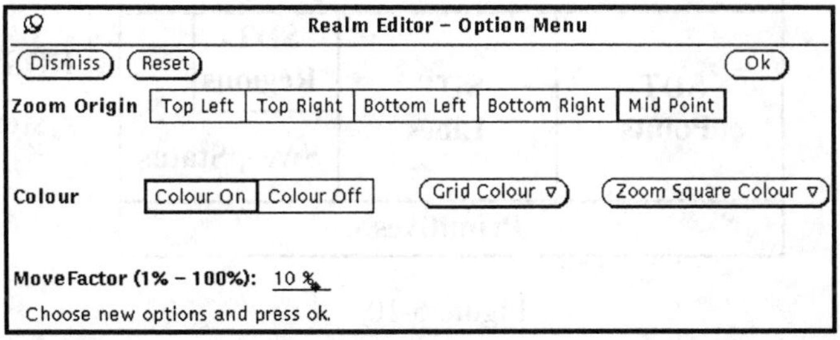

Realm Editor – Option Menu
Dismiss Reset Ok
Zoom Origin Top Left Top Right Bottom Left Bottom Right Mid Point
Colour Colour On Colour Off Grid Colour ▽ Zoom Square Colour ▽
MoveFactor (1% – 100%): 10 %
Choose new options and press ok.

Figure 6-9

determines either the mid point or one of the four corner points of the realm section as the reference point for zooming. The zoom origin is indicated in the control information area. The editor on its own examines if a colour monitor is connected to the computer. The colour mode can be switched on and off and has no effect on monochrome monitors. Two menus offer colour palettes for the grid colour and the zoom square colour. The move factor can have a value between 1% and 100%.

6.3 The ROSE System

6.3.1 Modularization and Architecture

The *ROSE system* implements the realm-based geometric algorithms and data types of the executable ROSE algebra and hence forms a kind of "geo-kernel". It is a collection of software modules and contains

- the modules *SDTPoints*, *SDTLines*, and *SDTRegions* which provide data structures for the realm-based spatial data types points, lines, and regions with corresponding construction and access operations according to Section 5.2,
- the module *ROSE* which realizes the realm-based geometric algorithms for the operators of the executable ROSE-Algebra according to Section 5.3,
- the module *SweepStatus* which manages the sweep line status for ROSE operations with regions operands according to Section 5.3.2, and
- the module *Primitives* which provides operations that are similar to those of the module *RobustGeoPrimitives* of the Realm system.

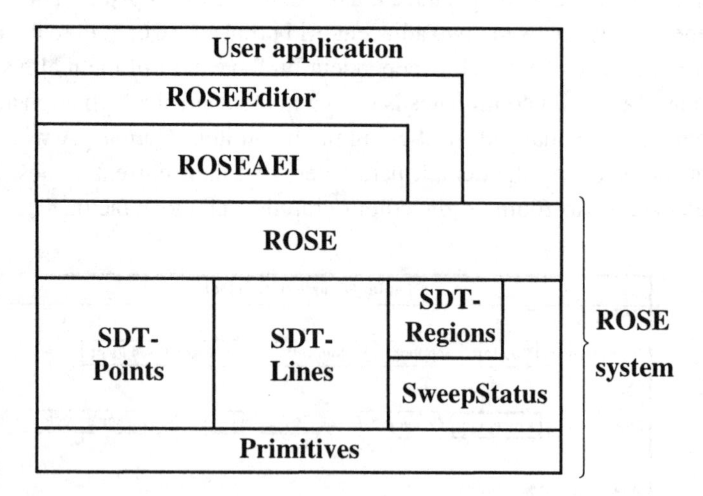

Figure 6-10

The module *ROSEAEI* realizes the *ROSE Application Environment Interface* (*ROSE-AEI*) which supports a better handling of the ROSE system. Operations are provided for database-independent input, output, storage, and retrieval of realm-based spatial objects. The implementation of the ROSE-AEI is based on the ROSE system.

The module *ROSEEditor* has so far not been designed and implemented. It can be viewed as a special user application on top of the ROSE system and the ROSE-AEI and is to allow to view and interactively manipulate ROSE objects.

Figure 6-10 shows the architecture of the ROSE system, the ROSE-AEI, and the ROSE editor.

6.3.2 Implementation Aspects

The three modules *SDTPoints*, *SDTLines*, and *SDTRegions* provide data structures for the realm-based spatial data types *points*, *lines*, and *regions* as opaque types with corresponding construction and access operations. As an example, in Figure 6-11 the representation of a *regions* object is shown (for *points* and *lines* objects it is similar). A *regions* object is given as (a pointer to) a record whose last component is an array *elem*. One can dynamically allocate storage to represent *regions* objects of any desired size. The array serves as a storage pool for three different kinds of nodes representing *halfsegments*, *faces*, or *holes*, respectively. Halfsegments are organized in an AVL-tree to allow updates in $O(\log n)$ time. Additional pointers connect all halfsegments within the object, within a face, and within a cycle (outer cycle or hole cycle) into linked lists ordered in halfsegment order. Additionally, all faces, and for each face its holes, are linked. Hence the complete structure of a *regions* object is explicitly represented and access operations are offered (in the module hiding this representation) to perform all kinds of scans in linear time. Furthermore, bounding boxes are stored for the object, each face, and each hole. The record contains general information about the object such as the root segment of the AVL-tree, fields for *perimeter*, *diameter* and *area*; the *attr* field tells which of these values have already been computed for this particular object.

The module *ROSE* implements the operators of the executable ROSE-Algebra and is based on traversal techniques, on the plane-sweep paradigm, and on graph theory. The module *SweepStatus* manages the sweep line status for all ROSE operations with *regions* operands and uses an AVL tree for its representation. In Section 5.3.3, the graph algorithms have been described from a conceptual point of view. What really happens is that the graph structure is analysed at the time of *closing* an object, that is, after all segments have been inserted. More precisely, the construction of a *regions* object consists of the following steps:

```
TYPE
  OBJATTRIBS   = (Closed, Perimeter, Diameter, Area);
  ATTRIBSET    = SET OF OBJATTRIBS;

  COMPATTRIBS = (InsideAbove, HoleSegment);
  COMPSET      = SET OF COMPATTRIBS;

  FIELDTYPE    = (HalfsegField, FaceField, HoleField);

  SELECTTYPE   = (RegionsSelected, FaceSelected, CycleSelected);

  REGIONSELEM =
      RECORD
        CASE kind : FIELDTYPE OF
          HalfsegField:
            h                 : HALFSEGMENT; (* Key-element. *)
            attrib            : COMPSET;     (* Element status.*)
            left              : CARDINAL;    (* AVL-tree. *)
            right             : CARDINAL;
            height            : CARDINAL;
            next_in_regions : CARDINAL;      (* In order lists.*)
            next_in_face      : CARDINAL;
            next_in_cycle     : CARDINAL;
        | FaceField:
            face_bbox       : BBOX;      (* Face bounding box. *)
            first_in_face : CARDINAL; (* First halfsegment. *)
            last_in_face  : CARDINAL; (* (Help pointer.) *)
            last_in_cycle : CARDINAL; (* (Help pointer.) *)
            first_hole      : CARDINAL; (* First hole in face. *)
            last_hole       : CARDINAL; (* (Help pointer.) *)
            next_face       : CARDINAL; (* Face list. *)
        | ELSE
            hole_bbox       : BBOX;      (* Hole bounding. *)
            first_in_hole : CARDINAL; (* First Halfsegment. *)
            last_in_hole  : CARDINAL; (* (Help pointer.) *)
            next_hole       : CARDINAL; (* Hole list. *)
        END;
      END;

  REGIONS      =
      POINTER TO RECORD
        attr        : ATTRIBSET;   (* The object's status. *)
        perimeter : REAL;          (* Length of Segments. *)
        diameter  : REAL;          (* Diameter. *)
        area        : REAL;        (* Area of object. *)
        bbox        : BBOX;        (* Bounding box. *)
        count       : CARDINAL;    (* Number of faces. *)
        holes       : CARDINAL;    (* Number of holes. *)
        free        : CARDINAL;    (* Number of free fields. *)
        first_idx : CARDINAL;      (* Idx of smallest halfseg. *)
        face_idx  : CARDINAL;      (* Idx of first face. *)
        root_idx  : CARDINAL;      (* Idx of root of AVL-tree. *)
        act_idx   : CARDINAL;      (* Idx of selected halfseg. *)
        act_face  : CARDINAL;      (* Idx of selected face. *)
        sel_kind  : SELECTTYPE; (* Kind of traversal. *)
        max_idx   : CARDINAL;      (* Idx of largest half-field. *)
        act_hole  : CARDINAL;      (* Idx of selected hole. *)
        elem        : ARRAY [1..MaxInRegions] OF REGIONSELEM
      END;
```

Figure 6-11

- Allocate storage, insert n halfsegments into an AVL-tree.

To close the object:

- Perform an inorder traversal of the tree to link all halfsegments of the object; compute the bounding box.
- Use algorithm *InsideAttribute* to compute *InsideAbove* attributes.
- Use algorithm *CycleClassification* (including algorithm *InitEdgeAndNodeArray*) to attach a unique *cycle number* to each segment.
- Use a plane sweep to determine for each hole segment the cycle number of the outer cycle of its surrounding face.
- In a final scan of the complete list of segments, link all segments within faces and cycles (this is possible since each segment has now an associated cycle number and face number) and compute the remaining information such as bounding boxes, links of faces and holes, etc.

Clearly, the whole construction takes no more than $O(n \log n)$ time and $O(n)$ space. An analogous strategy is used for the more simple *lines* and *points* objects. Because all this information is now explicitly available in the data structures, the algorithms and running times for some operations change: all algorithms for operation **no_of_components** perform a simple lookup in $O(1)$ time. The algorithm for **r_contour** simply scans the list of faces and for each face the list of segments of its outer cycle which requires only $O(k \log k)$ time (where k is the size of the result object). For operators computing **diameter**, **length**, **area** and **perimeter**, only the first call takes $O(n)$ time. The value is then stored with the object so that subsequent calls are lookups in $O(1)$ time. Furthermore, some operations use filter techniques and first compare bounding boxes of objects, some in a second step also component bounding boxes, in order to avoid running the more expensive algorithms on the actual halfsegments, whenever possible. Estimating the size of the result is necessary in all operations constructing new spatial objects to allocate the appropriate amount of storage for them.

The module *ROSEAEI* makes the use of the ROSE system easier for the user by providing operations concerning database-independent input, output, storage, and retrieval of realm-based spatial objects. The user's concept of a collection of homogeneous spatial objects is simply that of a list with the usual operations. Additionally, each object can be associated with a name and thus directly accessed. A scan is used to traverse a list of objects once from the beginning to the end, and its creation is always associated with the list. Several scans may exist simultaneously on a list. During a scan the spatial object of the current element can be accessed, a spatial object can be assigned to the current element, and a name which allows direct access can be associated with the current element. File operations allow persistent and database-independent storage and retrieval of a collection of spatial objects. Such collections can be loaded from a binary file into a list, stored from a list into a binary file, generated from a specially formatted text file representation into a list, and be mapped from a list to such a text file representation.

Some remarks regarding the ROSE editor, which has still to be implemented, are made in Section 6.3.3. For further details concerning the ROSE system, the study of [Ri95] is recommended.

6.3.3 The ROSE Editor

The *ROSE editor* has not yet been designed and implemented. We plan to construct the ROSE editor (like the realm editor) on the basis of the visualization kernel so that only the application-specific parts of the editor have to be realized. Planned tasks of the editor are, for instance, to display ROSE objects graphically, to interactively construct ROSE objects by selecting either a set of line segments or points of a given realm basis with the mouse, or to select two defined geometric objects and then apply an operator of the executable ROSE algebra to them.

6.4 Integration of Realm System and ROSE System

The coupling of ROSE system and Realm system to one integrated system (Figure 6-12) has the advantage that the functionalities of both systems can be used simultaneously. Spatial objects can now be constructed under the control of the Realm system which is responsible for the maintenance of the realm properties. For this purpose, a two-way linking of both systems is necessary. Realm changes (i.e., updates caused by insertion or deletion of realm objects) in the Realm system are propagated to the corresponding associated spatial objects in the ROSE system. In the

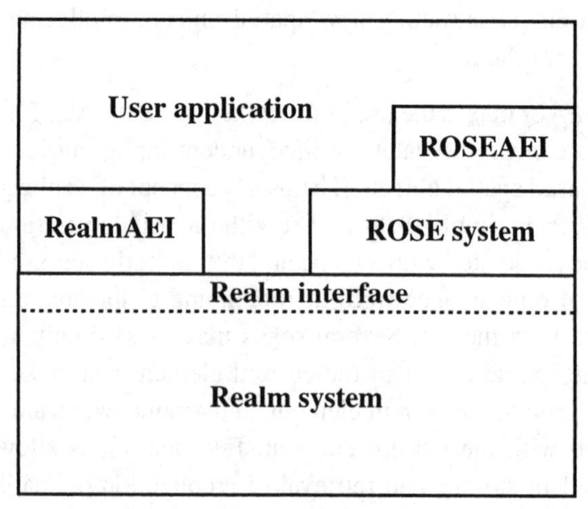

Figure 6-12

other direction, the ROSE system must inform the Realm system about the location of points and segments in spatial objects.

6.5 Related Work

The problem of the correct, robust, and efficient implementation of geometric algorithms is also topic of some other projects. Whereas in our context the topic is considered from the viewpoint of spatial database systems - spatial operations are implemented by methods of computational geometry - these projects entirely work in the field of computational geometry and are only partially about to approach the database side. Besides LEDA [MN89, MN95] and WOCG [EKMNS90] in particular the XYZ project [NSLAB91, Sc91a, Sc91b] has to be mentioned. The goal of this project is to develop correct, robust, and efficient software for a multitude of geometric standard problems (predominantly in two-dimensional space) which is applicable in practice, deals also with numerical errors and special algorithmic cases, and is designed as a collection of modules called the XYZ GeoBench. Furthermore, hints are given for increasing the reliability of a geometric library, and concepts are provided for the design of a graphical user interface supporting animation of geometric algorithms.

Chapter 7
Conclusions, Open Problems, and Future Work

7.1 Conclusions

This book started with the statement that the definition and implementation of spatial data types is probably the most fundamental issue in the development of spatial database systems. Spatial data types provide a fundamental abstraction for modelling the geometric structure of spatially-referenced objects and their spatial relationships, properties, and operations. After the formulation of general design criteria for the modelling of spatial data types, within the framework of this book (and three earlier publications) we in three steps introduced the ROSE algebra as a system of spatial data types and operations (i.e., as a spatial algebra) fulfilling these design criteria. The three steps are: (1) the concept of a *realm* [GS93] as a discrete geometric basis realized within the *Realm system*, (2) the formal definition of the *ROSE algebra* itself offering realm-based spatial data types and operations [GS95], and (3) the *ROSE system* [GRS95] as an implementation of the ROSE algebra, which realizes its types and operations by providing efficient data structures and algorithms defined over a discrete grid.

The realm concept solves several problems related to spatial data types for database systems and has been realized within the Realm system [Sc95]. In particular, this concept solves the problems of numerical robustness and topological correctness below and within the realm layer, enforces geometric consistency of related spatial objects, and enables one to formally define quite general spatial data types or algebras on top of it that enjoy nice closure properties not only in theory but also in computational practice. Starting from error-free integer arithmetic over robust geometric primitives to realms and realm-based structures and primitives, a precise formal framework has been developed that makes it easy to define spatial algebras and to implement them correctly.

The ROSE algebra has a number of interesting features: It (i) offers (values of) data types of a very general structure, (ii) has a complete formal definition of the semantics of types and operations, (iii) has realms as a discrete geometric basis which allows for a correct and robust implementation of types and operations in terms of integer arithmetic, (iv) treats consistency between distinct spatial objects with common parts, and (v) has a general object model interface which allows it to cooperate with different kinds of database systems.

In the following, some complementary remarks are made regarding these features. It may also be interesting to compare to the geo-relational algebra (in the sequel geo-algebra for short) [Gü88a] which was implemented in the Gral system [Gü89, BG92].

General types and operations. The ROSE algebra has very general data types to represent points, lines, and regions in the plane. For example, it is now possible to represent the whole area of a state including islands or separate land areas in a single *regions* object, or a complete highway network in a single *lines* object. On the one hand, this generality makes the spatial objects and operations conceptually more difficult, requires a quite elaborate system of definitions, and also needs more effort in the implementation. This is why in the geo-algebra a decision was made to deal only with simple polygons and single-component objects. On the other hand, the generality is needed in applications (in the Gral system this became obvious when the German state of Niedersachsen had to be represented which encloses - as a hole - the state of Bremen and has besides its main area several offshore islands in the North Sea). This complexity has been managed through the several definition layers of the ROSE algebra. Apart from the better capability to model spatial objects, an important benefit is that the types are now closed under set operations of the underlying point sets - for any type one can form union, difference (**plus**, **minus**) or aggregate over its values (**sum**) which makes the rather complex fusion operation (see Section 2.2.2) a simple by-product of grouping. Besides, all operations are now defined in the most general way (e.g., the **closest** operation is available for all spatial data types). In contrast, in the geo-algebra it was not possible to define a union or difference operator on regions since it would have led to holes, and intersection had to be defined as a relation operation because a resulting set of intersection objects could not be represented as a single spatial object.

Rigorous definition. The carrier sets of the types and the semantics of all operations have been defined completely, down to the level of simple arithmetic primitives on integers. As a result, there is no ambiguity for a programmer about the precise meaning of operations or about allowed structures. This is very important because when dealing with complex spatial structures inevitably questions about special cases come up like “Is it allowed that the boundary of a hole in a region touches the outer boundary?” or “Qualify two adjacent regions as intersecting?” The ROSE algebra definition gives precise answers to all such questions to an implementor and, if not to end users, at least to people writing manuals for end users.

Numerical robustness, finite resolution. The underlying realm provides the ROSE algebra with a discrete basis and shields it from all problems of numerical robustness. Integer coordinates are used for the representation of objects of spatial data types. Critical operations like testing whether points lie on the border of regions or whether two regions have segments in common become feasible. In contrast, in the geo-algebra operations like **meets** or **common_border** were omitted, because - with real numbers representing coordinates of spatial objects - it was not clear how these operations could be implemented in a numerically robust way.

Data model independence, clean object model interface. The ROSE algebra is not tied to any particular DBMS data model but can cooperate with many models and query languages. This might have been achieved in a trivial way by omitting all operations manipulating database objects (like **closest**, **overlay**) and not caring how the results of geometric operations can be used in the DBMS. Instead, we have defined an object model interface and investigated quite carefully the issues arising with the integration of the ROSE algebra into a query language. It has been demonstrated that an integration with, for example, an object-oriented model and query language can be achieved. To the author's knowledge, this is the first time that the problem of interfacing a general purpose query language with a complex application-specific sublanguage has been examined in some detail. Such interfaces will be important for cooperative database systems using external *computation services* [SW91, SW93].

Within the ROSE system [Ri95] the almost complete implementation of the first three groups of operators of the ROSE algebra (only the **dist** operator is missing) has been performed which deals with "atomic" objects (whereas the fourth group manipulates sets of database objects). The fact that ROSE objects are realm-based has led to relatively simple, efficient, and numerically robust algorithms. All manipulations of spatial objects are discrete (entirely based on integer arithmetic). Real numbers occur only to describe properties such as length or area of objects. The main algorithmic techniques used are simple or parallel traversal of objects, plane-sweep, and graph theory. A crucial concept is the use of ordered point or halfsegment sequences as a base representation of spatial objects. Manipulation of such sequences in parallel traversal or plane sweep implements most operations efficiently. On the other hand, we have also shown how the structure of objects (faces, holes, etc.) can be determined by graph algorithms and be represented in the data structures. The use of high-level primitives has made it possible to describe a relatively large number of algorithms in compact, precise notation. The author is not aware of any similar work - treating a whole algebra in a uniform way by giving precise algorithms including analysis of their complexity.

7.2 Open Problems

Some problems remain and need to be further investigated:

Topological correctness. Although it goes a long way, the approach of Greene and Yao does not completely guarantee topological correctness. As is also stated in [GY86], through the finite representation "... disjoint points and lines may collapse. However, aside from such degeneracies, we do guarantee that topology does not change." There has been a lot of work on numerical robustness and topological correctness for geometric computation (see Section 2.4). The approach of Greene and Yao has been selected because it fits well with the idea of realms as grid-based planar graphs underlying spatial data types. However, one might try to avoid the remaining anomalies by introducing integrity constraints. An example of such an integrity constraint is the rule that R-points must not lie on envelopes. The intuition behind this is that points that are so close to a segment are meant to lie on the segment. It would be the task of the update operations to maintain this constraint by redrawing a segment whenever a point is discovered to lie on the envelope (which can happen on point insertion or on segment insertion). But an examination of the consequences of this rule at the realm level and at the ROSE algebra level reveals that this rule poses many new and especially difficult problems. Further research is needed to find a set of consistent integrity constraints preserving the properties of the realm concept and the ROSE algebra and eliminating the remaining topological anomalies. The use of techniques from other approaches (e.g., symbolic reasoning) could also be examined.

Space overhead. By redrawing, many more segments may be created than were present in the original set of intersecting line segments. How many more is an interesting question that should be studied theoretically as well as in experiments with "real life" data. In any case, in the author's opinion one cannot trade correctness for space.

Invariance under redrawing. An unsatisfactory fact is that some of the numerical ROSE operations (**length**, **area**, etc.) yield slightly different results before and after a redrawing due to an update of the realm. Whereas slight numerical errors seem to be tolerable in contrast to topological errors, this may also lead to "discrete errors". For example, when a collection of spatially-referenced objects is sorted by area of its regions, the order may change through a realm update. Perhaps a definition of these operations can be found that is sufficiently consistent with the geometry of the objects, but invariant under redrawing.

Objects and operations violating realm closure, multiple realms. One is also interested in spatial objects that are not part of the given realm. For example, it should be possible to interactively draw a region and then to use it in a query. The new region cannot directly be compared with realm-based objects. One possible strategy might be to insert this region temporarily into the realm and to remove it again when the query has been processed. There may be other solutions. In this book, we have only discussed the case of a single realm underlying all spatial data of an application space, and we have restricted our attention to operations that are closed with respect to this underlying realm. But there are also interesting operations that refer to several realms and/or that leave the given realm, for example, the construction of a Voronoi diagram, a convex hull, or a buffer zone around a spatial object. One should study how these can be accommodated. One strategy might be to dynamically create a new realm containing just the new spatial objects, select a set of spatial data type objects in the database that might interact with these new geometries and create another "small" realm for them, and then use a "merge" operation on realms to compute all intersections correctly. This leads to multiple realms over the same area. Another reason why one might be interested in several realms is to reduce space overhead (by not intersecting spatial objects of different realms).

Implementation of the ROSE algebra. Algorithms for the **dist** operations and the set-manipulating operations of the fourth group of ROSE operators are still missing.

7.3 Future Work

7.3.1 Spatial Type Extension Packages

One focus will be on associating the ROSE algebra with different DBMS data models fulfilling some minimal requirements. The plan is to study, design, and implement general abstract "algebra interfaces" between extensible database systems and so-called external "extension packages". The ROSE algebra shall be equipped with such interfaces and be modified to a *Spatial Type Extension Package* (*STEP*). At the modelling level this corresponds to the concept of the object model interface (OMI) which enables the ROSE algebra to cooperate with different DBMS data models and query languages. A characteristic feature of current extensible database systems (e.g., [Gü89, SPSW90, SR86]) is that they have their own interfaces and extension mechanisms. Our goal is to define general interfaces which make as little assumptions as possible about the implementation of the DBMS intending to cooperate with the ROSE algebra. By pursuing this strategy, STEP could be coupled with different extensible systems.

Figure 7-1 illustrates two kinds of possible coupling mechanisms. The abbreviations in the figure have the following meaning: STIP = Spatial Type Integration Package, II = Integration Interface, GEOS = Geo-Server, NI = Network Interface.

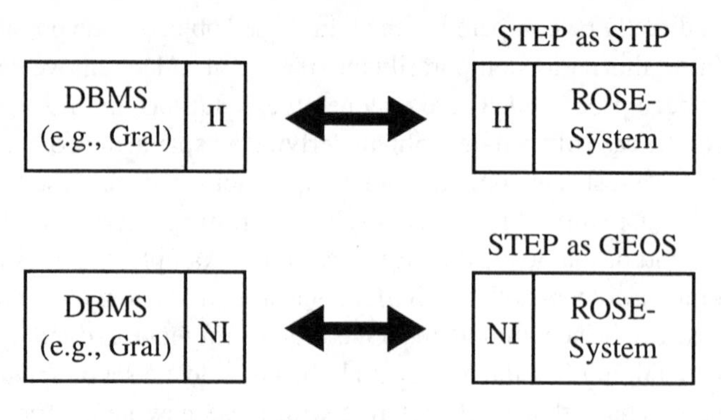

Figure 7-1

By applying the first kind of coupling mechanism, STEP as *STIP* (*Spatial Type Integration Package*), database system and ROSE system are linked together in a way similar to the integration and extension strategy as it is used for example in the Gral system. But while there the ROSE system would be embedded into and woven with the database system, here a general interface called the *integration interface* is used so that a clearer separation of both systems is achieved. Although within the Gral system extensions are organized very systematically, the addition of data types, operators, index structures, and optimization rules occurs at different places within the system.

The STIP coupling mechanism could now work as follows: The ROSE system provides a descriptive text file whose format is universally valid for extension packages. This description gives information for instance about existing types and operators and the names of corresponding implemented data types and procedures, about special index structures contained in the extension package, about cost functions for the operators, and perhaps about optimization rules. Hence, the description of extensions of any kind is now centralized in one text file. Within its integration interface the DBMS contains a procedure which can read such a descriptive file, which can make the corresponding entries into the system catalogues, which modifies the import lists of modules, etc. The database system itself must also provide a descriptive file which names the services that can be used by the extension package (e.g., providing memory pages for the construction of ROSE object representations). After the extension package has read this description and made corresponding entries into own management structures, both systems can be compiled and linked together.

The second kind of coupling mechanism, STEP as *GEOS* (*Geo-Server*), views STEP as an external, database-independent geo-server used by the DBMS as a computation service for performing geometric calculations. The database system itself is also understood as a service in the form of a database server whose functions can be called for instance by an application program. The approach of a geo-server represents the most general, most flexible, most consistent and at the same time most ambitious case of a cooperation between a geometric component and a database system, because it in principle allows heterogeneity of computer systems, system software, and programming languages. Application program, database system, and GEOS form autonomous subsystems that run as independent processes and are connected with each other by a communication network. Database system and GEOS communicate over a *network interface*.

In this context there are, of course, many open problems that have to be scrutinized. Some of them are enumerated and briefly sketched in the following:

- How must a geometric system like GEOS be structured and organized to allow a cooperation with a DBMS in a way which is as simple as possible (architecture of a geo-server)? How is extensibility of a geo-server achieved with respect to new types, operators, etc.?
- What are the communication mechanisms between GEOS and DBMS? What is the minimal information that database server and geo-server have to exchange for a cooperation? What does the network interface look like?
  Here the question is posed concerning fundamental communication and operation models for the planned network. Communication models describe mechanisms that are needed for the mutual understanding of different services with respect to the exchange of data. Operation models describe how these data are made available and processed, respectively, by a service. Within a communication network a protocol for the data exchange and for the call of service functions has to be designed. Its essential goal is to define fundamental communication structures and rules that are valid for all services participating in the network.
- Where are the spatial operations executed?
  The obvious answer seems to be that spatial operations are executed within the geo-server. But a principle of query processing is that data movements should be minimized because they are expensive. Besides, before a data movement frequently a restructuring of the data is necessary. Hence, a strategy might be not to transport the data to the geo-server but the programs containing operator implementations to the database server and to apply them there to the spatial data.
- Should functions of the geo-server be called directly by the application programs or always indirectly by interposing the DBMS?
- Where in the architecture of the whole system are spatial storage and index structures located? Are they part of STEP or part of the DBMS? What do generally applicable spatial storage and index structures look like? What does an interface for the integration of new spatial storage and index structures look like?
- Which role does parallelism play in the whole system? Can, for instance, query evaluations be executed in both systems partially in parallel or has the DBMS to wait for the answer of the geo-server? What does the architecture of the geo-serv-

er look like if at the same time different or equal functions of the geo-server are claimed by several database users or application programs?

To the author's knowledge, the approach of a loose coupling of application program, database system, and geometric computation service by three independent processes together with the sketched open problems has so far been examined only by a few researchers. The author is aware of the works presented in [SW91, SW93] and of an attempt to couple the XYZ GeoBench [NSLAB91, Sc91a, Sc91b] to a spatial database system on the basis of DASDBS [DSW90, PSSWD87, SPSW90].

7.3.2 Vague Spatial Objects

Nearly all publications in the field of spatial database systems and spatial data modelling (and also this book) implicitly assume that the extent and hence the *boundary* of spatial objects is precisely determined and universally recognized. This leads exclusively to *exact object models*. Spatial objects are represented by precisely described points, lines, and regions in a defined reference frame. The properties of the space at the points, along the lines, or within the regions are given by attributes whose values are assumed to be constant over the total extent of the objects. Examples are especially man-made spatial objects representing engineered artifacts (like highways, roads, houses, and bridges) and some predominantly immaterial spatial objects exerting social control (like countries and districts with their political and administrative boundaries or land parcels with their cadastral boundaries). We denote this kind of entities as *determinate spatial objects*.

Increasingly, researchers are beginning to realize that there are many spatial objects in reality which do not have sharp boundaries or whose boundaries cannot be precisely determined. Examples are natural, social, or cultural phenomena like land features with continuously changing properties (such as population density, soil quality, vegetation), oceans, biotopes, deserts, an English speaking area, or mountains and valleys. The transition between a valley and a mountain usually cannot be exactly determined so that the two spatial objects "valley" and "mountain" cannot be precisely separated and defined. Frequently, the indeterminacy of spatial objects is associated with temporal changes; for example, clouds and sandbanks dynamically change their shapes in the course of time. We denote this kind of entities as *vague* or *indeterminate spatial objects*.

Only little research has so far been performed for vague spatial concepts and vague spatial data types. In the sequel we will briefly give some reasons for this and outline the state-of-the-art in this research area.

So far, in spatial data modelling boundaries are considered as sharp lines that represent abrupt changes of spatial phenomena and that describe and thereby distinguish regions with different characteristic features. The assumption of crisp boundaries harmonizes very well with the internal representation and processing of spatial objects in a computer which requires precise and unique internal structures. Hence, in the past, there has been a tendency to force reality into determinate objects.

In practice, however, there is no apparent reason for the whole boundary of a region to be sharp or to have a constant degree of vagueness. There are a lot of geographical application examples illustrating that the boundaries of spatial objects can be indeterminate. For instance, boundaries of geological, soil, and vegetation units (see for example [Al94, Bu96, KV91, LAB96, WH96]) are often sharp in some places and vague in others; many human concepts like "the Indian Ocean" are implicitly vague.

The treatment of spatial objects with indeterminate boundaries is especially problematic for the computer scientist who is confronted with the difficulties how to model such objects in his database system so that they correspond to the user's intuition, how to finitely represent them in a computer format, how to develop spatial index structures for them, and how to draw them. He is accustomed to the abstraction process of simplifying spatial phenomena of the real world through the concepts of conventional binary logic, reduction of dimension, and cartographic generalization to precisely defined, simply structured, and sharply bounded objects of Euclidean geometry like points, lines, and regions.[20]

In reality, there are essentially two categories of *indeterminate boundaries*: sharp boundaries whose position and shape are unknown or cannot be measured precisely, and boundaries which are not well-defined or which are useless (for example, between a mountain and a valley) and where essentially the topological relationship between spatial objects is of interest.

Spatial objects with indeterminate boundaries are difficult to model and are so far not supported in spatial database systems. Only a few formal modelling approaches can be found in the literature. According to the two categories of boundaries, two kinds of vagueness or indeterminacy concerning spatial objects have to be distinguished: *Uncertainty* relates either to a lack of knowledge about the position and shape of an object with an existing, real border (*positional* uncertainty) or to the inability of measuring such an object precisely (*measurement* uncertainty). *Fuzziness* is an intrinsic feature of an object itself and describes the vagueness of an object which certainly has an extent but which inherently cannot have or does not have a precisely definable border.

The subject of modelling spatial vagueness has so far exclusively and only informally been treated by geographers but rather neglected by computer scientists. Coucle-

[20] Ironically speaking, this abstraction process itself mapping reality onto a mathematical model implicitly introduces a certain kind of vagueness and imprecision.

lis [Co96] has given a first attempt of a taxonomy of vague spatial objects. In the literature at least three alternatives are proposed as general design methods:

- *fuzzy models* [Al94, Ba93, Bu96, Du91, Ed94, HB93, KV91, LAB96, LGL92, LL93, Us96, Wa94, WH96, WHS90] which are all based on fuzzy set theory and predominantly model fuzziness,
- *probabilistic models* [Bl84, Bu96, Fi93, Sh93] which are based on probability theory and predominantly model positional and measurement uncertainty, and
- *exact models* [CF96, CG96, Sc96, ES97] which transfer data models, type systems, and concepts for spatial objects with sharp boundaries to spatial objects without clear boundaries and which predominantly model uncertainty but also aspects of fuzziness.

Fuzzy sets were first introduced by Zadeh [Za65] to treat imprecise concepts in a definable way. Fuzzy set theory is an extension or generalization of classical boolean set theory and deals only with fuzziness, not with uncertainty. Fuzziness is not a probabilistic attribute, in which the grade of membership of an individual in a set is connected to a given statistically defined probability function. Rather, it is an admission of the possibility that an individual is a member of a set or that a given statement is true. Examples of fuzzy spatial objects include mountains, valleys, biotopes, oceans, and many other geographic features which cannot be rigorously bounded by a sharp line.

Probability theory can be used to represent uncertainty. It defines the grade of membership of an entity in a set by a statistically defined probability function. Examples are the uncertainty about the spatial extent of particular entities like regions defined by some property such as temperature, or the water level of a lake.

The main difficulty of fuzzy and probabilistic models is that their use with spatial data is still a non-trivial application. On the one hand, our current computational technology does not allow efficient processing of uncertain and fuzzy spatial data. On the other hand, it is an open problem how to integrate and transform these models into the concept of spatial data types.

A benefit of the exact object model approach is that existing definitions, techniques, data structures, algorithms, etc., need not be redeveloped but only modified and extended, or simply used. The currently proposed exact methods for designing vague spatial objects model only *vague regions* and do not deal with *vague points* and *vague lines*. Three proposals use some kind of *zone* concept, either without holes [CF96, CG96] or with holes [Sc96]. The central idea is to consider determined zones surrounding the indeterminate boundaries of a region and expressing its minimal and maximal extension. The zones serve as a description and separation of the space that certainly belongs to the region, the space that possibly belongs to the region, and the space that is certainly outside.

While [CF96] and [CG96] are mainly interested in classifications of topological relationships between vague regions for which a simple model is assumed, [Sc96] proposes a model of complex vague regions with vague holes and focusses on their formal definition. Unfortunately, the three approaches are limited to "concentric" object models and have problems with geometric closure properties.

The model described in [ES97] is much more general and much simpler than the approaches suggested so far. It offers a formal definition of vague regions together with basic operations and predicates. Determinate regions are subsumed as a special case of vague regions. A *vague region* is modelled as a pair of disjoint (determinate) regions. The first region, called the *kernel,* describes the determinate part of the vague region, that is, the area which definitely and always belongs to the vague region. The second region, called the *boundary,* describes the vague part of the vague region, that is, the area for which we cannot say with any certainty whether it or parts of it belong to the vague region or not. *Maybe* the boundary or parts of it belong to the vague region, *maybe* this is not the case. Or we could say that this is *unknown*. The semantics of the boundary of a vague region is not fixed by this model but depends on the meaning the application associates with it.

The operations and predicates on vague regions allow a more fine-grained investigation of spatial situations than in the pure exact case. Vague spatial operations essentially comprise union, intersection, difference, and complement of vague regions. They fulfil geometric closure properties, that is, the result of each operation yields a vague region as well. Vague spatial predicates consider different kinds of indeterminate *intersect* and *inside* relationships.

Future research can relate to extensions along several different lines. First, the presented concept of vagueness could be extended to a formal definition of data types and operations for *vague points* and *vague lines.* This also implies operations concerning objects of different vague types and thus leads to a *vague spatial algebra.* Second, another direction of extension is the notion of vagueness itself. As yet, there is only one kind of vagueness, but there are many applications which can be best described by taking different degrees of vagueness (e.g. zones of decreasing pollution) into account. Third, there is the issue how vague spatial data types can be integrated into other data models and query languages. An integration might cause some trouble, since it either requires a redefinition of the data types or a redefinition and duplication of operations. The reason is that the vagueness of spatial objects affects the domains of standard data types like booleans and reals (for more details, see [ES97]). Fourth, it is certainly interesting to investigate fuzzy and probabilistic concepts for modelling vague spatial objects.

7.3.3 Spatiotemporal Objects

Current spatial applications increasingly pose the requirement of modelling and analysing the evolution of spatial phenomena through time. The reason for this is that dynamic changes of spatial objects over time are ubiquitous in reality. Examples are the trajectories of animals like birds and whales, orbits of satellites, radar monitoring of air traffic, changes of the world map, ecological applications like glacier or volcano observation, multimedia applications, and many more. Researchers have recognized for a long time that the concepts of space and time are deeply interconnected and that the perception of time to a large extent depends on spatial changes. But in database research this knowledge has so far not resulted in an integrated view and treatment of space and time.

Up to now, spatial database systems and geographical information systems have predominantly modelled only static but no dynamic (that is, temporal) aspects of spatial phenomena. Temporal database technology has only managed to offer appropriate temporal concepts for standard applications but failed to model changes in spatial applications as an example of non-standard applications. Hence, current efforts attempt to achieve an appropriate kind of interaction and synergy between both subareas of database research. The notion *spatiotemporal* is used to indicate the simultaneous support of space and time, in one or more dimensions.

A first glance reveals that it does not suffice to simply merge both technologies but that things are more complicated and that new concepts are indispensable. This conclusion necessitates new spatiotemporal data models, data types, operations, query languages, data structures, algorithms, storage structures, and indexing techniques.

The research on *spatiotemporal data types* is currently at the beginning. The aspect how to represent spatiotemporal objects and spatial changes conceptually and in data structures is an equally troublesome and open issue. In [Wo94] spatiotemporal objects are defined as simplicial complexes having both spatial and temporal extent. Furthermore, operations like union, intersection, difference, spatial projection and temporal projection are provided on these objects. Some few papers relate to implementation issues. [Kä94] and [RYG94] adopt the snapshot view and investigate the problem of representing spatial changes of regions over time. In [PD95] a time-based spatial data model and its realization are discussed; unfortunately, this approach has a low-level view of data. Relationships between spatiotemporal objects have so far not been considered.

Bibliography

[Ab87] R. Abler. The National Science Foundation Center for Geographic Information and Analysis. *Int. Journal of Geographical Information Systems*, vol. 1, no. 4, pp. 303-326, 1987.

[Ab89] D.J. Abel. SIRO-DBMS: A Database-Toolkit for Geographical Information Systems. *Int. Journal of Geographical Information Systems*, vol. 3, no. 2, pp. 103-116, 1989.

[AHU83] A.V. Aho, J.E. Hopcroft & J.D. Ullman. *Data Structures and Algorithms*. Addison-Wesley, 1983.

[Al61] P. Alexandroff. *Elementary Concepts of Topology*. Dover Publications, 1961.

[Al90] D. Alves. A Data Model for Geographic Information Systems. *Proc. of the 4th Int. Symp. on Spatial Data Handling*, vol. 2, pp. 879-887, 1990.

[Al94] D. Altman. Fuzzy Set Theoretic Approaches for Handling Imprecision in Spatial Analysis. *Int. Journal of Geographical Information Systems*, vol. 8, no. 3, pp. 271-289, 1994.

[An89] L. Anselin. What is Special about Spatial Data? Alternative Perspectives on Spatial Data Analysis. Technical Paper 89-4, National Center for Geographic Information and Analysis, 1989.

[Ap85] H.J. Appelrath. GEO – Concept of an Application-Neutral Geographical DB-System and its Implementation as INGRES Front-end. (In German). *Proc. of the GI-Fachtagung "Datenbanksysteme für Büro, Technik und Wissenschaft"*, Informatik-Fachbericht 94, Springer, pp. 476-486, 1985.

[Ar83] M.A. Armstrong. *Basic Topology*. Springer Verlag, 1983.

[AS90] W.G. Aref & H. Samet. An Approach to Information Management in Geographical Applications. *Proc. of the 4th Int. Symp. on Spatial Data Handling*, pp. 589-598, 1990.

[AS91] W.G. Aref & H. Samet. Extending a DBMS with Spatial Operations. *Proc. of the 2nd Int. Symp. on Advances in Spatial Databases (SSD'91)*, Springer-Verlag, LNCS 525, pp. 299-318, 1991.

[Ba89] F. Bancilhon. Query Languages for Object-Oriented Database Systems: Analysis and a Proposal. *Proc. of the BTW* (Datenbanksysteme in Büro, Technik und Wissenschaft), Informatik-Fachbericht 204, Springer-Verlag, pp. 1-18, 1989.

[Ba93] R. Banai. Fuzziness in Geographical Information Systems: Contributions from the Analytic Hierarchy Process. *Int. Journal of Geographical Information Systems*, vol. 7, no. 4, pp. 315-329, 1993.

[BBGST86] D.S. Batory, J. Barnett, J. Garza, K. Smith, K. Tsukuda, C. Twichell & T. Wise. GENESIS: A Reconfigurable Database Management System. University of Texas at Austin, Dept. of Computer Science, Report TR-86-07, 1986.

[BCD89] F. Bancilhon, S. Cluet & C. Delobel. A Query Language for the O_2 Object-Oriented Database System. *Proc. of the 2nd Workshop on Database Programming Languages*, 1989.

[BDK92] F. Bancilhon, C. Delobel & P. Kanellakis. *The O_2Book*. Morgan-Kaufmann, 1992.

[BDQV90] K. Bennis, B. David, I. Quilio & Y. Viemont. GeoTropics Database Support Alternatives for Geographic Applications. *Proc. of the 4th Int. Symp. on Spatial Data Handling*, pp. 599-610, 1990.

[Be84] G.B. Beretta. *An Implementation of a Plane-Sweep Algorithm on a Personal Computer.* Dissertation ETH Nr. 7538, Swiss Federal Institute of Technology Zürich, 1984.

[BFPWDA95] M.L. Barja, A.A.A. Fernandes, N.W. Paton, M.H. Williams, A. Dinn & A.I. Abdelmoty. Design and Implementation of ROCK & ROLL: A Deductive Object-Oriented Database System. *Information Systems*, vol. 20, no. 3, pp. 185-211, 1995.

[BG92] L. Becker & R.H. Güting. Rule-Based Optimization and Query Processing in an Extensible Geometric Database System. *ACM Transactions on Database Systems*, vol. 17, pp. 247-303, 1992.

[Bi67] G. Birkhoff. *Lattice Theory.* American Mathematical Society Colloquium Publications, vol. 25, 1967.

[BKK84] F.W. Burton, V.J. Kollias & J.G. Kollias. Consistency in Point-in-Polygon Tests. *The Computer Journal*, vol. 27, no. 4, pp. 375-376, 1984.

[Bl84] M. Blakemore. Generalization and Error in Spatial Databases. *Cartographica*, vol. 21, 1984.

[BO79] J.L. Bentley & T. Ottmann. Algorithms for Reporting and Counting Geometric Intersections. *IEEE Transactions on Computers*, vol. C-28, pp. 643-647, 1979.

[BS77] R.R. Berman & M. Stonebraker. GEO-QUEL: A System for the Manipulation and Display of Geographic Data. *Computer Graphics*, vol. 11, no. 2, pp. 186-191, 1977.

[Bu79] W. Burton. Logical and Physical Data Types in Geographic Information Systems. *Geo-Processing*, vol. 1, pp. 167-181, 1979.

[Bu86] P.A. Burrough. *Principles of Geographic Information Systems for Land Resources Assessment.* Oxford University Press, 1986.

[Bu89] L. Buisson. Reasoning on Space with Object-Centered Knowledge Representations. *Proc. of the 1st Int. Symp. on the Design and Implementation of Large Spatial Databases* (*SSD'89*), Springer-Verlag, LNCS 409, pp. 325-344, 1989.

[Bu96] P.A. Burrough. Natural Objects with Indeterminate Boundaries. *Geographic Objects with Indeterminate Boundaries*, GISDATA Series vol. 2, Taylor & Francis, pp. 3-28, 1996.

[CAR80] R.J. Cox, B.K. Aldred & D.W. Rhind. A Relational Data Base System and a Proposal for a Geographical Data Type. *Geo-Processing*, vol. 1, pp. 217-229, 1980.

[CCR93] Z. Cui, A.G. Cohn & D.A. Randell. Qualitative and Topological Relationships in Spatial Databases. *Proc. of the 3rd Int. Symp. on Advances in Spatial Databases* (*SSD'93*), LNCS 692, pp. 296-315, 1993.

[CDFGM86] M. Carey, D. DeWitt, D. Frank, G. Graefe, M. Muralikrishna, J. E. Richardson & E.J. Shekita. The Architecture of the EXODUS Extensible Database System. *Proc. of the IEEE/ACM Int. Workshop on Object-Oriented Database Systems*, pp. 52-65, 1986.

[CF80] N.S. Chang & K.S. Fu. Query-by-Pictorial-Example. *IEEE Transactions on Software Engineering*, vol. SE-6, pp. 519-524, 1980.

[CF81] N.S. Chang & K.S. Fu. Picture Query Languages for Pictorial Data-Base Systems. *Computer*, vol. 11, pp. 23-33, 1981.

[CF96] E. Clementini & P. di Felice. An Algebraic Model for Spatial Objects with Indeterminate Boundaries. *Geographic Objects with Indeterminate Boundaries*, GISDATA Series, vol. 2, Taylor & Francis, pp. 153-169, 1996.

[CFO93] E. Clementini, P. di Felice & P. van Oosterom. A Small Set of Formal Topological Relationships Suitable for End-User Interaction. *Proc. of the 3rd Int. Symp. on Advances in Spatial Databases (SSD'93)*, Springer-Verlag, LNCS 692, pp. 277-295, 1993.

[CG82] R.W. Claire & S.C. Guptill. Spatial Operators for Selected Data Structures. *Proc. of Auto-Carto 5*, pp. 189-200, 1982.

[CG96] A.G. Cohn & N.M. Gotts. The 'Egg-Yolk' Representation of Regions with Indeterminate Boundaries. *Geographic Objects with Indeterminate Boundaries*, GISDATA Series, vol. 2, Taylor & Francis, pp. 171-187, 1996.

[CL90] A. Choi A & W.S. Luk. A Bi-Level Object-Oriented Data Model for GIS Applications. *Proc. of the IEEE Compsac*, pp. 238-244, 1990.

[Cr78] F.H. Croom. *Basic Concepts of Algebraic Topology.* Springer-Verlag, 1978.

[Co96] H. Couclelis. Towards an Operational Typology of Geographic Entities with Ill-defined Boundaries. *Geographic Objects with Indeterminate Boundaries*, GISDATA Series vol. 2, Taylor & Francis, pp. 45-55, 1996.

[CW87] K.K.J. Chan & D. White. Map Algebra: An Object-Oriented Implementation. *Proc. of the Int. Workshop on Advanced Research in Geographical Information Systems*, vol. II, pp. 127-150, 1987.

[CZ94] E.P.F. Chan & R. Zhu. QL/G - A Query Language for Geometric Data Bases. Technical Report CS-94-25, Department of Computer Science, University of Waterloo, Canada, 1994.

[Da90] J. Dangermond. A Classification of Software Components Commonly Used in Geographic Information Systems. *Introductory Readings in Geographic Information Systems*, Taylor & Francis, pp. 30-51, 1990.

[Di86] K.R. Dittrich. Object-Oriented Database Systems: the Notion and the Issues. *Proc. of the IEEE/ACM Int. Workshop on Object-Oriented Database Systems*, pp. 2-4, 1986.

[DMBCG87] U. Dayal, F. Manola, A. Buchmann, U. Chakravarthy, D. Goldhirsch, S. Heiler, J. Orenstein & A. Rosenthal. Simplifying Complex Objects: The PROBE Approach to Modelling and Querying Them. *Proc. of the BTW* (Datenbanksysteme in Büro, Technik und Wissenschaft), Springer-Verlag, pp. 17-37, 1987.

[DP90] B.A. Davey & H.A. Priestley. *Introduction to Lattices and Order.* Cambridge University Press, 1990.

[DS88] D. Dobkin & D. Silver. Recipes for Geometry & Numerical Analysis - Part I: A Numerical Study. *Proc. of the ACM Symp. on Computational Geometry*, pp. 93-105, 1988.

[DS90] D. Dobkin & D. Silver. Applied Computational Geometry: Towards Robust Solutions of Basic Problems. *Journal of Computer and System Sciences*, vol. 40, pp. 70-87, 1990.

[DSW90] G. Dröge, H.-J. Schek & A. Wolf. Extensibility in DASDBS. (In German). *Informatik Forschung und Entwicklung*, vol. 5, no. 4, pp. 162-176, 1990.

[Du91] S. Dutta. Topological Constraints: A Representational Framework for Approximate Spatial and Temporal Reasoning. *Proc. of the 2nd Symp. on*

Advances in Spatial Databases (*SSD'91*), Springer-Verlag, LNCS 525, pp. 161-180,1991.

[ECF94] M.J. Egenhofer, E. Clementini & P. di Felice. Topological Relations between Regions with Holes. *Int. Journal of Geographical Information Systems*, vol. 8, no. 2, pp. 129-142, 1994.

[Ed94] G. Edwards. Characterizing and Maintaining Polygons with Fuzzy Boundaries in GIS. *Proc. of the 6th Int. Symp. on Spatial Data Handling*, pp. 223-239, 1994.

[EF91] M.J. Egenhofer & R.D. Franzosa. Point-Set Topological Spatial Relations. *Int. Journal of Geographical Information Systems*, vol. 5, no. 2, pp. 161-174, 1991.

[EF95] M.J. Egenhofer & R.D. Franzosa. On the Equivalence of Topological Relations. *Int. Journal of Geographical Information Systems*, vol. 9, no. 2, pp. 133-152, 1995.

[EFJ89] M.J. Egenhofer, A. Frank & J.P. Jackson. A Topological Data Model for Spatial Databases. *Proc. of the 1st Int. Symp. on the Design and Implementation of Large Spatial Databases* (*SSD'89*), Springer-Verlag, LNCS 409, pp. 271-286, 1989.

[Eg89a] M.J. Egenhofer. Spatial SQL: A Spatial Query Language. Report 103, Dept. of Surveying Engineering, University of Maine, 1989.

[Eg89b] M.J. Egenhofer. A Formal Definition of Binary Topological Relationships. *Third Int. Conf. on Foundations of Data Organization and Algorithms (FODO)*, Springer-Verlag, LNCS 367, pp. 457-472, 1989.

[Eg91a] M.J. Egenhofer. Extending SQL for Graphical Display. *Cartography and Geographic Information Systems*, vol. 18, no. 4, pp. 230-245, 1991.

[Eg91b] M.J. Egenhofer. Deficiencies of SQL as a GIS Query Language. *Cognitive and Linguistic Aspects of Geographic Space*, Kluwer Academic Publishers, pp. 477-491, 1991.

[Eg91c] M.J. Egenhofer. Reasoning about Binary Topological Relations. *Proc. of the 2nd Symp. on Advances in Spatial Databases* (*SSD'91*), Springer-Verlag, LNCS 525, pp. 143-160, 1991.

[Eg91d] M.J. Egenhofer. Categorizing Topological Relationships - Quantitive Refinements of Qualitative Spatial Information. Technical Report, Department of Surveying Engineering, University of Maine, 1991.

[Eg93] M.J. Egenhofer. What's Special about Spatial? Database Requirements for Vehicle Navigation in Geographic Space. *Proc. of the ACM SIGMOD Conf. on Management of Data*, pp. 398-402, 1993.

[EG94] M. Erwig & R.H. Güting. Explicit Graphs in a Functional Model for Spatial Databases. *IEEE Transactions on Knowledge and Data Engineering*, vol. 6, no. 5, pp. 787-804, 1994.

[Eg94] M.J. Egenhofer. Spatial SQL: A Query and Representation Language. Query Languages for Geographic Information Systems. *IEEE Transactions on Knowledge and Data Engineering*, vol. 6, no. 1, pp. 86-95, 1994.

[EH90] M.J. Egenhofer & J. Herring. A Mathematical Framework for the Definition of Topological Relationships. *Proc. of the 4th Int. Symp. on Spatial Data Handling*, pp. 803-813, 1990.

[EH91] M.J. Egenhofer & J. Herring. High-Level Spatial Data Structures. *Geographical Informations Systems: Overview, Principles and Applications*, 1991.

[EH92] M.J. Egenhofer & J. Herring. Categorizing Binary Topological Relationships between Regions, Lines, and Points in Geographic Databases. Technical Report, Department of Surveying Engineering, University of Maine, 1992.

[EKMNS90] P. Epstein, A. Knight, J. May, T. Nguyen & J. Sack. A Workbench for Computational Geometry (WOCG). Technical Report, Carleton University, 1990.

[ELNR87] H.-D. Ehrich, F. Lohmann, K. Neumann & I. Ramm. A Database Language for Scientific Map Data. *Geologisches Jahrbuch*, Sonderband, 1987.

[EM88] H. Edelsbrunner & E.P. Mücke. Simulation of Simplicity: A Technique to Cope with Degenerate Cases in Geometric Algorithms. *Proc. of the ACM Symp. on Computational Geometry*, pp. 118-133, 1988.

[Er94] M. Erwig. *Graphs in Spatial Databases*. Doctoral Thesis, FernUniversität Hagen, 1994.

[ES97] M. Erwig & M. Schneider. Vague Regions. *Proc. of the 5th Int. Symp. on Advances in Spatial Databases (SSD'97)*, 1997. To appear.

[FCKSA88] W.R. Franklin, N. Chandrasekhar, M. Kankanhalli, M. Seshan & V. Akman. Efficiency of Uniform Grids for Intersection Detection on Serial and Parallel Machines. New Trends in Computer Graphics, *Proc. of the Int. Symp. on Computational Geometry*, 1988.

[FDPW97] A.A.A. Fernandes, A. Dinn, N.W. Paton & M.H. Williams. Extending a Deductive Object-Oriented Database System with Spatial Data Handling Facilities. Draft version, 1997.

[Fi93] J.T. Finn. Use of the Average Mutual Information Index in Evaluating Classification Error and Consistency. *Int. Journal of Geographical Information Systems*, vol. 7, no. 4, pp. 349-366, 1993.

[FK86] A. Frank & W. Kuhn. Cell Graphs: A Provable Correct Method for the Storage of Geometry. *Proc. of the 3rd Int. Symp. on Spatial Data Handling*, pp. 411-436, 1986.

[Fo85] A.R. Forrest. Computational Geometry in Practice. *Fundamental Algorithms for Computer Graphics*, Springer Verlag, pp. 707-723, 1985.

[Fo87] A.R. Forrest. Computational Geometry and Software Engineering: Towards a Geometric Computing Environment. State-of-the-Art in Computer Graphics, *Techniques for Computer Graphics*, Springer Verlag, pp. 23-37, 1987.

[Fo89] S. Fortune. Stable Maintenance of Point Set Triangulations in Two Dimensions. *30th Annual Symp. on Foundations of Computer Science*, pp. 494-499, 1989.

[Fr75] J. Freeman. The Modelling of Spatial Relations. *Computer Graphics and Image Processing*, vol. 4, pp. 156-171, 1975.

[Fr82] A. Frank. MAPQUERY: Data Base Query Language for Retrieval of Geometric Data and their Graphical Representation. *Computer Graphics*, vol. 16, pp. 199-207, 1982.

[Fr84] W.R. Franklin. Cartographic Errors Symptomatic of Underlying Algebra Problems. *Proc. of the 1st Int. Symp. on Spatial Data Handling*, pp. 190-208, 1984.

[Fr87] A. Frank. Overlay Processing in Spatial Information Systems. *Proc. of the 8th Int. Symp. on Computer-Assisted Cartography, AUTOCARTO 8*, pp. 16-31, 1987.

[Fr90] A. Frank. Spatial Concepts, Geometric Data Models and Data Structures. *Computer and Geosciences*, 1990.

[Fr91] A. Frank. Properties of Geographic Data: Requirements for Spatial Access Methods. *Proc. of the 2nd Symp. on Advances in Spatial Databases (SSD'91)*, Springer-Verlag, LNCS 525, pp. 225-234, 1991.

[Ga64] S. Gaal. *Point Set Topology*. Academic Press, 1964.

[GCKPS89] G. Gardarin, J.P. Cheiney, G. Kiernan, D. Pastre & H. Stora. Managing Complex Objects in an Extensible Relational DBMS. *Proc. of the 15th Int. Conf. on Very Large Databases*, pp. 33-44, 1989.

[Gi77] P.J. Giblin. *Graphics, Surfaces and Homology.* Chapman and Hall, 1977.

[GNT91] M. Gargano, E. Nardelli & M. Talamo. Abstract Data Types for the Logical Modeling of Complex Data. *Information Systems*, vol. 16, no. 5, 1991.

[Go90] M.F. Goodchild. Keynote Address: Spatial Information Science. *Proc. of the 4th Int. Symp. on Spatial Data Handling*, vol. 1, pp. 3-12, 1990.

[Gr78] G. Grätzer. *General Lattice Theory.* Academic Press, 1978.

[GRS95] R.H. Güting, T. de Ridder & M. Schneider. Implementation of the ROSE Algebra: Efficient Algorithms for Realm-Based Spatial Data Types. *Proc. of the 4th Int. Symp. on Advances in Spatial Databases (SSD'95)*, LNCS 951, pp. 216-239, 1995.

[GS93] R.H. Güting & M. Schneider. Realms: A Foundation for Spatial Data Types in Database Systems. *Proc. of the 3rd Int. Symp. on Advances in Spatial Databases*, Springer Verlag, LNCS 692, pp. 14-35, 1993.

[GS95] R.H. Güting & M. Schneider. Realm-Based Spatial Data Types: The ROSE Algebra. *VLDB Journal*, vol. 4, pp. 100-143, 1995.

[GSS89] L. Guibas, D. Salesin & J. Stolfi. Epsilon-Geometry: Building Robust Algorithms from Imprecise Computations. *Proc. of the SIAM Conf. on Geometric Design*, pp. 208-217, 1989.

[Gu84] A. Guttman. R-Trees: A Dynamic Index Structure for Spatial Searching. *Proc. of the ACM SIGMOD Conf.*, pp. 47-57, 1984.

[Gü88a] R.H. Güting. Geo-Relational Algebra: A Model and Query Language for Geometric Database Systems. *Proc. of the Int. Conf. on Extending Database Technology*, pp. 506-527, 1988.

[Gü88b] R.H. Güting. Modeling Non-Standard Database Systems by Many-Sorted Algebras. Fachbereich Informatik, Universität Dortmund, Report 255, 1988.

[Gü89] R.H. Güting. Gral: An Extensible Relational Database System for Geometric Applications. *Proc. of the 15th Int. Conf. on Very Large Databases*, pp. 33-44, 1989.

[Gü91] R.H. Güting. Extending a Spatial Database System by Graphs and Object Class Hierarchies. *Proc. of the Int. Workshop on Database Management Systems for Geographical Applications*, pp. 103-124, 1991.

[Gü93] R.H. Güting. Second-Order Signature: A Tool for Specifying Data Models, Query Processing, and Optimization. *Proc. of the ACM SIGMOD Conf.*, pp. 277-286, 1993.

[Gü94a] R.H. Güting. GraphDB: A Data Model and Query Language for Graphs in Databases. FernUniversität Hagen, Report 155, short version at *Proc. of the 20th Int. Conf. on Very Large Databases*, pp. 297-308, 1994.

[Gü94b] R.H. Güting. An Introduction to Spatial Database Systems. *VLDB Journal (Special Issue on Spatial Database Systems)*, vol. 3, no. 4, pp. 357-399, 1994.

[GY86] D. Greene & F. Yao. Finite-Resolution Computational Geometry. *Proc. of the 27th IEEE Symp. on Foundations of Computer Science*, pp. 143-152, 1986.

[HB93] G.B.M. Heuvelink & P.A. Burrough. Error Propagation in Cartographic Modelling Using Boolean Logic and Continuous Classification. *Int. Journal of Geographical Information Systems*, vol. 7, no. 3, pp. 231-246, 1993.

[HC91] L.M. Haas & W.F. Cody. Exploiting Extensible DBMS in Integrated Geographic Information Systems. *Proc. of the 2nd Int. Symp. on Advances in Spatial Databases* (*SSD'91*), Springer-Verlag, LNCS 525, pp. 423-450, 1991.

[He89] J.R. Herring. The Category Theory of Spatial Paradigms. Languages of Spatial Relations: Initiative Two Specialist Meeting Report, NCGIA Technical Paper 89-2, pp. 47-52, 1989.

[He90] J.R. Herring. The Definition and Development of a Topological Spatial Data System. *Photogrammetry and Land Information Systems*, pp. 57-70, 1990.

[He91] J.R. Herring. The Mathematical Modeling of Spatial and Non-Spatial Information in Geographic Information Systems. *Cognitive and Linguistic Aspects of Space*, 1991.

[HEF90] J.R. Herring, M.J. Egenhofer & A.U. Frank. Using Category Theory to Model GIS Applications. *Proc. of the 4th Int. Symp. on Spatial Data Handling*, pp. 820-829, 1990.

[HHK88] C.M. Hoffmann, J.E. Hopcroft & M.S. Karasick. Towards Implementing Robust Geometric Computations. *Proc. of the ACM Symp. on Computational Geometry*, pp. 106-117, 1988.

[Ho89] C.M. Hoffmann. The Problems of Accuracy and Robustness in Geometric Computation. *Computer*, vol. 22, no. 3, pp. 31-42, 1989.

[HS93] Z. Huang & P. Svensson. Neighborhood Query and Analysis with GeoSAL, a Spatial Database Language. *Proc. of the 3rd Int. Symp. on Advances in Spatial Databases* (*SSD'93*), Springer-Verlag, LNCS 692, pp. 413-436, 1993.

[HSH92] Z. Huang, P. Svensson & H. Hauska. Solving Spatial Analysis Problems with GeoSAL, A Spatial Query Language. *Proc. of the 6th Int. Working Conf. on Scientific and Statistical Database Management*, 1992.

[HSW89] A. Henrich, H.-W. Six & P. Widmayer. The LSD Tree: Spatial Access to Multidimensional Point- and Non-Point-Objects. *Proc. of the 15th Int. Conf. on Very Large Data Bases*, pp. 45-53, 1989.

[Hu93] Z. Huang. *Design of GeoSAL, a Database Language for Spatial Data Analysis*. PhD Thesis, Royal Institute of Technology, Environmental and Natural Resources Information Systems, Stockholm, Sweden, 1993.

[IP87] K.J. Ingram & W.W. Phillips. Geographic Information Processing Using an SQL-Based Query Language. *Proc. of the 8th Int. Symp. on Computer-Assisted Cartography*, pp. 326-335, 1987.

[JC88] T. Joseph & A. Cardenas. PICQUERY: A High Level Query Language for Pictorial Database Management. *IEEE Transactions on Software Engineering*, vol. 14, pp. 630-638, 1988.

[Ka88] W. Kainz. Application of Lattice Theory in Geography. *Proc. of the 3rd Int. Symp. on Spatial Data Handling*, pp. 135-142, 1988.

[Ka89] W. Kainz. Order, Topology, and Metric in GIS. *Proc. of the ASPRS/ACSM Annual Convention*, vol. 4, pp. 154-160, 1989.

[Ka90] W. Kainz. Spatial Relationships - Topology Versus Order. *Proc. of the 4th Int. Symp. on Spatial Data Handling*, pp. 814-819, 1990.

[Kä94] T. Kämpke. Storing and Retrieving Changes in a Sequence of Polygons. *Int. Journal of Geographical Information Systems*, vol. 8, no. 6, pp. 493-513, 1994.

[KBS91a] H.-P. Kriegel, T. Brinkhoff & R. Schneider. The Combination of Spatial Access Methods and Computational Geometry in Geographic Database Systems. *Proc. of the 2nd Int. Symp. on Advances in Spatial Databases* (*SSD'91*), Springer-Verlag, LNCS 525, pp. 5-21, 1991.

[KBS91b] H.-P. Kriegel, T. Brinkhoff & R. Schneider. An Efficient Map Overlay Algorithm Based on Spatial Access Methods and Computational Geometry. *Int. Workshop on DBMS's for Geographical Applications*, pp. 56-73, 1991.

[KEG93] W. Kainz, M.J. Egenhofer & I. Greasley. Modelling Spatial Relations and Operations with Partially Ordered Sets. *Int. Journal of Geographical Information Systems*, vol. 7, no. 3, pp. 215-229, 1993.

[KGK93] W. Kim, J. Garza & A. Keskin. Spatial Data Management in Database Systems: Research Directions. *Proc. of the 3rd Int. Symp. on Advances in Spatial Databases* (*SSD'93*), Springer-Verlag, LNCS 692, pp. 1-13, 1993.

[KK81] J.M. Keil & D.G. Kirkpatrick. Computational Geometry on Integer Grids. *Proc. of the 19th Annual Allerton Conf. on Communication, Control, and Computing*, pp. 41-50, 1981.

[KM81] U.W. Kulisch & W.L. Miranker. *Computer Arithmetic in Theory and Practice*. Academic Press, 1981.

[KM83] P. Kornerup & D.W. Matula. Finite Precision Rational Arithmetic: An Arithmetic Unit. *IEEE Transactions on Computers*, vol. C-32, pp. 378-388, 1983.

[KM85] R.G. Karlsson & J.I. Munro. Proximity on a Grid. *Proc. of the 2nd Symp. on Theoretical Aspects of Computer Science*, Springer-Verlag, LNCS 182, pp. 187-196, 1985.

[Kn81] D.E. Knuth. *The Art of Computer Programming: Seminumerical Algorithms*. Vol. 2, Addison-Wesley, 1981.

[KO88a] R.G. Karlsson & M.H. Overmars. Scanline Algorithms on a Grid. *BIT*, vol. 28, pp. 227-241, 1988.

[KO88b] R.G. Karlsson & M.H. Overmars. Normalized Divide-and-Conquer: A Scaling Technique for Solving Multi-Dimensional Problems. *Information Processing Letters*, vol. 26, pp. 307-312, 1988.

[KP76] H. Koch & H. Pieper. Zahlentheorie. Ausgewählte Methoden und Ergebnisse. VEB Deutscher Verlag der Wissenschaften, 1976.

[KV91] V.J. Kollias & A. Voliotis. Fuzzy Reasoning in the Development of Geographical Information Systems. *Int. Journal of Geographical Information Systems*, vol. 5, no. 2, pp. 209-223, 1991.

[KVW89] G.H. Kirby, M. Visvalingam & P. Wade. Recognition and Representation of a Hierarchy of Polygons with Holes. *The Computer Journal*, vol. 32, no. 6, pp. 554-562, 1989.

[KW87] A. Kemper & M. Wallrath. An Analysis of Geometric Modeling in Database Systems. *ACM Computing Surveys*, vol. 19, no. 1, pp. 47-91, 1987.

[LAB96] P. Lagacherie, P. Andrieux & R. Bouzigues. Fuzziness and Uncertainty of Soil Boundaries: From Reality to Coding in GIS. *Geographic Objects with Indeterminate Boundaries*, GISDATA Series vol. 2, Taylor & Francis, pp. 275-286, 1996.

[LC79] B.S. Lin & S.K. Chang. Picture Algebra for Interface with Pictorial Database Systems. *Proc. IEEE COMPSAC - Comp. Software & Applications Conf.*, pp. 525-530, 1979.

[Le77] W.J. LeVeque. *Fundamentals of Number Theory*. Addison-Wesley, 1977.

[LGL92] Y. Leung, M. Goodchild & C.-C. Lin. Visualization of Fuzzy Scenes and Probability Fields. *Proc. of the 5th Int. Symp. on Spatial Data Handling*, pp. 480-490, 1992.

[LL93] Y. Leung & K.S. Leung. An Intelligent Expert System Shell for Knowledge-Based Geographical Information Systems: 1. The Tools. *Int. Journal of Geographical Information Systems*, vol. 7, no. 3, pp. 189-199, 1993.

[LN87] U. Lipeck & K. Neumann. Modelling and Manipulating Objects in Geoscientific Databases. *Proc. of the 5th Int. Conf. on the Entity-Relationship Approach*, pp. 67-86, 1987.

[LP84] D.T. Lee & F.P. Preparata. Computational Geometry - A Survey. *IEEE Transactions on Computers*, vol. C-33, no. 12, pp. 1072-1101, 1984.

[LPV93] T. Larue, D. Pastre & Y. Viémont. Strong Integration of Spatial Domains and Operators in a Relational Database System. *Proc. of the 3rd Int. Symp. on Advances in Spatial Databases* (*SSD'93*), Springer-Verlag, LNCS 692, pp. 53-72, 1993.

[Ma37] H.M. MacNeille. Partially Ordered Sets. *Transactions of the American Mathematical Society*, vol. 42, pp. 90-96, 1937.

[Ma82] W. R. Mallgren. Formal Specification of Graphic Data Types. *ACM Transactions on Programming Languages and Systems*, vol. 4, no. 4, pp. 687-710, 1982.

[MC80] P.E. Mantey & E.D. Carlson. Integrated Geographic Data Bases: The GADS Experience. *Data Base Techniques for Pictorial Applications*, Springer, pp. 173-190, 1980.

[Me84] K. Mehlhorn. *Data Structures and Algorithms 3: Multidimensional Searching and Computational Geometry*. Springer Verlag, 1984.

[MF89] D.M. Mark & A.U. Frank. Concepts of Space and Spatial Languages. *Int. Symp. on Computer-Assisted Cartography (Auto-Carto 9)*, pp. 538-556, 1989.

[Mi88] V. Milenkovic. Verifiable Implementations of Geometric Algorithms Using Finite Precision Arithmetic. *Artifical Intelligence*, vol. 37, pp. 377-401, 1988.

[Mi89] V. Milenkovic. Double Precision Geometry: A General Technique for Calculating Line and Segment Intersections Using Rounded Arithmetic. *Proc. of the 30th Annual Symp. on Foundations of Computer Science*, pp. 500-505, 1989.

[MK80] D.W. Matula & P. Kornerup. Foundations of Finite Precision Rational Arithmetic. *Computing, Suppl. 2*, pp. 85-111, 1980.

[MK84] S.P. Mudur & P.A. Koparkar. Interval Methods for Processing Geometric Objects. *IEEE Computer Graphics & Applications*, vol. 4, no. 2, pp. 7-17, 1984.

[MK85] D.W. Matula & P. Kornerup. Finite Precision Rational Arithmetic: Slash Number Systems. *IEEE Transactions on Computers*, vol. C-34, pp. 3-18, 1985.

[MN89] K. Mehlhorn & S. Näher. LEDA: A Library for Efficient Data Types and Algorithms. Preliminary version, Universität des Saarlandes, 1989.

[MN95] K. Mehlhorn & S. Näher. LEDA: A Platform for Combinatorial and Geometric Computing. *Communications of the ACM*, vol. 38, no. 1, 1995.

[MO86] F. Manola & J.A. Orenstein. Toward a General Spatial Data Model for an Object-Oriented DBMS. *Proc. of the 12th Int. Conf. on Very Large Data Bases*, pp. 328-335, 1986.

[MPFDW96] V. Müller, N.W. Paton, A.A.A. Fernandes, A. Dinn & M.H. Williams. Virtual Realms: An Efficient Implementation Strategy for Finite Resolution Spatial Data Types. *Proc. of the 7th Int. Symp. on Spatial Data Handling*, Taylor & Francis, 1996.

[Mü85] H. Müller. Rastered Point Location. *Proc. Workshop on Graphtheoretic Concepts in Computer Science*, Trauner Verlag, pp. 281-293, 1985.

[Ne88] K.-H.Neumann. *A Geoscientific Database Language for User-Definable Geometric Data Types*. PhD Thesis (in German), University Braunschweig, F.R. Germany, 1988.

[NME90] G. Nagy, M. Mukherjee & D.W. Embley. Making Do with Finite Numerical Precision in Spatial Data Structures. *Proc. of the 4th Int. Symp. on Spatial Data Handling*, pp. 55-65, 1990.

[NP82] J. Nievergelt & F.P. Preparata. Plane-Sweep Algorithms for Intersecting Geometric Figures. *Communications of the ACM*, vol. 25, pp. 739-747, 1982.

[NS88] J. Nievergelt & P. Schorn. Line problems with Supra-Linear Growth. (In German). *Informatik-Spektrum*, vol. 11, no. 4, 1988.

[NSLAB91] J. Nievergelt, P. Schorn, M. Lorenzi, C. Ammann & A. Brüngger. XYZ: A Project In Experimental Geometric Computation. *Proc. of the Int. Workshop on Computational Geometry*, Springer-Verlag, LNCS 553, pp. 171-186, 1991.

[NW79] G. Nagy & S. Wagle. Geographic Data Processing. *ACM Computing Surveys*, vol. 11, no. 2, pp. 139-181, 1979.

[$O_2$93] *The O_2 User's Manual*. Version 4.1. O_2 Technology, 1993.

[OH86] S.L. Osborn & T.E. Heaven. The Design of a Relational Database System with Abstract Data Types for Domains. *ACM Transactions on Database Systems*, vol. 11, no. 3, pp. 357-373, 1986.

[OM88] J. Orenstein & F. Manola. PROBE Spatial Data Modeling and Query Processing in an Image Database Application. *IEEE Trans. on Software Engineering*, vol. 14, pp. 611-629, 1988.

[Oo90] B.C. Ooi. *Efficient Query Processing in Geographic Information Systems*. LNCS 471, Springer Verlag, 1990.

[Or90] J.A. Orenstein. An Object-Oriented Approach to Spatial Data Processing. *Proc. of the 4th Int. Symp. on Spatial Data Handling*, pp. 669-678, 1990.

[OSM89] B.C. Ooi, R. Sacks-Davis & K.J. McDonell. Extending a DBMS for Geographic Applications. *Proc. of the IEEE Int. Conf. on Data Engineering*, pp. 590-597, 1989.

[OTU87] T. Ottmann, G. Thiemt & C. Ullrich. Numerical Stability of Geometric Algorithms. *Proc. of the 3rd ACM Symp. on Computational Geometry*, pp. 119-125, 1987.

[Ov88a] M.H. Overmars. Efficient Data Structures for Range Searching on a Grid. *Journal of Algorithms*, vol. 9, pp. 254-275, 1988.

[Ov88b] M.H. Overmars. New Algorithms for Computer Graphics. *Advances in Computer Graphics*, Eurographics Seminars, Springer-Verlag, pp. 3-19, 1988.

[Ov88c] M.H. Overmars. Computational Geometry on a Grid: an Overview. *Theoretical Foundations for Computer Graphics and CAD*, Springer-Verlag, pp. 167-184, 1988.

[PC87] D.J. Peuquet & Z. Ci-Xiang. An Algorithm to Determine the Directional Relationship between Arbitrarily-Shaped Polygons in the Plane. *Pattern Recognition*, vol. 20, no. 1, pp. 65-74, 1987.

[PD95] D.J. Peuquet & N. Duan. An Event-Based Spatiotemporal Data Model (ESTDM) for Temporal Analysis of Geographical Data. *Int. Journal of Geographical Information Systems*, vol. 9, no. 1, pp. 7-24, 1995.

[Pe84] D.J. Peuquet. A Conceptual Framework and Comparison of Spatial Data Models. *Cartographica*, vol. 21, no. 4, pp. 66-113, 1984.

[Pe86] D.J. Peuquet. The Use of Spatial Relationships to Aid Spatial Database Retrieval. *Proc. of the 2nd Int. Symp. on Spatial Data Handling*, pp. 459-470, 1986.

[PE88] D.V. Pullar & M.J. Egenhofer. Toward Formal Definitions of Topological Relations Among Spatial Objects. *Proc. of the 3rd Int. Symp. on Spatial Data Handling*, pp. 225-241, 1988.

[Pe88] D.J. Peuquet. Toward the Definition and Use of Complex Spatial Relationships. *Proc. of the 3rd Int. Symp. on Spatial Data Handling*, pp. 211-223, 1988.

[PS85] F.P. Preparata & M.I. Shamos. *Computational Geometry.* Springer Verlag, 1985.

[PS93] D. Papadias & T. Sellis. The Semantics of Relations in 2D Space Using Representative Points: Spatial Indexes. *European Conf. on Spatial Information Theory* (*COSIT'93*), Springer-Verlag, LNCS 716, pp. 25-35, 1993.

[PSSWD87] H.B. Paul, H.J. Schek, M.H. Scholl, G. Weikum & U. Deppich. Architecture and Implementation of the Darmstadt Database Kernel System. *Proc. of the ACM-SIGMOD Int. Conf. on Management of Data*, pp. 196-207, 1987.

[Pu88] D. Pullar. Data Definition and Operators on a Spatial Data Model. *Proc. of the ACSM-ASPRS Annual Convention*, pp. 197-202, 1988.

[RFS88] N. Roussopoulos, C. Faloutsos & T. Sellis. An Efficient Pictorial Database System for PSQL. *IEEE Trans. on Software Engineering*, vol. 14, pp. 639-650, 1988.

[Ri95] T. de Ridder. The ROSE System. Modula-2 Program System (Source Code). Fernuniversität Hagen, Praktische Informatik IV, Software Report 1, 1995. Available as a LaTeX file for printing and/or as a compressed collection of ASCII files.

[RNLE85] I. Ramm, K. Neumann, U.W. Lipeck & H.-D. Ehrich. An Interface for Geo-Scientific Databases. (In German). Report Nr. 85-08, Institut für Informatik, Abteilung Datenbanken und Informationssysteme, TU Braunschweig, 1985.

[Ro84] J.W. van Roessel. A Relational Approach to Vector Data Structure Conversion. *Proc. of the 1st Int. Symp. on Spatial Data Handling*, vol. 1, pp. 78-95, 1984.

[Ro87] J.W. van Roessel. Design of a Spatial Data Structure Using the Relational Normal Forms. *Int. Journal of Geographical Information Systems*, vol. 1, no. 1, pp. 33-50, 1987.

[RYG94] H. Raafat, Z. Yang & D. Gauthier. Relational Spatial Topologies for Historical Geographical Information. *Int. Journal of Geographical Information Systems*, vol. 8, no. 2, pp. 163-173, 1994.

[Sa85] A. Saalfeld. Lattice Structures in Geography. *Proc. of the 7th Int. Symp. on Computer-Assisted Cartography, AUTOCARTO 7*, pp. 482-489, 1985.

[Sc68] H. Schubert. *Topology.* Allyn and Bacon, 1968.

[Sc91a] P. Schorn. *Robust Algorithms in a Program Library for Geometric Computation.* PhD Dissertation 9516, ETH Zürich, 1991.

[Sc91b] P. Schorn. Implementing the XYZ GeoBench: A Programming Environment for Geometric Algorithms. Computational Geometry: Methods, Algorithms and Applications, *Proc. of the Int. Workshop on Computational Geometry*, Springer Verlag, LNCS 553, pp. 187-202, 1991.

[Sc94] P. Schorn. Degeneracy in Geometric Computation and the Perturbation Approach. *The Computer Journal*, vol. 37, no. 1, 1994.

[Sc95a] M. Schneider. *Spatial Data Types in Database Systems*. Doctoral Thesis, FernUniversität Hagen, 1995.

[Sc95b] M. Schneider. The Realm System. Modula-2 Program System (Source Code). Fernuniversität Hagen, Praktische Informatik IV, Software Report 3, 1995. Available as a LaTeX file for printing and/or as a compressed collection of ASCII files.

[Sc96] M. Schneider. Modeling Spatial Objects with Undetermined Boundaries Using the Realm/ROSE Approach. *Geographic Objects with Indeterminate Boundaries*, GISDATA Series vol. 2, Taylor & Francis, pp. 141-152, 1996.

[SCFLM86] P. Schwarz, W. Chang, J.C. Freytag, G. Lohman, J. McPherson, C. Mohan & H. Pirahesh. Extensibility in the Starburst Database System. *Proc. of the IEEE/ACM Int. Workshop on Object-Oriented Database Systems*, pp. 85-92, 1986.

[SH91] P. Svensson & Z. Huang. Geo-SAL: A Query Language for Spatial Data Analysis. *Proc. of the 2nd Int. Symp. on Advances in Spatial Databases* (*SSD'91*), Springer-Verlag, LNCS 525, pp. 119-140, 1991.

[Sh93] R. Shibasaki. A Framework for Handling Geometric Data with Positional Uncertainty in a GIS Environment. *GIS: Technology and Applications*, World Scientific, pp. 21-35, 1993.

[SI88] K. Sugihara & M. Iri: Geometric Algorithms in Finite-Precision Arithmetic. Research Memorandum RMI 88-10, Department of Mathematical Engineering and Information Physics, University of Tokyo, 1988.

[Sm90] C.S. Smyth. A Reference Data Model for Spatial and Geographic Applications. *Proc. of the 4th Int. Symp. on Spatial Data Handling*, pp. 869-878, 1990.

[SP92] T.R. Smith & K.K. Park. Algebraic Approach to Spatial Reasoning. *Int. Journal of Geographical Information Systems*, vol. 6, no. 3, pp. 177-192, 1992.

[SPSW90] H.-J. Schek, H.B. Paul, M.H. Scholl & G. Weikum. The DASDBS Project: Objectives, Experiences, and Future Prospects. *IEEE Transactions on Knowledge and Data Engineering*, vol. 2, pp. 25-43, 1990.

[SR86] M. Stonebraker & L. Rowe. The Design of Postgres. *Proc. of the ACM-SIGMOD Int. Conf. on Management of Data*, pp. 340-355, 1986.

[SRG83] M. Stonebraker, B. Rubenstein & A. Guttmann. Application of Abstract Data Types and Abstract Indices to CAD Databases. *Proc. of the ACM/IEEE Conf. on Engineering Design Applications*, pp. 107-113, 1983.

[SS85] M. Segal & C.H. Séquin. Consistent Calculations for Solid Modeling. *Proc. of the Symp. on Computational Geometry*, pp. 29-37, 1985.

[St86] M. Stonebraker. Inclusion of New Types in Relational Data Base Systems. *Proc. of the 2nd Int. Conf. on Data Engineering*, pp. 262-269, 1986.

[SV89] M. Scholl & A. Voisard. Thematic Map Modeling. *Proc. of the 1st Int. Symp. on the Design and Implementation of Large Spatial Databases* (*SSD'89*), Springer-Verlag, LNCS 409, pp. 167-190, 1989.

[SV91] M. Scholl & A. Voisard. Object-Oriented Database Systems for Geographic Applications: An Experiment with O_2. *Proc. Int. Workshop on Database Management Systems for Geographical Applications*, pp. 239-273, 1991.

[SW91] H.J. Schek & G. Weikum. Extensibility, Cooperation, Federation of Database Systems. (In German). *Proc. of the BTW (Datenbanksysteme in Büro, Technik und Wissenschaft)*, Informatik-Fachberichte 270, Springer, pp. 38-71, 1991.

[SW93] H.J. Schek & A. Wolf. From Extensible Databases to Interoperability between Multiple Databases and GIS Applications. *Proc. of the 3rd Int. Symp. on Advances in Spatial Databases (SSD'93)*, Springer-Verlag, LNCS 692, pp. 207-238, 1993.

[Ti80] R.B. Tilove. Set Membership Classification: A Unified Approach to Geometric Intersection Problems. *IEEE Transactions on Computers*, vol. C-29, pp. 874-883, 1980.

[To90] C.D. Tomlin. *Geographic Information Systems and Cartographic Modeling*. Prentice Hall, 1990.

[Us96] E. L. Usery. A Conceptual Framework and Fuzzy Set Implementation for Geographic Features. *Geographic Objects with Indeterminate Boundaries*, GISDATA Series, vol. 2, Taylor & Francis, pp. 71-85, 1996.

[Va91] C.R. Valenzuela. Data Analysis and Modeling. *Remote Sensing and Geographical Information Systems for Resource Management in Developing Countries*, pp. 335-348, 1991.

[Vo92] A. Voisard. *Bases de données géographiques: du modèle de données à l'interface utilisateur.* PhD Thesis, University of Paris-Sud (Centre d'Orsay), 1992.

[Wa88] D. Wagner. A Method of Evaluating Polygon Overlay Algorithms. *Proc. of the ACSM-ASPRS Annual Convention*, pp. 173-183, 1988.

[Wa94] F. Wang. Towards a Natural Language User Interface: An Approach of Fuzzy Query. *Int. Journal of Geographical Information Systems*, vol. 8, no. 2, pp. 143-162, 1994.

[WB93] M.F. Worboys & P. Bofakos. A Canonical Model for a Class of Areal Spatial Objects. *Proc. of the 3rd Int. Symp. on Advances in Spatial Databases (SSD'93)*, Springer-Verlag, LNCS 692, pp. 36-52, 1993.

[WH87] T.C. Waugh & R.G. Healey. The GEOVIEW Design - A Relational Data Base Approach to Geographic Data Handling. *Int. Journal of Geographical Information Systems*, vol. 1, no. 2, pp. 101-118, 1987.

[WH96] F. Wang & G.B. Hall. Fuzzy Representation of Geographical Boundaries in GIS. *Int. Journal of Geographical Information Systems*, vol. 10, no. 5, pp. 573-590, 1996.

[WHS90] F. Wang, G.B. Hall & Subaryono. Fuzzy Information Representation and Processing in Conventional GIS Software: Database Design and Application. *Int. Journal of Geographical Information Systems*, vol. 8, no. 2, pp. 143-162, 1994.

[Wo89] A. Wolf. The DASDBS GEO-Kernel: Concepts, Experiences, and the Second Step. *Proc. of the 1st Int. Symp. on the Design and Implementation of Large Spatial Databases*, Springer Verlag, LNCS 409, pp. 67-88, 1989.

[Wo92] M.F. Worboys. A Generic Model for Planar Geographical Objects. *Int. Journal of Geographical Information Systems*, vol. 4, no. 4, pp. 369-385, 1992.

[Wo94] M. Worboys. A Unified Model for Spatial and Temporal Information. *The Computer Journal*, vol. 37, no. 1, 1994.

[WSSH88] P.F. Wilms, P.M. Schwarz, H.-J. Schek & L.M. Haas. Incorporating Data Types in an Extensible Database Architecture. *Proc. of the 3rd Int. Conf. on Data and Knowledge Bases*, pp. 180-192, 1988.

[Ya88] C.-K. Yap. A Geometric Consistency Theorem for a Symbolic Perturbation Scheme. *Proc. of the ACM Symp. on Computational Geometry*, pp. 134-142, 1988.

[Ya92] F.F. Yao. Computational Geometry. Algorithms and Complexity, *Handbook of Theoretical Computer Science*, vol. A, Elsevier Science Publishers B.V., pp. 343-389, 1992.

[Za65] L.A. Zadeh. Fuzzy Sets. *Information and Control*, vol. 8, pp. 338-353, 1965.

Appendix A
Definition of Robust Geometric Primitives

This appendix presents the formal definitions of the robust geometric primitives (informally described in Section 3.2) in terms of error-free integer arithmetic and elementary analytical geometry. Furthermore, a condition is given which allows to decide whether an available computer integer arithmetic can be used for our purposes or not.

For an *N*-point *p*, *p.x* and *p.y* denote its first and second component, respectively. Two *N*-points *p* and *q* are <u>equal</u>,

$$p = q \quad :\Leftrightarrow \quad p.x = q.x \wedge p.y = q.y \,.$$

Two *N*-segments $s_1 = (p_1, p_2)$ and $s_2 = (q_1, q_2)$ are <u>equal</u>,

$$s_1 = s_2 \quad :\Leftrightarrow \quad (p_1 = q_1 \wedge p_2 = q_2) \vee (p_1 = q_2 \wedge p_2 = q_1)$$

Let $s_1 = (p_1, p_2) = ((x_{11}, y_{11}), (x_{12}, y_{12}))$ and $s_2 = (q_1, q_2) = ((x_{21}, y_{21}), (x_{22}, y_{22}))$ be two *N*-segments. For the calculation of a possible intersection point of the two *N*-segments we use the following matrix representation where λ, μ are rational numbers (to be represented by pairs of *Integer* values).

$$\begin{bmatrix} x_{11} \\ y_{11} \end{bmatrix} + \lambda \left(\begin{bmatrix} x_{12} \\ y_{12} \end{bmatrix} - \begin{bmatrix} x_{11} \\ y_{11} \end{bmatrix} \right) = \begin{bmatrix} x_{21} \\ y_{21} \end{bmatrix} + \mu \left(\begin{bmatrix} x_{22} \\ y_{22} \end{bmatrix} - \begin{bmatrix} x_{21} \\ y_{21} \end{bmatrix} \right)$$

This leads to the following inhomogeneous linear equation system in two variables:

$$\begin{aligned} x_{11} - x_{21} &= -\lambda\,(x_{12} - x_{11}) + \mu\,(x_{22} - x_{21}) \\ y_{11} - y_{21} &= -\lambda\,(y_{12} - y_{11}) + \mu\,(y_{22} - y_{21}) \end{aligned}$$

Let $a_{11} := x_{11} - x_{12}$, $a_{12} := x_{22} - x_{21}$, $b_1 := x_{11} - x_{21}$, $a_{21} := y_{11} - y_{12}$, $a_{22} := y_{22} - y_{21}$, and $b_2 := y_{11} - y_{21}$. Then

$$\begin{aligned} a_{11}\,\lambda + a_{12}\,\mu &= b_1 \\ a_{21}\,\lambda + a_{22}\,\mu &= b_2 \end{aligned} \quad \Rightarrow \quad \begin{aligned} \lambda\,(a_{11}\,a_{22} - a_{12}\,a_{21}) &= b_1\,a_{22} - b_2\,a_{12} \\ \mu\,(a_{11}\,a_{22} - a_{12}\,a_{21}) &= b_2\,a_{11} - b_1\,a_{21} \end{aligned}$$

With $D := a_{11}\, a_{22} - a_{12}\, a_{21}$, $D_1 := b_1\, a_{22} - b_2\, a_{12}$, $D_2 := b_2\, a_{11} - b_1\, a_{21}$, and $D \neq 0$ we get

$$\lambda = \frac{D_1}{D}, \mu = \frac{D_2}{D}. \qquad \text{(EQ 1)}$$

Two N-segments intersect if $D \neq 0$ and $0 < \lambda < 1$ and $0 < \mu < 1$. Note that the situation where an end point of one segment lies on the other segment is excluded. In particular no two end points are equal. An N-segment s_1 touches an N-segment s_2 if $D \neq 0$ and $(\lambda = 0 \vee \lambda = 1)$ and $0 < \mu < 1$. Two N-segments s_1 and s_2 touch each other if s_1 touches s_2 or s_2 touches s_1. Two N-segments are parallel if $D = 0$.

For an N-segment $s = ((x_1, y_1), (x_2, y_2))$, $x\text{-}ext(s) := \{min(x_1, x_2), ..., max(x_1, x_2)\} \subseteq N$ and $y\text{-}ext(s) := \{min(y_1, y_2), ..., max(y_1, y_2)\} \subseteq N$ denote the x- and y-intervals of its bounding box. The resulting intervals are called *N-intervals*. Two N-intervals I_1 and I_2 *overlap* if $card(I_1 \cap I_2) > 1$. They are *disjoint* if $I_1 \cap I_2 = \emptyset$. Two N-segments s_1, s_2 overlap if

(i) $D = 0$
(ii) $D_1 = D_2 = 0$
(iii) $x\text{-}ext(s_1)$ and $x\text{-}ext(s_2)$ overlap $\vee$ $y\text{-}ext(s_1)$ and $y\text{-}ext(s_2)$ overlap.

If condition (iii) does not hold and for $s_1 = ((x_{11}, y_{11}), (x_{12}, y_{12}))$ and $s_2 = ((x_{21}, y_{21}), (x_{22}, y_{22}))$ the condition

$$max(x_{11}, x_{12}) < min(x_{21}, x_{22}) \vee max(x_{21}, x_{22}) < min(x_{11}, x_{12}) \vee$$
$$max(y_{11}, y_{12}) < min(y_{21}, y_{22}) \vee max(y_{21}, y_{22}) < min(y_{11}, y_{12})$$

is fulfilled, the two N-segments are called aligned.

Two N-segments $s_1 = (p_1, p_2)$ and $s_2 = (q_1, q_2)$ meet if they have exactly one end point in common. Two N-segments are disjoint if they are neither equal nor meet nor intersect nor overlap nor touch. If two N-segments $s_1 = (p_1, p_2) = ((x_{11}, y_{11}), (x_{12}, y_{12}))$ and $s_2 = (q_1, q_2)$ intersect, then intersection(s_1, s_2) is the N-point $(\bar{x}, \bar{y}) := (round\_to\_N(x_0), round\_to\_N(y_0))$ where

(i)
$$x_0 = x_{11} + \lambda\,(x_{12} - x_{11})$$
$$y_0 = y_{11} + \lambda\,(y_{12} - y_{11}) \qquad \text{(EQ 2)}$$

x_0 and y_0 are two rational numbers resulting from solving the two equations in exact rational arithmetic (to be implemented through the integer primitives alone). λ is chosen as mentioned in (EQ 1).

(ii) the function *round_to_N* rounds a rational number to the "nearest" number in N.

For the function *round_to_N* we give a simple algorithm to show that integer arithmetic is sufficient to calculate the "nearest" number in N from a positive rational number $c = \frac{a}{b}$:

> **algorithm** *round_to_N*
> **input**: Two *Integer* numbers $a, b \geq 0$
> **output**: The value of the *Integer* number z which is nearest to the quotient a/b
> **begin**
>   **if** $a \geq b$ **then** $z := a$ *div* b; $a := a$ *mod* b **else** $z := 0$ **end-if**;
>   (* Now $a < b$ so that $0 < a/b < 1$ holds *)
>   **if** $a = 0$ **then return** z **end-if**;
>   **if** $2 * a \leq b$ **then return** z **else return** $z + 1$ **end-if**
> **end** *round_to_N*.

Let $s = (p_1, p_2) = ((x_1, y_1), (x_2, y_2))$ be an N-segment and let $p = (x, y)$ be an N-point. p lies on s, for short: p <u>on</u> s, if

(i) $(x_2 - x_1)(y - y_1) + (x - x_1)(y_1 - y_2) = 0$
(ii) $x \in x\text{-}ext(s) \wedge y \in y\text{-}ext(s)$

An N-point p lies within an N-segment s, for short: p <u>in</u> s, if additionally to (i) and (ii) holds

(iii) $x \notin \{x_1, x_2\} \vee y \notin \{y_1, y_2\}$.

One can observe that the largest numbers occur in the equations (EQ 2), namely

$$x_0 = \frac{x_{11}D + D_1(x_{12} - x_{11})}{D}, \; y_0 = \frac{y_{11}D + D_1(y_{12} - y_{11})}{D},$$

which leads to the requirement that numbers in the range $[-2n^3, 2n^3]$ with $n = |N|$ should be representable by an integer arithmetic in order to be error-free with respect to overflow. Either a computer integer arithmetic or integer type of a programming language can provide this range of numbers for an application, or one needs to implement a special integer type with corresponding operations.

Appendix B

Definition Layers for Realm-Based Spatial Data Types

ROSE Algebra Operations	Objects: *points*, *lines*, *regions* Operations: =, ≠, **inside**, **edge_inside**, **vertex_inside**, **area_disjoint**, **edge_disjoint**, **disjoint**, **intersects**, **meets**, **adjacent**, **encloses**, **on_border_of**, **border_in_common**, **intersection**, **plus**, **minus**, **common_border**, **vertices**, **contour**, **interior**, **count**, **dist**, **diameter**, **length**, **area**, **perimeter**, **sum**, **closest**, **decompose**, **overlay**, **fusion**
Spatial Data Types and Spatial Algebra Primitives	Objects: *points*, *lines*, *regions* Operations: ***union***, ***intersection***, ***difference***, (***area-***)***inside***, ***edge-inside***, ***vertex-inside***, ***area-disjoint***, ***edge-disjoint***, (***vertex-***)***disjoint***, ***meet***, ***adjacent***, ***intersect***, ***encloses***, ***on_border_of***, ***border_in_common***
Realms, Realm-Based Structures and Realm-Based Primitives	Objects: *R*-point, *R*-segment; *R*-cycle, *R*-face, *R*-unit, *R*-block Operations: *on*, *in*, *out*, (*area-*)*inside*, *edge-inside*, *vertex-inside*, *area-disjoint*, *edge-disjoint*, (*vertex-*)*disjoint*, *adjacent*, *meet*, *encloses*, *intersect*, *dist*, *area*
Robust Geometric Primitives	Objects: *N*-point, *N*-segment Operations: =, ≠, meet, overlap, intersect, disjoint, on, in, touches, intersection, parallel, aligned
Integer Arithmetic	Objects: integers in the range $[-2n^3, 2n^3]$ (n integer grid size) Operations: +, -, *, **div**, **mod**, =, ≠, <, ≤, ≥, >

Appendix C

Translation of the Descriptive ROSE Operators into Executable Operators

The structure of this table is explained in Section 5.1. In column "time complexity" (TC), n denotes the total size of the operand(s), m the size of the <u>regions</u> operand (only used if there is just one), and k the size of the result object.

Descriptive Operator		Executable Operator	PT	PS	G	TC
geo × *geo* → <u>bool</u>	=	**pp_equal, ll_equal, rr_equal**	x			$O(n)$
	≠	**pp_unequal, ll_unequal, rr_unequal**	x			$O(n)$
	disjoint	**pp_disjoint, ll_disjoint**	x			$O(n)$
		rr_disjoint		x		$O(n \log n)$
geo × <u>regions</u> → <u>bool</u>	**inside**	**pr_inside, lr_inside**		x		$O(n \log m)$
		rr_inside		x		$O(n \log n)$
<u>regions</u> × <u>regions</u> → <u>bool</u>	**area_disjoint**	**rr_area_disjoint**		x		$O(n \log n)$
	edge_disjoint	**rr_edge_disjoint**		x		$O(n \log n)$
	edge_inside	**rr_edge_inside**		x		$O(n \log n)$
	vertex_inside	**rr_vertex_inside**		x		$O(n \log n)$
ext_1 × ext_2 → <u>bool</u>	**intersects**	**ll_intersects**	x			$O(n)$
		lr_intersects, rl_intersects		x		$O(n \log m)$
		rr_intersects		x		$O(n \log n)$
	meets	**ll_meets**	x			$O(n)$
		lr_meets, rl_meets		x		$O(n \log m)$
		rr_meets		x		$O(n \log n)$

Descriptive Operator		Executable Operator	PT	PS	G	TC
	border_in_common	**ll_border_in_common, lr_border_in_common, rl_border_in_common, rr_border_in_common**	x			$O(n)$
area × *area* → *bool*	**adjacent**	**rr_adjacent**		x		$O(n \log n)$
area × *area* → *bool*	**encloses**	**rr_encloses**		x		$O(n \log n)$
points × *ext* → *bool*	**on_border_of**	**pl_on_border_of, pr_on_border_of**	x			$O(n)$
points × *points* → *points*	**intersection**	**pp_intersection**	x			$O(n + k \log k)$
lines × *lines* → *points*	**intersection**	**ll_intersection**	x			$O(n + k \log k)$
regions × *regions* → *regions*	**intersection**	**rr_intersection**		x		$O(n \log n)$
regions × *lines* → *lines*	**intersection**	**rl_intersection**		x		$O(n \log m + k \log k)$
geo × *geo* → *geo*	**plus**	**pp_plus, ll_plus**	x			$O(n + k \log k)$
		rr_plus		x		$O(n \log n)$
geo × *geo* → *geo*	**minus**	**pp_minus, ll_minus**	x			$O(n + k \log k)$
		rr_minus		x		$O(n \log n)$
$ext_1 \times ext_2$ → *lines*	**common_border**	**ll_common_border, lr_common_border, rl_common_border, rr_common_border**	x			$O(n + k \log k)$
ext → *points*	**vertices**	**l_vertices, r_vertices**	x			$O(n + k \log k)$
regions → *lines*	**contour**	**r_contour**			x	$O(n \log n)$ / $O(k \log k)$
lines → *regions*	**interior**	**l_interior**			x	$O(n \log n)$
geo → *int*	**no_of_components**	**p_no_of_components**	x			$O(n)$ / $O(1)$
		l_no_of_components, r_no_of_components			x	$O(n \log n)$ / $O(1)$
$geo_1 \times geo_2$ → *real*	**dist**	**pp_dist, pl_dist, pr_dist, lp_dist, ll_dist, lr_dist, rp_dist, rl_dist, rr_dist**				
geo → *real*	**diameter**	**p_diameter, l_diameter, r_diameter**	*special algorithm*			$O(n)$ / $O(1)$

Descriptive Operator		Executable Operator	PT	PS	G	TC
lines → *real*	**length**	**l_length**	x			$O(n)$ / $O(1)$
regions → *real*	**area**	**r_area**	x			$O(n)$ / $O(1)$
	perimeter	**r_perimeter**	x			$O(n)$ / $O(1)$

Index

A

above predicate 35
adjacent predicate 34, 63
adjacent predicate 154
<u>*adjacent*</u> primitive 133, 135
adjacent primitive 143
algebraic topology 11, 50, 68
<u>aligned</u> primitive 89, 260
application environment interface 215, 216
arc 28
area data type 32, 45
area operation 35, 63
area operation 157
<u>*area*</u> primitive 139
area_disjoint predicate 154
<u>*area-disjoint*</u> primitive 133, 135
area-disjoint primitive 143
<u>*(area-)inside*</u> primitive 133, 135, 139
(area-)inside primitive 143
atom 61
attribute function 149
auxiliary point 93

B

base area 61
base segment 94
behavioural object orientation 47
behind predicate 35
below predicate 35
bent line 96
best approximation 98
blocks function 138
border operation 37
border polygon 93
border_in_common predicate 154
border_in_common primitive 143
boundary 240
  indeterminate 241
  sharp 240
boundary operation 37, 60, 63, 70, 71, 72
boundary representation 30, 55
bounding point 93
boundingFaces operation 72
box operation 37
buffer zoning 37

C

cardinality operation 36, 63
cell complex, *k*-cell complex 71, 72
cell, *k*-cell 53, 71
cellular decomposition 68, 70
cellular topology 68
center operation 37
centroid operation 63
chain 29
  simple 29
choose operation 37
clipping operation 41
close predicate 35
closed ball 53
closed disc 53
closed set 52
closest operation 42
closest operation 160
closure operation 60
closure properties. *See* design criteria
combinatorial topology 50, 68
common_border operation 155
compl operation 58
completely hooked segments 95
completeness of a spatial type system 16, 67
completeness principles 71
  completeness of incidence 71
  completeness of inclusion 72
complex object. *See* spatial object
component bounding box 210
components operation 36, 37, 56, 63
compose operation 42, 59
composition of topological relationships 65
computation service 235, 238, 240
Computational Geometry 4, 6, 31, 44, 45, 77
conceptual level/model. *See* spatial data type
connected predicate 63
connectedness 17, 56, 60
connectivity structure 14
contains predicate 34, 35, 65
context level 11, 12
continued fraction 80, 97, 99

continued fraction algorithm 100
continued fraction expansion 99
continuous topological transformation 49
contour operation 155
convergent 99
convex hull 68
convex_hull operation 36
cover operation 40
covered_by predicate 34
covers predicate 65
cross predicate 66
crossing difference vector 107
curve 28
cut operation 36
cut predicate 34
cycles function 137

D

data model independence. *See* design criteria
data normalization 81
decompose operation 42, 59
decompose operation 160
descriptive algebra 169
design criteria for spatial data types
  closure properties 3, 44, 58, 67, 72, 87, 176, 233
  convenience 16
  data model independence 4, 46, 235
  efficiency 4, 45, 176
  extensibility 4, 33, 45
  finite resolution 4, 44, 48, 77, 234
  formal definition 3, 4, 5, 43, 44, 48, 233
  generality and versatility 3, 44, 67, 73
  geometric consistency 4, 6, 45, 72, 88, 233
  minimality 16
  numerical robustness 4, 5, 11, 44, 48, 49, 73, 77, 87, 176, 233, 234
  rigorous definition 4, 44, 234
  robust implementation 3, 5
  topological correctness 4, 5, 7, 11, 44, 48, 49, 68, 73, 77, 87, 176, 233
designer's level/model. *See* spatial data type
diameter operation 35
diameter operation 157
difference operation 36, 63
difference primitive 142
difference vector 105
dimension 65, 69
dimension-extended method 66
diophantine approximation 97, 98
direction operation 36
directional relationship. *See* spatial predicate
discrete segment intersection problem 90
discrete space 83
disjoint predicate 34, 65, 66
disjoint primitive 90, 260
disjoint primitive 138
disjoint primitive 143
dist operation 36
dist operation 157
dist primitive 139
distance algorithm 208
dominating point 174
double precision geometry 82

E

edge cracking 81
edge_disjoint predicate 154
edge_inside predicate 154
edge-disjoint primitive 133, 135
edge-disjoint primitive 143
edge-inside primitive 133, 135
edge-inside primitive 143
efficiency. *See* design criteria
embedding 61
encloses predicate 154
encloses primitive 135
encloses primitive 143
envelope 82, 92, 93, 94
  left, right 93
  proper 93
envelope point 93
ep operation 35
ε-arithmetic 80
epsilon geometry 82
equal predicate 34, 55, 63, 65
equal primitive 89, 259
Euclidean geometry 68, 77
Euclidean metric 53
Euclidean space 51, 53, 68, 83
event point 183
event point schedule 183
exact object model 240
executable algebra 169
extend operation 37
extensibility
  of spatial database systems 2

See design criteria
extensible database system 45
exterior operation 60

F

face 69
face thickness 81
faces function 136, 138
far predicate 35
Farey sequence, Farey series 80, 97, 98
filter technique 209
finite precision computational geometry 11, 83, 214
finite resolution computational geometry 5, 6, 83, 214
finite resolution. *See* design criteria
finite-precision arithmetic 78
formal definition method 48
formal definition. *See* design criteria
four-intersection model, 4-intersection model 64, 73
fusion operation 39, 59
fusion operation 160
fuzziness 241
fuzzy relationship. *See* spatial predicate

G

gap polygon 96
general topology 50, 59
generality. *See* design criteria
generalization 28
generalization operation 39
generalized region 67
generic area 62
geo operation 58
geographical data 2
geographical information system 2, 28
geometric algorithm 77, 88
  correctness of a 79, 80
  efficiency of a 79
  robustness of a 79, 80
  stability of a 79, 80
geometric consistency. *See* design criteria
geometric data 1
geometric data type 2
geometric database system 1
geometric modelling 13
geometric primitive 5, 79
  See also robust geometric primitive
Geometry data type 58
geo-relational algebra 3, 32, 42, 55
geo-server model 238
Gral system 3, 48
graph algorithm 170, 235
greatest element 75
greatest lower bound 75
grid point 92

H

half plane 27
half plane segment 27
halfsegment 174, 227
  dominating point of a 174
  left, right 174
  ordered sequence of halfsegments 174
Hasse diagram 74
Hausdorff space 52
hidden variable method 81
homeomorphic space 52
hook 94
  induced 96
  original 95
hooked segment 94

I

implementation level/model. *See* spatial data type
in predicate 34, 66
in primitive 89, 261
in primitive 132
in_circle predicate 34
in_front_of predicate 35
in_window predicate 34
infinite-precision arithmetic 78
inside operation 36
inside predicate 34, 35, 55, 60, 65
inside predicate 154
integer arithmetic 7, 85, 89, 100, 233
integration interface 238
interior operation 37, 60, 70, 72
interior operation 155
interiorFaces operation 72
intersect predicate 34, 55
intersect primitive 90, 260
intersect primitive 138, 139
intersect primitive 143
intersection operation 36, 63
intersection operation 155

intersection primitive 90, 260
intersection primitive 142
intersects predicate 154
interval arithmetic 80
into predicate 35

K

kind 144

L

lattice 74, 76
  complete 76
least element 75
least upper bound 75
left predicate 35
length operation 35
length operation 157
length primitive 139
line 28
  vague 242
line segment 29, 55, 92
linear order. *See* order
lines algebra 175
lines data type 5, 7, 87, 131, 138, 141, 153, 227
lower bound 75

M

many-sorted algebra 144
many-sorted signature 144
maxdist operation 36
maximal best vector 104
mediant 98
meet predicate 34, 56, 60, 65
meet primitive 89, 260
meet primitive 133, 135, 138, 139
meet primitive 143
meeting point 138, 178
meets predicate 154
merge operation 39
metric 52
metric relationship. *See* spatial predicate
metric space 51, 52, 53
metric topology 53
mindist operation 36
minimal best vector 104
minimum feature separation 81
minimum feature size 81
minus operation 155

N

near predicate 35
neighbour predicate 63
neighbourhood 52, 53, 63
neighbouring predicate 34
network 14
network interface 239
next_to predicate 35
nine-intersection model, 9-intersection model 65
N-interval 260
no_of_components operation 157
nodes operation 35
non-standard application 45
non-standard database system 1
normal completion 76
north predicate 35
N-point 7, 89, 122
N-rectangle 124
N-segment 7, 89, 122
numerical robustness. *See* design criteria

O

O_2SQL/ROSE 161
object aggregation function 149
object bounding box 209
object construction operation 36
object extension function 149
object model interface 4, 8, 141, 148, 234, 235
object modelling 13, 43
object transformation operation 37
on primitive 89, 261
on primitive 131
on_border_of predicate 154
on_border_of primitive 143
open ball 53
open disc 53
open line segment 92
open set 52
order
  linear 74
  partial 50, 74
  strict 50, 74
  total 74
order relation 74
order theory 11, 50, 73
ordered set

linearly 74
partially 74
strictly 74
totally 74
out primitive 132
outside predicate 34, 55
over predicate 35
overlap number 191
overlap predicate 34, 60, 63, 65, 66
overlap primitive 89, 260
overlay operation 38
overlay operation 160

P

parallel primitive 89, 260
parallel traversal algorithm 170, 177, 235
partial order. *See* order
partial quotient 99
partition 8, 14, 32, 146
perimeter operation 35, 63
perimeter operation 157
perturbation 79
perturbation approach 81
perturbation-free approach 80
plane-sweep 235
plane-sweep algorithm 6, 170, 182
plus operation 155
point 29, 52, 55
vague 242
point set 26, 49
point set theory 11, 49, 55
point set topology 11, 50, 59
points 153
points algebra 173
points data type 5, 7, 87, 141, 153, 227
polygon 28
closed 30
simple 30
polygon operation 37
polygon with holes 29, 31, 44, 55, 67, 73, 131, 133
poset. *See* order
properties of spatial data
connectedness 17
geometric, spatial 17
metric, locational 17
non-geometric, non-spatial, aspatial, thematic 17
reference frame, reference system 17
spatial dependence 17
temporal change 17
topological 17

R

rational arithmetic 80
R-block 7, 131, 138
R-cycle 7, 131
realm 5, 7, 85, 86, 122, 153, 233
representation of a 123
stored 140
viewed as a spatial index structure 88, 130
viewed as a spatially embedded planar graph 86, 122, 196
virtual 140
realm application environment interface 216
Realm editor 9, 215, 216
realm interface 7, 85, 122, 130
realm object identifier (roid) 123
Realm system 9, 215, 216, 233
realm-based
computational geometry 6, 176
geometric algorithm 6, 169, 176
primitive 85, 131, 233
spatial algebra 153
spatial data type 87, 141
spatial object 5, 86
structure 7, 85, 131, 233
redrawing 5, 7, 81, 85, 92, 94, 103
redrawing algorithm
generalized 116
restricted 103
reduction 28
region 56
generalized 67
spatial 65
vague 242, 243
region expansion 37
regions algebra 174
regions data type 5, 7, 87, 131, 141, 153, 227
regions function 137
regular closed set 44, 60
regular set 57
regular set operation 44, 57, 61
regularity 60
regularization function 61
restriction type 147
R-face 7, 131, 134
right predicate 35

rigorous definition. *See* design criteria
robust geometric primitive 7, 85, 89, 233
robust implementation. *See* design criteria
ROSE algebra 5, 6, 8, 141, 153, 233
  descriptive 8
  executable 8, 226
ROSE application environment interface 227
ROSE editor 9, 215, 216, 230
ROSE system 9, 171, 215, 226, 233
rotate operation 37
rounding error 78
R-point 7, 122
R-segment 7, 122
R-unit 7, 131, 135

S

scalar product 80
second-order signature 144, 171
segment 92
segment classification 185, 191
segments operation 35
separation 60, 63
set-theory 58
SHAPES algebra 58
signature specification 145
simple traversal algorithm 177, 235
simplex, *k*-simplex 68
  oriented 70
simplicial complex 69, 72
  oriented 70
simplicial decomposition 68, 72
simplicial topology 68
simulation of simplicity 82
skeleton 61, 72
south predicate 35
sp operation 35
space 50
spatial algebra 6, 8, 233
spatial algebra primitive 8, 141, 143
spatial closeness 52
spatial component identifier (scid) 123
spatial concept 11, 18, 22
  entity-oriented 19
  feature-based 19
  position-based 19
  space-oriented 19
spatial data 1, 22
  See also properties of spatial data; spatial data type; spatial object
spatial data model 3, 11, 20, 22, 29
spatial data modelling 11, 12, 240
spatial data structure 11, 21, 49
  raster-oriented 22
  vector-oriented 22
spatial data type 2, 3, 22, 29, 85, 86, 148, 233
  application-independent 45
  application-specific 45
  as abstract data type 47
  as opaque type 48
  based on half planes 27
  based on point sets 26
  classification 23
  conceptual model for a 15, 22, 43
  designer's model for a 15, 22, 43, 48
  flat view of a 141
  formal definition method for a 48
  implementation model for a 15, 48, 49
  semantics of a 48
  standardization 48
  structured view of a 142
  universal 25
  user's model for a 15, 22, 43, 48
  See also realm-based spatial data type
spatial data type 25
spatial database system 1, 3, 22, 29, 233
spatial dependence 17
spatial information theory 79
spatial join 42
spatial modelling 13, 14, 22
spatial object 2, 13, 22, 55
  determinate 240
  extended structure 29
  high-level view of a 24, 25, 29, 33, 46
  indeterminate 240
  simple structure 29
  vague 240
  viewed as complex object 18, 22, 46
spatial operation 22, 29, 33, 48, 78, 233
  on sets of spatially-referenced objects 38, 153, 158
  returning a number 35, 153, 156
  returning a spatial object 36, 153, 155
  returning constituents of a spatial object 35
spatial order relationship 34
spatial predicate 34, 42
  expressing a directional relationship 35
  expressing a fuzzy relationship 35

expressing a metric relationship 34
expressing a spatial order relationship 34
expressing a strict order relationship 34
expressing a topological relationship 34, 153
spatial query language 3, 22, 29, 78
spatial region 65
spatial relationship 34, 233
spatial selection 42
spatial strict order relationship 34
spatial type extension package 4, 8, 237
spatial type integration package 238
spatial type system 29
spatially-referenced object 2, 13, 32, 38, 233
spatiotemporal data type 244
sphere 53
strict inclusion 63
strict order relationship 34
strict order. *See* order
strongly-connected predicate 63
structural object orientation 47
structure modelling 14, 38, 43
sum operation 160
superimposition operation 39
surrounding operation 37
sweep line 183
sweep line status 183

T
Thiessen diagram 41
through predicate 35
topological correctness. *See* design criteria
topological invariant 49, 50, 64
topological relationship. *See* spatial predicate
topological space 51, 52
topologically equivalent spaces 52
topology 49, 51, 52
total order. *See* order
touch predicate 34, 66
touch primitive 89, 260
touches primitive 89, 260
translate operation 37
type constructor 144
type mapping function 160
type variable 145

U
uncertainty 241
measurement 241
positional 241
under predicate 35
unequal predicate 34, 55
unequal predicate 154
union operation 36, 63
union primitive 142
unit segment 92
units function 136, 137
upper bound 75
user's level/model. *See* spatial data type

V
vector 54
geometrically independent vectors 68
linearly independent vectors 54
vector space 51, 54
vertex shifting 81
vertex_inside predicate 154
(vertex-)disjoint primitive 133, 135
(vertex-)disjoint primitive 143
vertex-inside primitive 133, 135
vertex-inside primitive 143
vertices operation 155
visibility 15, 22, 46
visualization kernel 221
Voronoi diagram 41
voronoi operation 41

W
windowing operation 41

X
xc operation 35

Y
yc operation 35

Z
zone 242

Lecture Notes in Computer Science

For information about Vols. 1–1229

please contact your bookseller or Springer-Verlag

Vol. 1230: J. Duncan, G. Gindi (Eds.), Information Processing in Medical Imaging. Proceedings, 1997. XVI, 557 pages. 1997.

Vol. 1231: M. Bertran, T. Rus (Eds.), Transformation-Based Reactive Systems Development. Proceedings, 1997. XI, 431 pages. 1997.

Vol. 1232: H. Comon (Ed.), Rewriting Techniques and Applications. Proceedings, 1997. XI, 339 pages. 1997.

Vol. 1233: W. Fumy (Ed.), Advances in Cryptology — EUROCRYPT '97. Proceedings, 1997. XI, 509 pages. 1997.

Vol 1234: S. Adian, A. Nerode (Eds.), Logical Foundations of Computer Science. Proceedings, 1997. IX, 431 pages. 1997.

Vol. 1235: R. Conradi (Ed.), Software Configuration Management. Proceedings, 1997. VIII, 234 pages. 1997.

Vol. 1236: E. Maier, M. Mast, S. LuperFoy (Eds.), Dialogue Processing in Spoken Language Systems. Proceedings, 1996. VIII, 220 pages. 1997. (Subseries LNAI).

Vol. 1238: A. Mullery, M. Besson, M. Campolargo, R. Gobbi, R. Reed (Eds.), Intelligence in Services and Networks: Technology for Cooperative Competition. Proceedings, 1997. XII, 480 pages. 1997.

Vol. 1239: D. Sehr, U. Banerjee, D. Gelernter, A. Nicolau, D. Padua (Eds.), Languages and Compilers for Parallel Computing. Proceedings, 1996. XIII, 612 pages. 1997.

Vol. 1240: J. Mira, R. Moreno-Díaz, J. Cabestany (Eds.), Biological and Artificial Computation: From Neuroscience to Technology. Proceedings, 1997. XXI, 1401 pages. 1997.

Vol. 1241: M. Akşit, S. Matsuoka (Eds.), ECOOP'97 – Object-Oriented Programming. Proceedings, 1997. XI, 531 pages. 1997.

Vol. 1242: S. Fdida, M. Morganti (Eds.), Multimedia Applications, Services and Techniques – ECMAST '97. Proceedings, 1997. XIV, 772 pages. 1997.

Vol. 1243: A. Mazurkiewicz, J. Winkowski (Eds.), CONCUR'97: Concurrency Theory. Proceedings, 1997. VIII, 421 pages. 1997.

Vol. 1244: D. M. Gabbay, R. Kruse, A. Nonnengart, H.J. Ohlbach (Eds.), Qualitative and Quantitative Practical Reasoning. Proceedings, 1997. X, 621 pages. 1997. (Subseries LNAI).

Vol. 1245: M. Calzarossa, R. Marie, B. Plateau, G. Rubino (Eds.), Computer Performance Evaluation. Proceedings, 1997. VIII, 231 pages. 1997.

Vol. 1246: S. Tucker Taft, R. A. Duff (Eds.), Ada 95 Reference Manual. XXII, 526 pages. 1997.

Vol. 1247: J. Barnes (Ed.), Ada 95 Rationale. XVI, 458 pages. 1997.

Vol. 1248: P. Azéma, G. Balbo (Eds.), Application and Theory of Petri Nets 1997. Proceedings, 1997. VIII, 467 pages. 1997.

Vol. 1249: W. McCune (Ed.), Automated Deduction – CADE-14. Proceedings, 1997. XIV, 462 pages. 1997. (Subseries LNAI).

Vol. 1250: A. Olivé, J.A. Pastor (Eds.), Advanced Information Systems Engineering. Proceedings, 1997. XI, 451 pages. 1997.

Vol. 1251: K. Hardy, J. Briggs (Eds.), Reliable Software Technologies – Ada-Europe '97. Proceedings, 1997. VIII, 293 pages. 1997.

Vol. 1252: B. ter Haar Romeny, L. Florack, J. Koenderink, M. Viergever (Eds.), Scale-Space Theory in Computer Vision. Proceedings, 1997. IX, 365 pages. 1997.

Vol. 1253: G. Bilardi, A. Ferreira, R. Lüling, J. Rolim (Eds.), Solving Irregularly Structured Problems in Parallel. Proceedings, 1997. X, 287 pages. 1997.

Vol. 1254: O. Grumberg (Ed.), Computer Aided Verification. Proceedings, 1997. XI, 486 pages. 1997.

Vol. 1255: T. Mora, H. Mattson (Eds.), Applied Algebra, Algebraic Algorithms and Error-Correcting Codes. Proceedings, 1997. X, 353 pages. 1997.

Vol. 1256: P. Degano, R. Gorrieri, A. Marchetti-Spaccamela (Eds.), Automata, Languages and Programming. Proceedings, 1997. XVI, 862 pages. 1997.

Vol. 1258: D. van Dalen, M. Bezem (Eds.), Computer Science Logic. Proceedings, 1996. VIII, 473 pages. 1997.

Vol. 1259: T. Higuchi, M. Iwata, W. Liu (Eds.), Evolvable Systems: From Biology to Hardware. Proceedings, 1996. XI, 484 pages. 1997.

Vol. 1260: D. Raymond, D. Wood, S. Yu (Eds.), Automata Implementation. Proceedings, 1996. VIII, 189 pages. 1997.

Vol. 1261: J. Mycielski, G. Rozenberg, A. Salomaa (Eds.), Structures in Logic and Computer Science. X, 371 pages. 1997.

Vol. 1262: M. Scholl, A. Voisard (Eds.), Advances in Spatial Databases. Proceedings, 1997. XI, 379 pages. 1997.

Vol. 1263: J. Komorowski, J. Zytkow (Eds.), Principles of Data Mining and Knowledge Discovery. Proceedings, 1997. IX, 397 pages. 1997. (Subseries LNAI).

Vol. 1264: A. Apostolico, J. Hein (Eds.), Combinatorial Pattern Matching. Proceedings, 1997. VIII, 277 pages. 1997.

Vol. 1265: J. Dix, U. Furbach, A. Nerode (Eds.), Logic Programming and Nonmonotonic Reasoning. Proceedings, 1997. X, 453 pages. 1997. (Subseries LNAI).

Vol. 1266: D.B. Leake, E. Plaza (Eds.), Case-Based Reasoning Research and Development. Proceedings, 1997. XIII, 648 pages. 1997 (Subseries LNAI).

Vol. 1267: E. Biham (Ed.), Fast Software Encryption. Proceedings, 1997. VIII, 289 pages. 1997.

Vol. 1268: W. Kluge (Ed.), Implementation of Functional Languages. Proceedings, 1996. XI, 284 pages. 1997.

Vol. 1269: J. Rolim (Ed.), Randomization and Approximation Techniques in Computer Science. Proceedings, 1997. VIII, 227 pages. 1997.

Vol. 1270: V. Varadharajan, J. Pieprzyk, Y. Mu (Eds.), Information Security and Privacy. Proceedings, 1997. XI, 337 pages. 1997.

Vol. 1271: C. Small, P. Douglas, R. Johnson, P. King, N. Martin (Eds.), Advances in Databases. Proceedings, 1997. XI, 233 pages. 1997.

Vol. 1272: F. Dehne, A. Rau-Chaplin, J.-R. Sack, R. Tamassia (Eds.), Algorithms and Data Structures. Proceedings, 1997. X, 476 pages. 1997.

Vol. 1273: P. Antsaklis, W. Kohn, A. Nerode, S. Sastry (Eds.), Hybrid Systems IV. X, 405 pages. 1997.

Vol. 1274: T. Masuda, Y. Masunaga, M. Tsukamoto (Eds.), Worldwide Computing and Its Applications. Proceedings, 1997. XVI, 443 pages. 1997.

Vol. 1275: E.L. Gunter, A. Felty (Eds.), Theorem Proving in Higher Order Logics. Proceedings, 1997. VIII, 339 pages. 1997.

Vol. 1276: T. Jiang, D.T. Lee (Eds.), Computing and Combinatorics. Proceedings, 1997. XI, 522 pages. 1997.

Vol. 1277: V. Malyshkin (Ed.), Parallel Computing Technologies. Proceedings, 1997. XII, 455 pages. 1997.

Vol. 1278: R. Hofestädt, T. Lengauer, M. Löffler, D. Schomburg (Eds.), Bioinformatics. Proceedings, 1996. XI, 222 pages. 1997.

Vol. 1279: B. S. Chlebus, L. Czaja (Eds.), Fundamentals of Computation Theory. Proceedings, 1997. XI, 475 pages. 1997.

Vol. 1280: X. Liu, P. Cohen, M. Berthold (Eds.), Advances in Intelligent Data Analysis. Proceedings, 1997. XII, 621 pages. 1997.

Vol. 1281: M. Abadi, T. Ito (Eds.), Theoretical Aspects of Computer Software. Proceedings, 1997. XI, 639 pages. 1997.

Vol. 1282: D. Garlan, D. Le Métayer (Eds.), Coordination Languages and Models. Proceedings, 1997. X, 435 pages. 1997.

Vol. 1283: M. Müller-Olm, Modular Compiler Verification. XV, 250 pages. 1997.

Vol. 1284: R. Burkard, G. Woeginger (Eds.), Algorithms — ESA '97. Proceedings, 1997. XI, 515 pages. 1997.

Vol. 1285: X. Jao, J.-H. Kim, T. Furuhashi (Eds.), Simulated Evolution and Learning. Proceedings, 1996. VIII, 231 pages. 1997. (Subseries LNAI).

Vol. 1286: C. Zhang, D. Lukose (Eds.), Multi-Agent Systems. Proceedings, 1996. VII, 195 pages. 1997. (Subseries LNAI).

Vol. 1287: T. Kropf (Ed.), Formal Hardware Verification. XII, 367 pages. 1997.

Vol. 1288: M. Schneider, Spatial Data Types for Database Systems. XIII, 275 pages. 1997.

Vol. 1289: G. Gottlob, A. Leitsch, D. Mundici (Eds.), Computational Logic and Proof Theory. Proceedings, 1997. VIII, 348 pages. 1997.

Vol. 1290: E. Moggi, G. Rosolini (Eds.), Category Theory and Computer Science. Proceedings, 1997. VII, 313 pages. 1997.

Vol. 1292: H. Glaser, P. Hartel, H. Kuchen (Eds.), Programming Languages: Implementations, Logigs, and Programs. Proceedings, 1997. XI, 425 pages. 1997.

Vol. 1294: B.S. Kaliski Jr. (Ed.), Advances in Cryptology — CRYPTO '97. Proceedings, 1997. XII, 539 pages. 1997.

Vol. 1295: I. Prívara, P. Ružička (Eds.), Mathematical Foundations of Computer Science 1997. Proceedings, 1997. X, 519 pages. 1997.

Vol. 1296: G. Sommer, K. Daniilidis, J. Pauli (Eds.), Computer Analysis of Images and Patterns. Proceedings, 1997. XIII, 737 pages. 1997.

Vol. 1297: N. Lavrač, S. Džeroski (Eds.), Inductive Logic Programming. Proceedings, 1997. VIII, 309 pages. 1997. (Subseries LNAI).

Vol. 1298: M. Hanus, J. Heering, K. Meinke (Eds.), Algebraic and Logic Programming. Proceedings, 1997. X, 286 pages. 1997.

Vol. 1299: M.T. Pazienza (Ed.), Information Extraction. Proceedings, 1997. IX, 213 pages. 1997. (Subseries LNAI).

Vol. 1300: C. Lengauer, M. Griebl, S. Gorlatch (Eds.), Euro-Par'97 Parallel Processing. Proceedings, 1997. XXX, 1379 pages. 1997.

Vol. 1302: P. Van Hentenryck (Ed.), Static Analysis. Proceedings, 1997. X, 413 pages. 1997.

Vol. 1303: G. Brewka, C. Habel, B. Nebel (Eds.), KI-97: Advances in Artificial Intelligence. Proceedings, 1997. XI, 413 pages. 1997. (Subseries LNAI).

Vol. 1304: W. Luk, P.Y.K. Cheung, M. Glesner (Eds.), Field-Programmable Logic and Applications. Proceedings, 1997. XI, 503 pages. 1997.

Vol. 1305: D. Corne, J.L. Shapiro (Eds.), Evolutionary Computing. Proceedings, 1997. X, 313 pages. 1997.

Vol. 1308: A. Hameurlain, A M. Tjoa (Eds.), Database and Expert Systems Applications. Proceedings, 1997. XVII, 688 pages. 1997.

Vol. 1310: A. Del Bimbo (Ed.), Image Analysis and Processing. Proceedings, 1997. Volume I. XXI, 722 pages. 1997.

Vol. 1311: A. Del Bimbo (Ed.), Image Analysis and Processing. Proceedings, 1997. Volume II. XXII, 794 pages. 1997.

Vol. 1312: A. Geppert, M. Berndtsson (Eds.), Rules in Database Systems. Proceedings, 1997. VII, 213 pages. 1997.

Vol. 1314: S. Muggleton (Ed.), Inductive Logic Programming. Proceedings, 1996. VIII, 397 pages. 1997. (Subseries LNAI).

Vol. 1315: G. Sommer, J.J. Koenderink (Eds.), Algebraic Frames for the Perception-Action Cycle. Proceedings, 1997. VIII, 395 pages. 1997.